Introduction

On our way home to our Triune GOD, in loving kindness, they reveal *their* ways and beliefs we are all called to emulate. And by our love *for* GOD, we hear, remember, and live in all their created ways. Our good works and eternal love are found in knowing and living within their ways and beliefs. And if we are blessed, that our eyes appreciate not the creation's beauty more, but *theirs*; that our ears hear not this world's words, but *theirs*; that our souls believe not this world's truths, but their*s*; that our understanding is not ours, but *theirs*; and our belief is as *theirs,* and not as our own. It is then, as with JESUS, our eternal joy to be one in GOD, and our eternal blessing is to do FATHER'S will. We do write about the eternal love from GOD. But more often we write in pain and sorrow about the dire need for an eternal love *from* all children towards GOD. A love made manifest and seen by GOD in our desirous loving obedience. There are truths in this book many don't want to hear; but through the eternal perspective and our righteous holy understanding, we share that one well-defined call of eternal life. We honestly deal with all commands, statutes, and science, in Scripture, as we hold them up in all strength, love, and fear, of offending and losing GOD.

Author, Lawrence Michael Stack. Theist, Dualist, Theorist, and Philosopher.

ISBN 978-0-578-74768-2 (Universal / Custom)

LCCN 2020915609

Our other books available…

The Theory of Created Conscious Perception

The Eternal Perspective

All Conscious Formulations

Lower case italicized font indicates mild emphasis added

.

Chapters

Book 4 How might we know FATHER

There is a choice given to all who have ever existed. Which is to exist in the ways and beliefs of shared eternal life, FATHER has created. Or choose our own temporary ways that lead to eternal condemnation and separation. Walk with GOD, seeking and finding FATHER'S created ways of life, or be separated forever. All souls knowingly or un-knowingly, admittedly or in denial, quickly or gradually, we all make our choice; and there are only these two choices. Life eternal rests in our becoming consciously and physically one in FATHER; while death rests in being an individual. It is we who must choose, and we each do choose. Eternal life is in knowing that FATHER has needs and desires. In loving and giving us his all, he needs to be loved by us in all the same ways. FATHER is the greatest of all wonders, and the designer, author, and creator of all lesser wonders. He is the author of the one pure truth and belief, the source of life and eternal love. FATHER hates all that is individual and divisive; and we aspire to his likeness, his substance, his attributes, his desires, to become one in his eternal understanding of belief and truth. But how dare we aspire to know the heights of FATHER'S mind, ways, and his will? We do so by his Holy command to become his children. And this, we only accomplish through his ordained steps of our sanctification, and the ordered approach FATHER has instituted; and in his ways, he makes us his promise. These ordained steps being we come at his call, hearing and living the teaching of HOLYSPIRIT. Scripture says GOD calls at seeing in us the things of promise towards eternal love. Being a compassionate, humble, contrite, fearful, believing obedient heart towards GOD'S family. The second step FATHER must see is a true full repentance, which must come by our true love for and to the one name given by which it can be received. Being JESUS, the MESSIAH. Repentance means to see and accept that in our life, we have not lived by FATHER'S created eternal ways. It is accepting that our life has been lived through our own *individual* desires and beliefs, and we have played a god over our own lives. Repentance is seeing how we have lived as a physical worldly creature, and not as a growing spiritual child of GOD. Repentance is losing *all* former false understanding, belief, and desire, and to adopt those of GOD. Repentance is more than being and saying we are sorry; it is turning to remove every ungodly aspect. This is of course a process, and the length of this process will depend upon the depths to which we have fallen, and the growing strength of our love for GOD. The stronger our current love for GOD, the faster and easier our change will be. And if CHRIST sees our love for GOD is true, and our repentance is real, he has the power to forgives those he finds worthy. Never think that if you bow, cry, and pray, you are by that act alone, forgiven. Forgiveness is not taken by us in prayer; it is only given by GOD. Within your nurtured hope of forgiveness and growth, there will be a growing sense of HOLYSPIRIT moving within. After our initial forgiveness, and GOD trying our souls, some may receive the indwelling of HOLYSPIRIT as early as their well contemplated burial Baptism; being our *death* to this world, being *born* again unto GOD'S promise of eternal love and life. Keeping all of this faithfully within our heart, we strive on to learn, strengthen, and live in all the ways and belief of GOD. We must study FATHER'S words and his perceived Creation. And by his ways of love, GOD'S beliefs and ways become ours. We are warned to *carefully* discern between that which is of GOD and eternal love, and that which is not. *We* are responsible for wanting, seeking, hearing, and loving, the one true belief of GOD. *Only* one interpretation of the WORD can be true. In our knowing this, we come humbled, exhibiting a natural fear of GOD. Malachi 2:5, "*MY COVENANT WAS WITH HIM OF LIFE AND PEACE; AND I GAVE THEM TO HIM FOR THE FEAR WHEREWITH HE FEARED ME, AND WAS AFRAID BEFORE MY NAME*". Our only hope rests in

HOLYSPIRIT leading us in our natural fear of losing GOD. We must sense a needful urgency, and in our strongest desire, seek and live the true ways and belief of GOD. HOLYSPIRIT reveals our true purpose for being gifted and created. Yet everywhere, we see a weakening desire for GOD'S pure truth, and a growing desire for the error of individual truths. From Adam on, mankind has sought to alter the pure truth of GOD. Such souls have become strong in the flesh, and weak in the spirit. These are souls not wanting the one pure truth, who then led others who don't want the one pure truth. Preachers betray GOD in being more accepting of mankind's fallen physical desires. Many believe GOD'S truth changes with the world; and in this, they bring false teachings, and empty their selves of GOD. The strong correct fear of GOD in this world, has all but disappeared, and its place in most souls is now filled with fear of physical death; this is the natural progression. Minus GOD, there is nothing left but physical death and the great fear of it. A child of GOD'S fear is not of physical death, but of losing our LORD who holds all love and life. How can one believe in FATHER, and not fear losing him? Can JESUS fear losing FATHER? Heb. 5:7-9, "*WHO IN THE DAYS OF HIS FLESH, WHEN HE HAD OFFERED UP PRAYERS AND SUPPLICATIONS WITH STRONG CRYING AND TEARS*

UNTO HIM THAT WAS ABLE TO SAVE HIM FROM DEATH, AND WAS HEARD IN THAT HE FEARED; THOUGH HE WAS A ***SON****, YET LEARNED HE OBEDIENCE BY THE THINGS WHICH HE SUFFERED; AND BEING MADE PERFECT, HE BECAME THE AUTHOR OF ETERNAL SALVATION UNTO ALL THEM THAT OBEY HIM;*" My LORD, who do you see *me* to be, do you see me at all? Is my love for you true? I long for your eternal ways, love, and truth, to reign over my life. FATHER, how can I being so low, speak to you, being so exalted on high. FATHER, show me what I truly am, and lead me to you. I know that my truest love and fear for you are my only hope, strength, and compass. If my love for you is true, I will be found according to your immutable promise. Teach and show me strength, assure me FATHER. LORD have eternal mercy upon all I am before you, that I should put away all offence from my existence. HOLYSPIRIT, forgive my afront to your Holiness, I know this body is created to be a shared Holy Temple of GOD, and you are the master, and I the student. Amen. To what relationship, has FATHER called us. The standards of life taught by most Christian Denominations, are not GOD'S eternal standard. Most *professed* Christians, believe FATHER wants them to enjoy physical life, being a part of this world's ways and traditions around them. They think GOD wants us to seek things like a godless education to get a high paying job. Most believe FATHER wants them to be a good person according to the modern standards of mankind. Most believe FATHER is accepting of their sinful nature, as long as they *try* to be caring, truthful, and honest; while never perfectly being any of these. Most believe in coming together in a form of worship on a fairly regular basis to praise, and pray for the physical needs they witness in this world. Most believe this is what we are called to do. Yet, minus Church attendance and Bible reading, the *so-called* secular good citizens of this world, live pretty much the same way. The world is full of godless secular humanitarians, generous with their free time and excess money when they see hunger, sickness, and poverty. And this charity among the secular follows the pattern seen in professed Christians, in giving according to the wealth and free time available in the physical life of the individual giver. We see over-all from rich to poor, very few give till they themselves are in any physical want. By this purely physical example, the only visible difference is, one group professes the name of CHRIST. And today even this is done less and less outside Church walls for fear of embarrassing one's self, or offending others. We do not condemn physical charity, or any desire to ease physical suffering, as we are commanded to do so. But we state that this is GOD'S second physical command. The modern church body in its social life and practice, has reduced or abandoned the first spiritual commandment. Being to love the LORD our GOD with *all* our heart, mind, and soul. In practice, most elevate the second physical commandment above the first, being to physically love thy neighbor as thyself; this is not Spiritual oneness with GOD. If this is all we are called to, where's the evidence of our *spiritual* holiness, righteousness, and our perfect love for GOD? So, what higher relationship are we called to? For example, consider how long ago the Nation of Israel was enslaved by Egypt. Yet they had good food, medical care, and shelter, and could believe in GOD inside their mind where no one can stop them. Why did FATHER then bring the house of Israel out into a barren desert wilderness, away from the provision of every basic physical need they had? And did his people want to be there in the desert? No. They quickly cried for the life and things they had before, and cried about what FATHER provided for their bodies. What was FATHER trying to instill between him and his children? He truly wants and needs to be our FATHER, to be responsible and *trusted* for our every good provision. Does FATHER want us to look to him for our daily bread, healing, and protection? Or does he want us to go out into this godless world, and compromise his ways and love to get them for ourselves? If we want to know what GOD wants for us, all we need do is look at the Garden of Eden before the fall. This is where and how GOD created us to be, and to live. All sin rests in mankind's refusal to surrender control for their provision to FATHER as a child should; and instead, just want to go after whatever *they* desire. FATHER sees this in the choices that darken a soul's life. JESUS said in Matt. 6:25 *"THEREFORE I SAY UNTO YOU, TAKE NO THOUGHT FOR YOUR LIFE, WHAT YE SHALL EAT, OR WHAT YE SHALL DRINK; NOR YET FOR YOUR BODY, WHAT YE SHALL PUT ON. IS NOT THE LIFE MORE THAN MEAT, AND THE BODY RAIMENT?* Our LORD has said that a sinful fleshly nature, lives and rules only by our Spiritual weakness. If any soul says they cannot quit sinning, and cannot be perfect before the LORD; they admit they are of the flesh, and not the SPIRIT. In his mercy, our LORD declares that by all the powers in Heaven and Creation, he has suffered, paid for, and will give to all who truly desire it; not just true forgiveness, but a Godly strength, born of the indwelling HOLYSPIRIT. These we receive only by our great desire and work to overcome the flesh and all sin. We are commanded to walk in righteousness.

Scripture says all who say otherwise, or deny his commands of promise, may have a form of Godliness, but in weakness do deny the power there of. Our search for GOD'S truth and love is life-long. It is not reading the Bible in the flesh, one or a hundred times. It is our growth in the truth that starts at our true rebirth, being a proper foundation in love and humility, which brings ever-increasing spiritual wisdom, patience, and faith. Impatient souls accept false doctrines, or fall away, having no core desire born of true love to patiently endure. If you truly give your heart to GOD, and truly receive the Baptism of HOLYSPIRIT, there are three ways to lose what you have been given. One is to commit the unforgivable sin. Two, is to *fall* out of love with GOD. And three is to *fall* into a false doctrine that allows and enables continual sin. GOD and his WORD will Judge us all. Remember that GOD needs and desires to love, *and* be loved, equally. GOD cannot lie; and has declared that by faith love and repentance, (turning from *all* sin) we will be forgiven. JESUS said in John 5:14, "*BEHOLD THOU ART MADE WHOLE: <u>SIN NO MORE,</u> LEST A WORSE THING COME UPON THEE*". 1st John 5:2-3 "*BY THIS WE KNOW THAT WE LOVE THE CHILDREN OF **GOD**, WHEN WE LOVE **GOD** AND <u>KEEP</u> HIS COMMANDMENTS. FOR <u>THIS IS THE LOVE OF **GOD**,</u>*

THAT WE KEEP HIS COMMANDMENTS: AND HIS COMMANDMENTS ARE NOT GRIEVIOUS" Grievous, means very hard. John 14:15, "*IF YOU LOVE ME KEEP MY COMMANDMENTS*". John 14:21, "*HE THAT HATH MY COMMANDMENTS, AND KEEPETH THEM, HE IT IS THAT LOVETH ME: AND HE THAT LOVETH ME SHALL BE LOVED OF MY* ***FATHER****, AND I WILL LOVE HIM, AND MANIFEST MYSELF TO HIM*". Matthew 19:17, "*WHY CALLEST THOU ME GOOD? THERE IS NONE GOOD BUT ONE, THAT IS* ***GOD****: BUT IF THOU WILT ENTER INTO LIFE, KEEP THE COMMANDMENTS*". Are we called to be perfect, can we be perfect. And what is the definition of this perfection in mankind. Genesis 17:1, "*I AM THE* ***ALMIGHTY GOD****, WALK BEFORE ME, AND BE THOU PERFECT*". 1st Kings 8:61, "*LET YOUR HEART THEREFOR BE PERFECT WITH THE* ***LORD*** *OUR* ***GOD****, TO WALK IN HIS STATUTES, AND TO KEEP HIS COMMANDMENTS, AS AT THIS DAY*". 1st Corinthians1:10, "*NOW I BESEECH YOU, BRETHEREN, BY THE NAME OF OUR* ***LORD*** *JESUS CHRIST, THAT YE ALL SPEAK THE SAME THING, AND THAT THERE BE NO DIVISION AMONG YOU; BUT THAT YE BE PERFECTLY JOINED TOGETHER IN THE SAME MIND AND IN THE SAME JUDGMENT*". Philippians 3:15, "*LET US THEREFOR*

AS MANY AS BE PERFECT, BE THUS MINDED: AND IF IN ANY THING YE BE OTHERWISE MINDED, ***GOD*** *SHALL REVEAL EVEN THIS UNTO US*". Matthew 5:48, "*BE YE PERFECT, EVEN AS YOUR FATHER WHICH IS IN HEAVEN IS PERFECT*". Luke 8:14, "*AND THAT WHICH FELL AMONG THORNS ARE THEY, WHICH WHEN THEY HAVE HEARD, GO FORTH, AND ARE CHOKED WITH CARES AND RICHES AND PLEASURES OF THIS LIFE, AND BRING NO FRUIT TO PERFECTION*". Hebrews, 6:1, "*THEREFORE LEAVING THE PRINCIPLES OF THE DOCTRINE OF CHRIST, LET US GO ON UNTO PERFECTION; NOT LAYING AGAIN THE FOUNDATION OF REPENTANCE FROM DEAD WORKS, AND OF FAITH TOWARDS GOD*". Why is it a teaching in Churches today, that we can't be perfect, thus surrendering to sin, and admitting defeat without a fight. Being in FATHER, CHRIST, and HOLYSPIRIT, we are far more than flesh, we are more than mankind; and we are far more than sin. John 1:12-13, "*BUT AS MANY AS RECEIVED HIM, TO THEM GAVE HE POWER TO BECOME THE SONS OF* ***GOD****, EVEN TO THEM THAT BELEIVED ON HIS NAME: WHICH WERE BORN, NOT OF BLOOD, NOR OF THE WILL OF MAN, BUT OF* ***GOD***". James 2:24, "*YE SEE THEN HOW THAT BY*

WORKS A MAN IS JUSTIFIED, AND NOT BY FAITH ONLY'. What is our very first *work* of GOD? It is to love the LORD GOD first above all things combined, with all our heart mind and soul. To love JESUS, is to love him for who and what he is. This means to love GOD'S ways, and to want to be just like JESUS in his Holiness and Righteousness, and to in love, obey every command. A soul who loves GOD more than their self without fail, can never sin or believe a false doctrine, because a soul is deceived and turned to commit sin, out of selfish desires in their heart. This makes our true eternal love for GOD, according to the first greatest commandment, an all-powerful and infallible force. GOD'S great love and mercy, are always found in our struggles born of love for him. Our good first work is to please GOD; and in this is found our perfection. Not being the fullness of FATHER'S perfection, who leads; but *our* following in his perfection. FATHER'S perfection is full, complete, and pure. And from CHRIST and HOLYSPIRIT on down, our perfection rests in our lovingly following and obeying FATHER. Our perfection is thereby true and pure; yet never FATHER'S fullness. Ours is always before us in our every moment and choice of desire. So, as long as we strive to learn and stay within our understanding of GOD'S perfection, and don't stray, we are without sin. This is the less than partial perfection of mankind, that comes in our not having FATHER'S full perfection, yet no imperfections. GOD'S love and mercy are not found in surrendering to accept or embrace imperfection; they are not found in claiming we cannot be perfect, or in denying his commands and powers are given to help us be perfect. All such teaching is a lie of eternal purpose of life, love, and our creation. Our hope for perfection dwells in our desired effort to emulate and please GOD out of love. In these is found our perfection GOD seeks. *Im*perfection takes root when the desires of the flesh control the soul. An evil soul surrenders to flesh and sin and accepts imperfection. Did not FATHER say to Satan, see my servant Job, a *perfect* man. All the great men and women in the Bible, have one thing in common, they loved the perfect created ways of GOD. They strived to please GOD in those perfect ways out of love; not for what GOD gives or can take away, but for who and what GOD is. When Martha asked JESUS to tell Mary, who sat at his feet learning, to help with the worldly work of preparing food for their flesh. JESUS said lovingly "*MAR'THA MAR'THA, THOU ART CAREFUL AND TROUBLED ABOUT MANY THINGS: BUT ONE THING IS NEEDFUL: AND MARY HATH CHOSEN THAT GOOD PART, WHICH SHALL NOT BE TAKEN AWAY FROM HER*". We must feed the Spirit first. Scripture declares over and over, that by CHRIST'S sacrifice, and the Godly understanding taught us by the indwelling HOLYGHOST, we have GOD'S power to overcome all things, if we believe, love, and desire to do so. Believe with all you are and know this in your heart, and you can attain this perfection of Spirit. This is not only possible, if this is your true great desire, it *is* easy. To do this, Scripture says flee all temptations, troubles, and controls, of this world. We must separate ourselves from the ungodly, and remove ourselves from the worldly attacks upon our soul. So, if we are persecuted, it should be only for our faithful practice of FATHER'S ways, and not for hypocrisy or worldliness. GOD commands many to go and share his good news, but not everyone is called to do this in every way. And for all who are called to venture out, Scripture says go led by HOLYSPIRIT, cautiously two by two, and never alone. We are told often in the Old and New Testament to come out of this fallen world, and touch not its unclean things. In this, we do not partake of their sins. If we sin, it is because of weakness in our love and our spirit towards GOD; we must nourish our love and spirit. And if we are truly honest before GOD, we must admit that if we sin, it is because of the strength in our flesh. Meaning we are still afflicted by the flesh, and we must crucify the flesh. If we stumble it is because our love for GOD is weaker than our desire for what causes us to sin. We must accept this truth and strengthen the spirit being weak, and weaken the flesh which is strong. A true love for FATHER, enables our enlightened understanding of his plan, his creation, his reason and purpose, and his eternal desires. These become mysteries to individuals, hidden from their view by individual desires, purposes, and reasons. If trapped in the ways of life that are of one's own creation, one must come out of their little kingdom they have made, and as commanded by our LORD, crucify themselves daily. If one walks in Satan's Kingdom and ways, they cannot also walk in the righteous ways of FATHER. Where and how in Satan's kingdom can we live, that allows us to lift each other up, perfecting in ourselves and in each other, the loving willful force of GOD. If this we desire and seek, we will pray, and search the Scriptures, and hear and know the answer. It is there in GOD'S many words, but wisdom needs to be heard from HOLYSPIRIT. If we truly hear this, we are on a path with GOD in all things. If one does not desire this above all things, but wishes to keep one foot in this world, having desire for its pleasures, then one will hear what *their* heart wants. I trust in GOD alone, looking out through the physical darkness, to see his spiritual brightness. FATHER, I strive to see and avoid temptations in this world. I wait for your blessing, and praise you in times of both want and fullness. I know the things of this world come against all love for you; I know both suffering *and* comfort are trials of our flesh and Spirit in this fallen world. All worldly loves are based on shared compromise and change. But I know *eternal* love is based on our changing, not GOD. I know our perfection and eternal existence are defined by so many broad, deep, and complex, qualities and qualia. Mankind's simplistic defining of the eternal, as never ending; and perfect, as being flawless, do not begin to define spiritual eternal perfection. We see all trials hold one common value, being what they teach us of GOD. Amen. We know GOD holds all accountable, who teach others false interpretation of reality and truth. We add not one word to Holy Scripture, and will not silently stand-by as one word is weakened, altered, or taken out of GOD'S loving royal commands. If one bears witness to that which they have seen, they are responsible for the account they give, and for all that comes upon others, by their account. That account being their solemn testimony, weighed, balanced, and considered, by others. So, concerning the words of GOD, I speak in great cautious fear for my soul. My GOD how can this world desire such false interpretations of your Holy Scriptures? GOD is not the author of confusion.

When GOD speaks, each word holds *their* meaning and intent, and *their* original meaning and intent are what we must seek; not our, or another's. If we are not profoundly sure of what we say, we should be silent. For if we are, even if unintentionally wrong, we still spread the destruction of lies. If one misleads another with false teaching, and that soul is turned astray in their search for the truth of GOD, both souls are accountable together. If one leads another astray, they are accountable for their part in all the evil that follows. We are all accountable for what we say, do, and choose to believe. What price is required of those who lead searching souls to weakness and eternal loss. It's *as* easy to deceive another unintentionally, as intentionally; and both bring the same result, a deceived weakened soul. How are souls eagerly deceived? All failures of mankind are in their desires for the flesh. Such souls for the most part, can't or won't believe in absolutes. If you remove absolutes, you are left with a divisive confused reality. Please name one subject that you, having full knowledge of, believe doesn't have an absolute. How about love and truth? Satan and his children say love and truth are not absolutes, they are what each soul makes of them. Can this be true, can we each play a god, and have our own definitions of life's great ideals and ways. FATHER will never bow to individual truths, contrary to his; only his one authored absolute understanding of eternal love and truth, are accepted. We see there can*not* be two absolute truths, or three. We also see one absolute eternal love; it too, being according to the definition and standard set by GOD. We believe in one perfect order, and one straight and narrow path, away from this world and towards GOD. These are absolute constants, and we will all find this same path if we search in eternal love through HOLYSPIRIT. 1st John 2:27, "*BUT THE ANOINTING WHICH YE HAVE RECEIVED OF HIM ABIDETH IN YOU, AND YE NEED NOT THAT ANY MAN TEACH YOU: BUT AS THE SAME ANOINTING TEACHETH YOU ALL THINGS, AND IS TRUTH, AND IS NO LIE, AND EVEN AS IT HAS TAUGHT YOU, YE SHALL ABIDE IN HIM.*" If you believe in GOD, and turn from, and do not trust in mankind's many beliefs, but stand in GOD'S belief alone; you can never be deceived. Once we see the absolutes of love and truth do exist, it brings a thirst in our soul for these to be written upon our heart and mind. We all must believe in something, and many are content with a made up self-supplied truth; but the fewer choose to lose ours, and hear GOD'S. A soul not nourished by GOD'S one pure truth, cannot live in peace with others eternally; they can only wither within their self forever. We all must be honest in not just giving all true answers, but in asking all true spiritual questions. In each of us, when it comes to GOD'S truth, at our *first* turning, there exists a measured difference between what we say we believe, and what we believe strong enough to live. We see strength of belief, varies in each of us. How honest can we be about the strength of our core beliefs. It takes serious consideration to give a close to truthful answer. If you hear a question like this, and the first thought that comes to mind is, I am pretty honest with myself, you answered to fast. Souls run from correction, and are afraid to confront this issue. Most souls instinctively avoid, and do not want to see their true state of belief. To see our uncovered bare soul, as GOD sees us, can be a painful mirror to look into, but it is vitally necessary. Being truly honest with ourselves takes effort. Scripture says the heart, meaning the core depths of our soul, is deceitful above all things, and who can know it. This speaks to levels of dishonesty within souls who hold individual beliefs of understanding, and do not have HOLYSPIRIT to teach them. We can only see ourselves truly, through the belief and understanding of GOD. Do you know all the ways and desires of GOD? Or do you only know what you want to know? If the one path to GOD is found only in GOD'S true eternal love and ways of life, this requires our continued sacrifice of growth in their ways, knowledge, and belief. In knowing GOD'S ways, more and more, we see where we have been, and still are destructive, and desire to change. Things we once thought ok, we now see as divisive in not aligning to GOD'S ways. At some point, does one's *resistance* to change arise? At some point, is it too much for one to change, too much worldly desire to give up and walk away from? And will you, as many then do, think you can settle in at *your* comfortable level of compromise? And by *your* ongoing verbal profession of belief alone, is it in *your* power, to affirm *your* adoption and election as a child of GOD? Today this is the predominant accepted belief and teaching, being a fearful breakage from the truth. The Church howls that our good works are not our salvation, and this is truth. Salvation is truly a gift, lest any should boast. Yet we are *commanded* by our love for GOD, to work and live in the true ways of GOD. Modern Churches praise GOD'S commandment of physical charitable works towards others. Yet the Church lost its hearing and voice, concerning the commandment to do the first good spiritual work, which is to love GOD high above all, and to go and sin no more. If we love GOD, we must surely love who GOD is. And if we love who GOD is, we must surely want to live lust like GOD. Many do not see this, but the fewer having opened eyes, see a vile spreading sickness within the Church. GOD and their words will judge. In the beginning there was the perfect order in this world created by GOD that had no death or suffering. It was kept in physical perfection by FATHER'S forceful desiring presence, and the loving obedience of the two souls placed in his Garden. These first two free-willed self-aware conscious entities called mankind, in loving emulation, walked perfect before GOD. And JESUS came daily and walked with, and taught them about their eternal Spiritual and physical need. For how long, we can't be sure, it could have been for tens, hundreds, or thousands of orbits around the sun. We read that JESUS in love, came each day walking in FATHER'S Garden, teaching, loving, and seeing only their perfect order. Then temptation came to mankind by Satan, and lustful desires arose to bring mankind's first act of rebellion against GOD'S pure truth. Sin was first born in Lucifer, and spread by him to 1/3 of the angels; and then he spread his confusion to mankind. Adam and Eve had always accepted the perfect truth from JESUS, up until the day Eve desired Satan's lie to be true. The horrible destructive power of a lie, searches for any weakness in a free-willed soul. Here we see the best example of a temptation to sin, *and* its results. This free-willed aspect has not stopped or changed in souls. And every tiny sin, attacks

FATHER'S love and Creation, as he cannot abide the presence of sin, which are all imperfections. Hearing this horrid news, FATHER recoiled and pulled his hands away from their imperfect sin, stretching the creation out from its dynamic perfection. FATHER'S hands that hold the universe, now pull the universe out of perfection faster than the speed of photons, towards complete dissolution, just as he warned. If we replace our ways of love towards another soul, with betrayal and unfaithfulness, does it not push them away. Isaiah 59:2, "*BUT YOUR INIQUITIES HAVE SEPERATED BETWEEN YOU AND YOUR* ***GOD****, AND YOUR SINS HAVE HID HIS FACE FROM YOU, THAT HE WILL NOT HEAR.*" Scripture tells us how Satan brought the power of death into the Earth. What is this power he used? It is simple deception to lure and tempt souls out of GOD'S perfect ways and care, and into all lusts for self and the flesh. Throughout history, FATHER has created and invested in stronger and stronger ways for us to redeem ourselves back to him and his ways. He's tried harder and harder, giving more and more. And finally, FATHER gave all he has, and promises he will give his all, up until the last second possible, trying to save all who can hear what his SON has said and done. But those who lust for the creation and the flesh, never hear FATHER'S full truth, it is perverted in their hearts to become their own desired truth. We offer a simple story to demonstrate the current state of self-deception that lies between love of self, and all true love for FATHER and his ways. There were once found among the creations of a loving FATHER, many gifted souls who "said" they loved him. And these souls were taught that if they *truly* loved him, they would follow his commands of love to be perfect and not introduce destruction into his perfect creation. By this, FATHER being perfect, pure, holy, and righteous, could have a loving eternal relationship with them, and they with him. FATHER told the people who professed they loved him, that any rejection of his eternal created ways is sin; and sin introduces the imperfection of destruction to his creations that sustain them. These people who *said* they love GOD more than all that the self-indulging pleasures that exists, tried to be perfect; but most often forgot, because of their love of self and their desires for the world and the flesh. FATHER saw and said these souls are weak in spirit and strong in their flesh, I will teach them so they can defeat the flesh. So, he gave to those who said they loved him, a book of chronicled history of GOD'S laws and ordained worship, that they might remember. The people who said they loved GOD tried *again* to be perfect in his loving eternal ways, yet most often they forgot, because of their greater love of self and desires for the world and the flesh. FATHER gave to these people who said they loved him, much love and help *many* times; yet each time the people who *said* they loved GOD and tried to be perfect in their love, often failed because of their stronger love for self and desires for the world and the flesh. There was great pain, loss, and grief, in the heart of FATHER. So, he said, I will give this people who *say* they love me, all the love, and all the spirit I have. And surely this gift of my everything, being the willing sacrifice of my only begotten SON, *my* greatest love who will suffer and bare *their* shame before *their* own eyes, that they may seek to truly repentant and receive my indwelling HOLYSPIRIT. By these *final* gifts of love, they can and must hold an eternal loving obedience for the omniscience of GOD; and remain perfect in the ways of eternal love and life. FATHER did then, and has always told mankind not to boast in finding and receiving this true love and perfection. As eternal life, and the knowledge of eternal love are a gift and a saving work of GOD, and never born of our self. But as before, the people who *said* they loved GOD, tried to be perfect in his love. But again, most still forgot by their love of self and their desires for the world and the flesh. This revealing a great hardness in their hearts, as they quickly tired of trying to be one with GOD. Some then subtly lied to those who still strived to be one with GOD, saying has not GOD said of perfection, it is a gift and not of your work. Cease your striving and enjoy this world with us, and you will still receive the gift. And many who *said* they loved GOD accepted this deceit in their heart, and espoused they no longer needed to live perfect to receive this gift. Believing in their lust and love of self, they will still receive this gift; for this they say is a loving gift that comes *from* GOD. While we, *who* do truly love FATHER in our core, continue perfecting our souls in seeking to please him. Which is not only that eternal love *from* FATHER, but *our* eternal love *for* FATHER. Do we receive salvation through the love from GOD alone? No. We are saved by CHRIST'S suffering, and the indwelling HOLYSPIRIT, *and* our desire to be one with GOD. How much do we each love GOD? How much do we each need to love GOD? Do we study ourselves and ponder these questions? Do we quickly say, I am not of this world; yet not want to talk about how and why we are not of this world? Which reveals we are more of this world than we think. GOD and their words will judge us. Jeremiah 2:11-13, "*HATH A NATION CHANGED THEIR GODS, WHICH ARE YET NO GODS? BUT MY PEOPLE HAVE CHANGED THEIR GLORY FOR THAT WHICH DOTH NOT PROFIT. BE ASTONISHED, O YE HEAVENS,*

AT THIS, AND BE HORRIBLY AFRAID, BE YE VERY DESOLATE, SAITH THE ***LORD****. FOR MY PEOPLE HAVE COMMITED TWO EVILS; THEY HAVE FORSAKEN ME THE FOUNTAIN OF LIVING WATERS, AND HEWED THEM OUT CISTERNS, BROKEN CISTERNS, THAT CAN HOLD NO WATER.*" We are told to be peculiar, and separate ourselves from Satan's world. Titus 2:14, "*WHO GAVE HIMSELF FOR US, THAT HE MIGHT REDEEM US FROM ALL INIQUITY, AND PURIFY UNTO HIMSELF A PECULIAR PEOPLE, ZEALOUS OF GOOD WORKS*". 1st Peter 2:9, "*BUT YE ARE A CHOSEN GENERATION, A ROYAL PRIESTHOOD, AN HOLY NATION, A PECULIAR PEOPLE; THAT YE SHOULD SHOW FORTH THE PRAISES OF HIM WHO HATH CALLED YOU OUT OF DARKNESS INTO HIS MARVELOUS LIGHT.*" 2nd Corinthians 6:17, "*WHEREFORE COME OUT FROM AMONG THEM, AND BE YE SEPERATE, SAITH THE* ***LORD****, AND TOUCH NOT THE UNCLEAN THING; AND I WILL RECEIVE YOU, AND WILL BE A* ***FATHER*** *TO YOU, AND YE SHALL BE MY SONS AND DAUGHTERS, SAITH THE* ***LORD*** *ALMIGHTY.*" 1st John 2:15-16, "*LOVE NOT THE WORLD, NEITHER THE THINGS THAT ARE IN THE WORLD. IF ANY MAN LOVE THE WORLD, THE LOVE OF THE* ***FATHER*** *IS NOT IN*

HIM. FOR ALL THAT IS IN THE WORLD, THE LUST OF THE FLESH, AND THE LUST OF THE EYES, AND THE PRIDE OF LIFE, IS NOT OF THE ***FATHER****, BUT IS OF THE WORLD.*" Do you wonder why the many divisions of Christian belief, center on how much FATHER gives to those who ask, and how much GOD asks in return; and how much GOD is involved in the little everyday things that happen to us? How much do you think FATHER changes reality for you? We must know that if GOD intervenes in our lives, this will, as a relative result, alter the whole world in endless ways. We must know this means reality is consciously deterministic. We must know that if GOD intervenes in our life, it must affect others, including believers. If one thinks GOD saved them from a ticket by making a traffic light stay yellow longer than normal; what would that involve? Who can believe FATHER shuffles creation and changes other people's lives and destiny to save them from a ticket, or so they won't be late for work. If FATHER did this, what would happen to the other people, the lost and saved going the other way? What might happen to them in their lives, now that they all had to wait for even a moment, for you? All their destinies have been changed forever. Many of us have heard a Preacher in a sermon say GOD brought someone to that Church that day for a reason, meaning some are there by divine intervention. Does FATHER over-ride free-will? We can hear HOLYSPIRIT'S call, but that is all. So, let all preachers be careful and explicit. FATHER created one shared reality, and if we truly love him, we must know he surely wants to let us know he is there and loves us. But it is not Scriptural or true, or within the laws and boundaries of all free-will reality set by FATHER himself, to say we see him going about changing and planning our life in this world. Yes, GOD speaks to a few in dreams and visions, but not often; and FATHER gives divine healing and Spiritual intervention to a few more to advance the Kingdom. But he does not offer, or give these to any who do not truly seek and love GOD; and surely not to all who ask, or profess CHRIST'S name in this horrid world. When I experience FATHER'S intervention, it is rare, big, undeniable, and for the Kingdom, and never just about my life here in this world. If I experience dreams, healings, or interventions, I know it is for my work towards FATHER'S Kingdom. Throughout Scripture we see his great divine interventions from time to time to help and increase his Kingdom, and guide the course of GOD'S people as a whole. But can we ask on a personal daily basis, to make our lives prosperous and comfortable? FATHER does not go in front of me or you, preparing a way for physical prosperity and comfort. In this world, FATHER went to great lengths to create chance, accountable choice, and free-will. Every time we have any gain of chance, we cannot in selfish dishonesty see or claim the intervention of GOD when there was none. This is the founding aspect of the false pre-destination and prosperity doctrines. In yielding to such confusion, one will never recognize GOD'S touch in reality. The rapid spreading of the Prosperity Doctrine, is telling in both its content and message, catering to the love of self and this world. Consider that if GOD warns us so often in Scripture, that prosperity and comfort in this world bring temptation and danger, why would GOD then give us these in abundance. JESUS went to great lengths to teach this in every word and act of his short life here in the flesh of man. Do any believe, that if you came from a well to do family who had the means to give great time and physical care for you; to guide you through *this* world, with patient understanding, comfort, and loving rewards. Was that FATHER'S blessing over you, or was it the chance of life, and a multitude of profitable choices made by your ancestors. If it was FATHER who gave you these as a blessing out of love for you; what do you tell the tiny child he loves, born into cruelty, want, and horrid existence of looking for food in the trash? We must understand, that though FATHER can perceive any physical manifestation he wills; he has for righteous reasons, self-imposed limits on what he *will* do. We must see that physical life is by far, the least important aspect, and that at a soon certain point of GOD'S intervention "free-will" will be diminished, altered, and overcome; so, to what extent will GOD intervene in our lives. Before answering our prayers, FATHER will search the desires, love, and obedience in our heart. Two, GOD will consider if our prayer benefits the Kingdom of GOD. And three, will our request strengthen our love and Spirit, and not our flesh, while not hurting another's love and Spirit. The side effects of GOD helping one, that harms others, is a great consideration. And there are surely other things beyond our privy GOD might consider in being omniscient. We see FATHER will not clear us a path, free of physical adversity and pain. But we can surely know FATHER is deep in the Spiritual rescue business of his children, and will not allow any Spiritual trial by devils we cannot bare without GOD'S help. When any of GOD'S true Children are at the edge of deep despair, or in true Spiritual danger; then as we do for our children in great and desperate need, GOD is there. But GOD *and we* know, it is good for children to learn to stand and battle in this world on our own. This is part of theirs, and our climbing out of Satan's world, and into FATHER'S. How much are we each not just in, but of this world? How far have we each come out of this world for GOD? How far must we come out? The answers to all questions are in Scriptures Godly interpretation. But the thing is, GOD says his wisdom and truth are hidden to the blind, who don't truly and deeply want to see its fullness. One can never find the great spiritual depths they do not seek. While in the flesh, one can read the Bible a hundred times, and never hear it properly, its truth is lost in a false interpretation, by the hardness of their heart. Not to eternally judge, but to open eyes, we see pulpits across the world, filled with souls who are weaker and in more need of GOD'S truth, than those they try to lead. These were taught in seminaries by weak men, handing down weak partial laws and truth, mixed with traditions and contradictions. These were taught the business acumen of not offending; as attendance and donations are a required engrained parts of modern church business. They sow seeds that for the most part, minimize or remove peoples fear of accountability, sin, and GOD. They preach individual paths, holding a small false measure of GOD. Most ministers are unknowing and unable to deal with much of anything of a true mature spiritual need. And in grave error, encourage all who listen to be prosperous, successful, and a lukewarm part of this world. These deliver watered-down selective aspects of FATHER'S words. Modern Pastors and professed Christians, gladly help with the physical wants of this world in their community; but such works are meant to be the second *lesser* flowers of our faith, not the root and stem of it. Our love for FATHER is revealed in forgiving, suffering, and giving; but far above all else, it is in our self-denial. It can never be our measure of giving worldly things first. We have all seen the child who was lavished by wealthy parents, yet this child does not feel or know how to spiritually love. Love only lives and grows in our Soul, not the body. Establishment Churches get lost in their work of the flesh, when spiritual growth towards purity must be our priority; and they are not one in the same. Many Preachers are modern-day Scribes and Pharisees, trying to keep a Church financially sound, filling pews as they go through lukewarm motions. They fill minds and hearts with error as to how far we are to go, and how much we are to give and surrender. False preachers try to not offend their customers about the worldly ways they live their lives. Do not be led by Preachers who wish to appease people more than GOD; we must give and surrender all to GOD. Eternal love and life are not gifts of honor without cost; and eternal love is the highest of all desires, witnessed as we strive in our faithful responsibility towards all the lively ways of GOD.

Consider how much we each fear GOD; do not think you can eternally love GOD any more, or less, than you eternally fear losing them. For that is impossible and at odds with eternal life and love's meaning. We must feel both equally, or feel neither in a *real* way. We can say and believe we love GOD, a hundred times every day, but until GOD sees and feels *our* eternal obedient love, our words and *belief* appear untrue to them. Love is a word most always misinterpreted, and individually defined. GOD seeks what *they* define and feel to be eternal love. How can we know we have this same great form of love for GOD? Scripture says, the best evidence, that we have this love GOD seeks, is when we strive in true desire to obey every word in Scripture, and walk perfect in all FATHER'S declared created ways of life, and not our own. We offer another parable to distinguish those who truly fear FATHER, from those who are currently vainly religious, lacking a natural sufficient fear of GOD... There was a great creation, being a beautiful vast garden, full of every bright color and beautiful flower. There were every small and large tree bearing shade, and all manner of bright pleasing fruits. There were all manner of nuts, berries, fragrances, and nectars. And there was the One Tree of Eternal Life. There were also bright glistening cool waters everywhere, and a great warm loving light upon it all. There were all perfect provisions of spiritual and physical life to enable mankind's senses before GOD. There was then placed in this vast perfect garden, the *first* two perfect souls of mankind, having their free-will choices. But there was also *one* tree placed in the midst (or middle) of the Garden, that they were warned, was not for them to eat from. They were told eating from it would mean disobeying and choosing a rebellious imperfection that would separate them from GOD; and that separation was called death. Yet they did eventually take and eat of it, and as warned, they become imperfect, separate, and individuals. So, the perfect Garden and Tree of Life, were taken from them. The Great Creator then tried and tried to find a way for them to return and find their perfection. And lastly in great suffering, a pouring out of his only begotten SON to be a great new beautiful Tree of life, being in the midst of, not a garden, but the whole world. This Tree was so great, high, and perfect, that it cast its warm shadow over the entire world; reaching up in strength, beauty, and majesty, as a ladder towards the Heavens. Reaching up to the very Throne of the Great Creator. Mankind was told they may climb and eat of the fruits, found in the very heights of this most perfect tree of life. And the higher we climb, the stronger and more secure its great spiritual branches became. The Great Creator, then wrote a Holy Book about the fallen souls and his Garden, and his Great perfect Tree of repentance. In this book, the Creator said if any soul, in a desire of love, climbs, lives, and eats, in the heights of this Great Tree. They may once again see, know, and reconnect, to all the great and perfect spiritual things far off that mankind had long ago lost; things they have never seen or knew before. But the Great Creator warned again, saying if any will desire this Great Tree, they must live in and love the heights of *this* Tree more than anything. To always climb and look up towards its perfect spiritual heights, and not love anymore the falling physical corruption below, being defiled and lost from perfection. This is our resurrection back to the heights of eternal love. The Creator said, if the people on the ground did these things, they could once again be part of the Great Spirit of the Great Tree. But the Creator warned them many times, to *fear* him in his sayings, that they must believe he will surely keep and uphold every word in his book. So, the people who did desire, climbed up into the Great Tree, some higher than others, some lower. Some loving the Great Tree more, some loving the Great Tree less. Some climbed high enough to behold marvelous far-off wonders they had never seen or known before. But soon, many of the souls who climbed up, *looked back down*, as they missed and desired and loved the physical pleasures, tastes, smells, and comforts, on the ground. Some who looked down, climbed back down to the lower weaker branches and settled in where they believed they still looked up, as they reached down to sample the pleasures of the ground. Soon many who looked down were back on the ground, but they gathered once a week under the tree to behold and reach up, thinking to touch and for a few moments hold onto the lowest weakest branches of the Great Tree. These no longer saw or knew those far-off marvelous wonders, for they gave up on the perfect heights of the Great Tree. These now behold only the physical things around them. Many who stay in the world, deceive themselves and say, we can surely enjoy the pleasures of this world, yet often return to touch the Great Tree of Life in the midst of the world. Subtly deceiving themselves, saying the Great Creator surely knows we love the perfect spiritual heights of his Great Tree, more than the physical pleasures and comforts found on the ground. And *he* will surely in his great mercy, welcome us when the Spirit brings his loving family to Heaven. And GOD has said it is a gift, and not of our works. So, those who looked down, perpetuated this false belief. Those who look down, did not hear or fear the warnings anymore; they denied the ordained ways to return to GOD'S perfection. Many in their weakness of belief, left the Tree completely, and returned no more. On that day of Judgment, the Great Spirit found from the root to the very top of that Tree, will begin naming his children, as those on the ground stand around it. But the SON declares to them, and all who did not climb, "*I never knew you*" because of their disbelief and lack of fear for GOD'S commands. Their great love was for the things of the world, and they dishonored GOD'S loving suffering Sacrifice. We see among the many Christian Denominations, that very few expectations are being preached; even as GOD says our perfection is the evidence of our love for GOD and eternal life. False denominations, hold false beliefs for how far GOD goes, or has gone to save each soul. Many Pastors preach against fearing GOD. Now as for fearing GOD, we are shown in certainty GOD has no desire or need for us to fear him. But GOD knows *we need to fear them*, or we will destroy ourselves and each other. The fear of losing love is a necessity for all souls. FATHER said to Lucifer, "*THOU WAST PERFECT IN ALL THY WAYS UNTIL INIQUITY WAS FOUND IN THEE*". FATHER inquired of Lucifer, how are thou fallen from perfection? His chosen imperfection and falling, came when Lucifer no longer loved or feared GOD enough to follow them; so, he refused to believe FATHER'S warnings. Flee any denomination that over emphasizes the merciful gifts of GOD, and or, under emphasizes FATHER'S warnings as to what is sin, what is commanded, and what is expected of us. There is a vile sound, having voices added continually as an evil chorus echoes throughout the mountains of this world, louder and louder. We hear a *false* mantra of have no fear, and simply say you believe in the name of the SON, and little to no more, and all imperfect roads lead to Heaven. Is GOD accepting of never-ending sin? That is a great lie that GOD will judge. Know and remember always, that FATHER needs and desires to love, and be loved equally. GOD help us see the darkness of this world, and leave it all behind. To always look up and see only your bright truth. Amen. Beware of preachers who entwine the lure of physical comfort, pleasure, beauty, and excitement of this world, with a loving search for GOD. I have seen many souls with a true hope, desire, and strong belief that a loving GOD does exist, having within themselves, all fertile promise. Yet I have seen them turned aside, weakened and set on paths of failure by false teachers. Over 250 false Christian Denominational Doctrines currently flood the earth; Matthew 23:15, "*WOE UNTO YOU, SCRIBES AND PHARISEES, HYPOCRITES! FOR YE COMPASS SEA AND LAND TO MAKE ONE PROSELYTE, AND WHEN HE IS MADE, YE MAKE HIM*

TWOFOLD MORE THE CHILD OF HELL THAN YOURSELF." We give only the pure truth, in that GOD prepares a place called Heaven for those who eternally love and seek GOD'S ways; but not for people who believe there is a god, and don't want to go to hell while living in rebellion. The devils *believe* in GOD also, and tremble in disobedience at hearing GOD'S names. Is your belief and worship, focused on the loving grace and mercy GOD has promised for all eternity. Or, is your belief focused on forsaking this fallen world and its ways, as commanded in a fervent love for GOD. We must return in full, the same faith and love shown us in the sacrifice of MESSIAH. Are we giving our greatest love to GOD in the fear and admonition commanded in Scripture. We need never worry about GOD'S love towards us if we live in loving obedience to eternal life; because GOD'S love can never fail. But it is *our* love *for* GOD that fails in our every disobedience to eternal life. In our every sin, the *circular* flow of GOD'S love through us, back to them, has failed; we have not loved GOD in our sinful moments. In this world filled with sin and sinners and temptation all around us, a great love for GOD is rare. But thankfully, in the New Heaven, temptation will not exist. HOLYSPIRIT calls all who err to a new start; to a Spiritual approach that will help us grow as fast and as far as we desire toward GOD without limit. Please understand, that if you hold a false truth or doctrine, this creates limitations and weakness in your walk, so you can never reach GOD. A deep and true love for all GOD'S ways, is our only guide. If that love is not in us, then by what other force can we be drawn and led straight to GOD. As the aligning pull of magnets draws each one together, so, it is with all true love for GOD'S ways. But if you reverse your magnets polarity, to not align with GOD'S ways, you cannot come to GOD, against the repelling force pushing every denying soul to one side or the other. If we obey GOD'S commandments and statutes, it is because we believe, love, *and* fear GOD, more than this world. If we don't, it is because we *don't* believe, love, or fear GOD more than this world. We all do those things we deeply desire, what other forceful reasons could there be. If one soul can't obey, yet another can, what does that mean? If your desires are different than those of GOD, how can you say you love the ways of GOD. And if you cannot say you desire and love the ways of GOD, how can you say you love GOD accept in selfish desire for what GOD might give you. How can many believe in different levels of emulating GOD, and lesser levels of denying the flesh. How can individuals find different beliefs and paths to GOD, when Scripture says there is one narrow straight way, and one pure truth and belief. We cannot accept any individual understanding of GOD'S ways and commands. In an average Church today, if ten people are put in a circle and asked a deep spiritually mature question about the truth and will of GOD. You can not only hear several, but often ten different beliefs to *some* certain degree. There can be surface agreement, but there will also be the deeper disagreement, as they politely agree to disagree. What is worse than these divisions of belief in the Church, is that most preachers say it's ok for the Church to be this way. Modern preachers are taught to stay away from areas of belief that reveal these divisions within their congregation. Division of truth in the Body of CHRIST, cannot be; we must be of one mind, one accord, one spirit, and under one GOD. Satan always divides to concur, and I see vast growing divisions of belief. Satan divides mankind's belief in GOD and Creation with confusions of hell. And if souls are greatly confused, they cannot see or hear FATHER'S created order any more. They will see unnatural orders, believing gender is up to individual interpretation, that sex is recreation and sport for pure physical pleasure. GOD'S created nature and purpose is a marital bond for *procreation* between a husband and wife alone, it is *not* for recreation! Accept even one tiny belief outside GOD'S created ways and purpose, and you thereby, call GOD a liar. There is one created forceful alignment that perpetuates eternal life; and countless unnatural abominable ways of hell's physical pleasures. We must teach and warn each other that minus GOD'S true foundation, our understanding of reality will always be wrong. Yes, many false beliefs can appear to stand on their own in a temporary perverted physical way, but a Child of GOD must build their spiritual foundation to withstand all evil. I see unprepared professed believers, stumble and fall when storms of confusion come against them. We must build our willful resolve, not just against the tragic storms of affliction we all experience, such as our loss of loved ones and the horror we see coming upon the young, weak, and old. We must build our loving belief to withstand the great twisting storms, and the quaking terrible days of GOD'S Wrath and Prophecy. We must recognize and withstand Satan's great deceptions as we pass by, knowing they are everywhere. What we accept as the first pure truth, is what we must build ourselves upon, and rely on to sustain us. How then should we preach and believe in each moment? Do preachers laxly preach the thoughts and belief of a continuing stability and comfort for the generation to come, praying for physical life to get better, and for GOD to restore physical order. Or do we preach we are in the last days, and must prepare for GOD'S coming cleansing; and to stand against a spiritual evil that comes against us all with such a deception of Spirit, that JESUS says in Matthew 24:22-25, "*EXCEPT THOSE DAYS BE SHORTENED, THEIR SHOULD NO FLESH BE SAVED: BUT FOR THE ELECTS SAKE THOSE DAYS SHALL BE SHORTENED. THEN IF ANYMAN SHALL SAY UNTO YOU, LO, HERE IS* ***CHRIST****, OR THERE: BELEIVE IT NOT. FOR THERE SHALL ARISE FALSE CHRISTS, AND FALSE PROPHETS, AND SHALL SHEW GREAT SIGNS AND WONDERS, INSO MUCH THAT, IF IT WERE POSSIBLE, THEY SHALL DECEIVE THE VERY ELECT.*" We are warned of a coming deception, only those *strong* in GOD'S love and truth will endure. To endure such things, requires sound Doctrine. When Scripture says *sound* Doctrine, that means unsound Doctrines exist. I have never read a single word in Scripture that declares a pre, or mid-tribulation rapture of the Church, not one word. I have read only the opposite in Scripture, and it is a most necessary thing in the tribulation for the elect to be there and endure it. One thing is for certain, those who believe in a premature rapture, will be unprepared and disillusioned when hit with the great deception of the embodied Antichrist, and the wrath of GOD. Most, if not all of these souls will fall away. 2nd Thessalonians

2:1-4, "*NOW WE BESEECH YOU, BRETHEREN, BY THE COMING OF OUR* ***LORD JESUS CHRIST****, AND BY OUR GATHERING TOGETHER UNTO HIM, THAT YE BE NOT SOON SHAKEN IN MIND, OR BE TROUBLED, NEITHER BY SPIRIT, NOR BY WORD, NOR BY LETTER AS FROM US, AS THAT THE DAY OF* ***CHRIST*** *IS AT HAND. LET NO MAN DECEIVE YOU BY ANY MEANS: FOR THAT DAY SHALL NOT COME, EXCEPT THERE COME A FALLING AWAY FIRST, AND THAT MAN OF SIN BE REVEALED, THE SON OF PERDITION; WHO OPPOSETH AND EXALTETH HIMSELF ABOVE ALL THAT IS CALLED GOD, OR THAT IS WORSHIPPED; SO THAT HE AS GOD SITTETH IN THE TEMPLE OF GOD, SHOWING HIMSELF THAT HE IS GOD.*" We see in the first sentence of the above verse, the word *Beseech*, meaning "urgently plead" that we be not turned to fables concerning the second coming of CHRIST, and his gathering of us to himself; or raptured if you prefer that term. In the second sentence we are told that day of gathering comes only at the very end, and that way before that, there comes a great falling away from GOD. Meaning, before and during the Tribulation, most all of the Church will follow false doctrines. Scripture says the true children will not be gathered to be with CHRIST, until after the Antichrist is revealed at the end of the

Tribulation, and after Satan enters the temple in Jerusalem, indwelling that image to stand and say he is GOD. It is such a hard thing to grasp, that so much of the Church now follow such false doctrines as a pre, or mid-trib. rapture, when every word in Scripture denies that. Scripture says the Church will be here, enduring trials until the end. Many will choose weak unsound doctrines that please their ears. Scripture describes these weak believers as half-hearted, lukewarm, and wanting no more than the merciful milk of truth for babes. I fear I too can fall away, if I lose my loving eternal perspective. We all should never underestimate the power and potential for evil around and within us. One third of the Angels in Heaven, being eye witnesses to the glory and power of GOD, still deceived their selves and succumbed to contrary core desires. We are conscious souls, gifted and created in hope that we choose the purpose and pleasure of GOD. Our physical side of life comes eternally from GOD. Yet being born in sin, we are born of temptation and ignorance of the lawful ways of love, governing all things of GOD. At conception, we are governed by a weak nature of the flesh to survive, and we are born as a probable offence, and a probable destructive force against GOD'S perfect order. Now a few souls are sent and received HOLYSPIRIT in the womb, for GOD'S specific purpose; but the rest, are born not having HOLYSPIRIT, across the line of life and in the hands of death. Search and learn by the teaching HOLYSPIRIT that our obedience to GOD is our love and goodness. Only in this can we be restored to his elect. Scripture says at the close of GOD'S desire for this creation, the Trinity will have done all they righteously can. They are all 3, right now up to that line of free-will, working to draw all who love GOD. Isaiah 5:4, "*WHAT COULD HAVE BEEN DONE MORE TO MY VINEYARD, THAT I HAVE NOT DONE IN IT? WHEREFORE, WHEN I LOOKED THAT IT SHOULD BRING FORTH GRAPES, BROUGHT IT FORTH WILD GRAPES*?" We know that though GOD reaches down, as far as we will reach up; yet we must never *expect* CHRIST to reach down to us more than he has, as his grace, mercy, and giving, are already sufficient; meaning GOD has already put all his truth and sacrificed goodness within our reach, when we deserved death. GOD has already gifted us our soul, and created a body for us, and created a world to live in. CHRIST has already suffered and suffered, and even died to show his love for us. What more should they do before we devote our eternal love to GOD. We must in our greatest love, take it upon ourselves to choose truth; and strive in love and run and fight with prayer and fasting to endure until the end. Do we heed in fear, the warnings of our LORD in Matthew 22:14, "*FOR MANY ARE CALLED, BUT FEW ARE CHOSEN.*" And 1st Peter 4:18, "*AND IF THE RIGHTEOUS SCARCELY BE SAVED, WHERE SHALL THE UNGODLY AND THE SINNER APPEAR.*" Matthew 7:13-16, "*ENTER YE IN AT THE STRAIT GATE: FOR WIDE IS THE GATE, AND BROAD IS THE WAY, THAT LEADETH TO DESTRUCTION, AND MANY THERE BE WHICH GO IN THEREAT: BECAUSE STRAIT IS THE GATE, AND NARROW IS THE WAY, WHICH LEADETH UNTO LIFE, AND FEW THERE BE THAT FIND IT.*" This last verse speaking of the wide and the narrow gate, does not speak of believers and non-believers. All those trying to enter Heaven through the wide gate, are those GOD calls lukewarm believers. These are widely compromised, and need a wide individual inclusive gate, to fit into heaven. But how can a soul go left and right at the same time? And if so, how is their path changed? Scripture does not say hold onto, and live our life professing JESUS. Scripture says profess JESUS and give our life to him, allowing his righteousness to reign in us. Do not say we are perfect by professing JESUS. Profess JESUS by being perfect in him. Are we called to simply occupy as simple souls, having a simple confession and profession of belief, faith, and a hope, of trying to live a little better more thankful life than the lost. Or are we called to build in us a temple of HOLYSPIRIT'S great strength, being the great love and power of FATHER himself. There are hundreds of warnings in Scripture, but what does each soul hear. How can many hear differing messages; only when their flesh interprets Scripture to fit their core desires. Our beliefs can never align with GOD'S, until all our desires align. We can never know enough about FATHER, from whom life, love, truth, and all blessings flow. Within FATHER'S earliest spoken words, there was enough forceful purposed desire and singularities coming forth, for him to perceive and create the entire universe and all physical life. When GOD speaks his desires, countless worlds are born, and all creation are set in their motions by GOD'S desires. All motion of spin, rotation, and orbits, are set by his force of will. One can call it gravity or space curvature, but it is the forceful entangled projected will of FATHER upon his perceptions of creation. When JESUS spoke into the elements that made up the wind clouds and Sea of Galilee, in the midst of huge waves and fierce winds to be still, immediately the wind was calm and the sea was flat. This was seen by those with him, to be the greatest show of our LORD'S force, in all Scripture. They saw healings, and the dead again breathe and praise GOD; yet, at the sight of reality as far as their eyes could see, bowing to his spoken word, Scripture says his disciples became fearful of him. They had just then, barely glimpsed the conscious force that stood before them. How can we not feel fear at every thought of GOD. If we know with all our heart, that such a projected force of will and understanding does exist, who orchestrated the whole of the universe; we then know we must learn of him. And by this knowledge, we grow to love and fear him. We must appreciate that though we will never be fully omniscient like FATHER, it is our good work to always seek and live in all his ways. To know that GOD in their love, will then show us such things and wonders, as to strain our ability to understand. All praise worship and love for you FATHER almighty. In modern godless cultures, nothing is valued more than the hell and death of individuality and Satan's equity. How do you see yourself? Do you see an individual? And what does this mean? Do you enjoy having your own personality, and different beliefs, hopes, desires, styles, and all the other attributes, including your personal possessions that make you unique and separate. Do you think everyone being different and separate, makes life better in agreeing with the old adage, variety is become the spice of life? Do you want to be your own individual person, to such an extent forever? Do you feel resistance to the idea of believing, knowing, desiring, and living, in the eternal unchanging one absolute truth and perfect order of how GOD made us, and all things to be? JESUS is perfect, and has never once failed to be perfect in the eyes of FATHER. JESUS in love, *always* does those things that please FATHER. Scripture calls us to emulate CHRIST in all ways, being our example. But is this the life you want? Do you love, live, and honor the Seventh Day Sabbath, as set in perpetuity by FATHER? We have to understand that we cannot be an individual, we must put away every offence of this imperfect world. Are professed Christians striving to break free from this world, or are most living in denial as they hang on to it in love? Letting go is not an easy thing to do if doubt tells you that this might be all there is; but letting go is our true act of faith. It takes true faith to not look back, to keep going away from all the wrongs we were and knew; and keep going towards all we want and need to be and know. Our Messianic worship and life is meant to be other-worldly; and not of the flesh, but of GOD. A different path than all the powers and pleasures this world offers, encourages, and indulges. Are we called to reject this world and the things therein? Or has GOD told us to take all the things of this world, and put Christian labels on them, thereby falsely sanctifying them? Can we hold Christian lusts, and have Christian mansions, yachts, Christian corporations, theme parks, and cruises? Can we have Christian restaurants, karate, aerobics, maybe even massages? What if someone opens a Christian nightclub with alcohol, dancing, flirting? *All* of these are lies of pandering, adorning, and appeasing, the flesh while in the world, yet claiming to be spiritual. Jeremiah 3:10, "*AND YET FOR ALL THIS HER TREACHEROUS SISTER JUDAH HATH NOT TURNED UNTO ME WITH HER WHOLE HEART, BUT <u>FEIGNEDLY,</u> SAITH* ***LORD***." The eyes of the world are closed, their ears don't hear, their heart is cold to GOD, their Spirit is dying or dead. How great is all deception? Remember that Satan, the father of lies and deception, never works alone, Satan always finds his willing accomplice. Deception and temptation need someone wanting to be tempted and deceived. Each time a soul is deceived, they joined with Satan in their own deception. Can a soul live for all fleshly comfort and desires, while simply professing the name of the SON, and be found good before FATHER? Or by helping the poor out of one's abundance? That is the level of obedience and love for FATHER, JESUS accused the Scribes and Pharisees of living. Do any know what will be in Heaven? The Bible tells us, yet most seem to hold a personal picture of heaven in their mind. And as with all other things, their visuals are born of their physical desires. Can you truly say that if every one of your current desires for what Heaven will be like, are totally wrong, do you still want to go there for eternity? Or will you need a few worldly things to be happy? I assure you, that Heaven will be eternally provisioned, structured, ordered, and ritualized. Remember GOD'S children were brought out of Egypt? Being led by the great force of GOD, and being fed manna from Heaven which they rejected, wanting the foods of this world and their former bondage. A Heaven one imagines, is a heaven that soul would create if *they* were GOD. Heaven cannot be found by *anyone* wanting to be GOD, or coveting anything physical. It is in trusting GOD to provide what is spiritually good; because we can never know what is eternally good. Heaven is in our great trust, love, and non-expectations of a little child. And this love is reflected in our living by all GOD'S commands here and now. The flesh wants so much to be its own master, and most souls want so much to be their own ruler. And such souls want to define their own Heaven. Do you in your mind have such a picture of Heaven? Does this picture please you? Can you trust and give that up completely, holding to nothing you have ever seen, heard, tasted, touched, desired, or known, of this world? Can you trust FATHER prepares us a place of spiritual love, peace, and joy, through a more ethereal body that will physically sense far less, with a great spiritual love being held under one existential belief. Living with GOD means our existing, filled with *their* higher conscious feelings and understandings not of this world. Many visualize Heaven as filled with physical pleasures and all the people they ever loved. Most do not realize that in their grand imaginations, GOD is not even with them in their heaven. Are you disillusioned and disappointed by the temporal temporary things of this world? If so, why continue to have hope in these? Dear GOD, I believe you love us as we love you, and whatever you prepare is for our well-being and eternal peace. Amen. We can be sure Heaven is not what any of us would physically create; and this is good, for in our lack of omniscience, no matter how hard we try, we could not design or ever sustain a Heaven that is perfect and eternal. Why do you want to be with GOD in Heaven? Do you clearly see why a love for GOD'S one shared spiritual life, must be Heavens pure foundation? Do you want GOD for your imagined promise of an eternal quality of physical life; or for an eternal spiritual life? What would you want if GOD took you and set you in the farthest blackest vacuum of space, with no sound, sight, feeling of cold or hot, nothing but your soul's weak thoughts that you still exist with a hope that GOD loves and will return for you? How long could you endure this barest existence in knowing only your own limited thoughts, born of, and built from your memories, just waiting with nothing more than knowing GOD exists. Is this minimal existence of life, still a precious thing? Or would you wish you ceased to exist? Being set there in the lonely dark, is it the qualities of physical pleasure, comfort, and security you will miss; and not eternal spiritual love, being the true great gift from GOD? Do we show we trust GOD loves us by accepting his will? Do we accept the one common spiritual path set out for us all in Scripture, even if it leads to an unknown place? Do we accept GOD'S determined end to our existence on earth, or the life of another, when we want them to continue? Do we see life here is but a thought and a blink of the eye, and what we physically endure or receive here, or how long we are here, matters not? The only value taken, is what we learn about GOD from it all. Matthew 6:31-34, "*THEREFORE TAKE NO THOUGHT, SAYING, WHAT SHALL WE EAT? OR, WHAT SHALL WE DRINK? OR, WHEREWITHAL SHALL WE BE CLOTHED? (FOR AFTER ALL THESE THINGS DO THE GENTILES SEEK:) FOR YOUR HEAVENLY* ***FATHER*** *KNOWETH THAT YE HAVE NEED OF ALL THESE THINGS. <u>BUT SEEK YE FIRST</u> THE KINGDOM OF* ***GOD****, AND HIS RIGHTEOUSNESS; AND*

ALL THESE THINGS SHALL BE ADDED UNTO YOU. TAKE THEREFORE NO THOUGHT FOR THE MORROW: FOR THE MORROW SHALL TAKE THOUGHT FOR THE THINGS OF ITSELF. SUFFICIENT UNTO THE DAY IS THE EVIL THEREOF." Do you know all the material things of this world, are GOD'S? Life here, is a tiny flicker in the whole of all eternal realities. How can we place such importance on the qualities and quantities of something so short; something that without doubt will be gone forever as if it never was? Why is it so hard for most believers to live towards eternal life instead of being consumed by their short temporary physical life? The true answer rests in one's desires and lust for this world. We cannot have eternal trust in GOD without eternal faith; or any eternal faith without eternal trust. And we cannot have either without eternal love for GOD. So, love, trust, and faith, are entangled as one. These must be our foundation and strength for the soul. How much comfort and assurance in your life comes from your job, home, family, friends, health insurance, vehicle, 401k, social security, and a few months groceries in your pantry. If you lost *all* those things, not knowing where you would sleep, or what you would eat or wear. Would you still trust in GOD and know they are there? Is GOD literally your greatest comfort, love, trust, and faith. JESUS warned of being possessed by worldly possessions. Do we truly mean it when we say, all that exists in our provision is GOD'S; or is this just lip service to GOD. We are to be Godly Stewards and nothing more. GOD wants to, and knows our heart, mind, and soul. Do you remember being a child on the edge of a dividing wall or a pool; and your dad saying jump, I will catch you. FATHER says jump, but as long as we have not jumped, we remain in and of this world. Is it the same to only believe FATHER will catch us, and take joy in that? Or does it mean more to jump? What if our belief is strong, but our desire to stay where we are is stronger? Can we try to say we have both? How great is the comfort of staying on the edge, and how great is the joy when we jump and are caught safely in the arms of FATHER above? Does our *not* jumping, indicate a weak love for FATHER? We are now in an evil tempting physical world; and if we don't understand that, how can we come out and avoid it? We hear people ask so often, that if GOD exists, why doesn't he stop all evil and pain? Such a worldly unknowing question to ask. We can never know the finer line of the self-imposed limits upon GOD, concerning their intervention in our free-will lives, for the reasons and purpose of allowing our desires and love to flourish. GOD'S self-imposed restrictions, extend from the depths of his wrathful destructive fire, to his

Holy Throne of mercy. In all of this, we know FATHER will preserve the perfect balance of justice for all that have ever made their choice. GOD very rarely stops a soul from fulfilling the free-will evil desires of their heart. These are bright lines GOD will never cross except to enhance and protect their Kingdom. And from the other side, to what extent does GOD permit Satan and all evil to accuse and try our faith and love, and interrupt our life? But a more interesting question is, does Satan even bother? If you are not under true spiritual attack, you must wonder why? We are not talking about every bad physical thing that happens to us all in this world full of chance, greed, and pride, which bring us worldly problems. Spiritual attacks are different, often having little to do with physical problems. A spiritual attack is when devils seek to insert doubt, confusion, and weakness, in our relationship with GOD. So, if you do not currently know and accept FATHER'S full truth, there is no current need for Satan to attack you, as you are currently defeated. If you cannot see the fact that FATHER created all things for *his* loving eternal purpose, you are confused. FATHER did not create anything physical for mankind's *fallen* temporary desires. And if we love GOD above all else, their joy is our joy also. So, in not being drawn into rebellion; and looking beyond this, we see only a loving

Holy GOD we trust, and we have no worries. If a soul doubts GOD, it is toward their self-destruction. We can know that all GOD has done, and *will* do, is for spiritual wellbeing first and most, and our physical platform, last and least. We see in Scripture a multitude of miracles at the Sacrificial coming of our KING. These were to show the world that it was GOD who had come down to correct and restore them. Those miracles proclaimed JESUS to be FATHER'S SON, and heralded FATHER'S intended mercy toward all souls who would and could receive it. But do not assume such physical intervention are available to everyone at all times. GOD heals or intervenes at times for eternal reasons and purpose. We do not say, don't pray for the sick and the lost. We do say that in the current apostate Church, following false doctrines, GOD'S purpose is scarcely seen; and so is GOD'S intervention. A sick Church goer with health insurance, runs in *fear* to the modern doctors, and when those doctors, surgeons, and chemicals, trade health for money, and fix their problem; they thank GOD and claim the mercy of GOD was upon them. This, as the poor around the world who profess GOD, minus doctors and medicines, die of the same disease. We cannot sin in this fallen world, and claim GOD is making our life unfold. I heard a minister from a pulpit say something unusual happened to him; he then said, but we all know there are no such thing as accidents or coincidences. He lied, standing there under, and wearing GOD'S Mantle of authority, teaching that all good and evil things happen by the hand of GOD. We hope his confused soul might see, and stop leading others into hell. All creations came from FATHER, and all physical guidance of life flows from him. But the runaway world is now full of chance and free-will conscious agents, who make their own choices for GOD, or this world. Let us choose truth, and strive on to perfection; do not stop where you are, never get comfortable. If we have faithfully and responsibly run towards GOD for a hundred years, we are not there yet. Not until we are sealed and hear; enter thou good and faithful servant. We must always see ourselves as not being a great deal closer than the day we first cried out to GOD for forgiveness. Never think you are safe from yourself and a desire for evil that may arise within you. If we have walked in perfection for a long while, Satan is laying pit falls, trials, and temptations, before us. But Satan cannot defeat us without our surrender to doubt or temptation. We must know GOD, and truth, and we must know what is sin. When we say *"know"* what does this mean? Some people think if they have something memorized, they know it. But we are sharing here the strongest possible meaning. We mean it's entire quality and quantity are part of us down to our core. We mean we can no longer be separated from this knowing in any way, as it is now a part of who we are. It is a part of life itself, being learned thru trials and spiritual revelation over and over, which can never be denied by any means. Do you know GOD like this? And when we say *GOD*, what do we mean? For there are beliefs in many false gods. And if you don't like any of those, you can make up and finely tune your own god to suit your desires, as many souls do. But here we are talking only about the one true GOD of the Holy Scriptures, who most do not know intimately. To *know GOD*, we must be introduced through HOLYSPIRIT in long learning, we must be shown by HOLYSPIRIT the great vast entangled qualities and quantities that declare the sum of who and what is GOD. We must worship FATHER, but only according to his will. And then add to that, all the true physical understanding of creation we are willing to see. FATHER is the over-arching protector of all who are precious to him. FATHER is the final destroyer of all forces that threaten his family and perfect order. FATHER is the author of eternal love and truth, the absolute ultimate. FATHER is the seat of all truth and wisdom, the unchanging origin of Righteousness and Holiness. To truly know a person, we live a great moment looking at what they do, their work, and what they love. We consider also all they have done, and add to that, their words, actions, and evident Spirit. And by this we come to know a soul as best we can. So, study all the evidence of FATHER'S works and deeds, and all revelations from HOLYSPIRIT who interprets and teaches. Through these we can know GOD more, each of us to the best of our desires, and efforts. And how can we know and interpret sin? Do we look among each other for a worldly consensus? GOD forbid. And by what standard do we see that upon our being tempted, we have also lusted? Lust, meaning to truly want. Well, if a temptation returns often, we are surely lusting for or coveting that thing in our heart. But in this world, the tenth commandment has endless weak interpretations. How does your soul interpret lust, when you read; "*thou shalt not covet*"? We all know it has to do with wanting, but wanting what and how much is where the number of interpretations increase. And if mankind holds differing interpretations, it is because they hear their interpretation through their flesh, and not from GOD. Can we know what is sin by looking among each other for a consensus? Of course not. But this is what is being done in this world; and in the modern godless Church. We are given from above that in the life of JESUS, we see the ways of Righteousness to emulate. We define sin, as the things

CHRIST would not do. If JESUS wanted it, it was not covetous. And if JESUS refused it, it is sinful. In JESUS, we saw not only did he not want the things of this world, he did not *once* take to himself the power to choose what he needed. FATHER provides all his Children's needs, including his only begotten SON. To desire one thing more, is taking that right to decide what we need, when that is the rightful responsibility of FATHER. Scripture says, with simple food and raiment, be content, 1st Tim 6:5-12, "*PERVERSE DISPUTINGS OF MEN OF CORRUPT MINDS, AND DESTITUTE OF THE TRUTH, SUPPOSING THAT GAIN IS GODLINESS: FROM SUCH WITHDRAW THYSELF. BUT GODLINESS WITH CONTENTMENT IS GREAT GAIN. FOR WE BROUGHT NOTHING INTO THIS WORLD, AND IT IS CERTAIN WE CAN CARRY NOTHING OUT. AND HAVING FOOD AND RAIMENT LET US BE THEREWITH CONTENT. BUT THEY THAT WILL BE RICH FALL INTO TEMPTATION AND A SNARE, AND INTO MANY FOOLISH AND HURTFUL LUSTS, WHICH DROWN MEN IN DESTRUCTION AND PERDITION. FOR THE LOVE OF MONEY IS THE ROOT OF ALL EVIL: WHICH WHILE SOME COVETED AFTER, THEY HAVE ERRED FROM THE FAITH, AND PIERCED THEMSELVES THROUGH WITH MANY SORROWS. BUT*

THOU, O MAN OF GOD, FLEE THESE THINGS; AND FOLLOW AFTER RIGHTEOUSNESS, GODLINESS, FAITH, LOVE, PATIENCE, MEEKNESS. FIGHT THE GOOD FIGHT OF FAITH, LAY HOLD ON ETERNAL LIFE, WHEREUNTO THOU ART ALSO CALLED, AND HAST PROFESSED A GOOD PROFESSION BEFORE MANY WITNESSES." Scripture conveys the will of GOD, that this is the way we should live. And if one should say, I *need* more than this, their sin is that they do just as Satan did, and sit in the seat of GOD. Let us never contend with each other over what is sin, look only to our example in JESUS and Scripture as to what is sin. GOD and their words will judge. We need to purify ourselves as commanded. Think of a chunk of ore from the ground, containing about a tenth of one percent copper. Copper is a good conductor of electricity, but no current passes through the ore because of its ninety-nine-point-nine percent impurity. So, can HOLYSPIRIT dwell in and work through an impure vessel? Do *you* in all your sincerity have any desire or expectation of becoming perfect and sinless in this world? If not, how is this not your great purpose? Lose the part of you that is covetous and impure, and be filled with HOLYSPIRIT. GOD cannot be pleased in our simply saying they are the most important thing in our life, but only in seeing that their and our ways and beliefs are become the same. This is the eternal anchoring loving purpose of the Holy Sabbath to hold us all together as one. We see FATHER must do many provisional works towards all of his Children in love. In all his Children's created weakness, FATHER is watching, caring, and doing all he can within his holiness. FATHER sanctified the Seventh day to be a day in which we, showing our love for him, are most diligent that we allow him to rest, as we rest in FATHER. Have you not read how FATHER is wearied, grieved, and wroth with trying to gather his Children. And on the Holy Sabbath, FATHER sees who his true Children are, as we most diligently turn from the world. And as a little child we climb into his lap, secured, resting, and gazing outward. In this, we are not just allowing FATHER his greater rest for that day, we enter into his rest with him. A day of solace, when FATHER calls his children by name, and holds close those who love him most. All his children wanting him to rest, and wanting to rest in his love and promise from their battle and struggles in this world. Love the seventh day of rest, for FATHER is there, and he has said *he* is *not* found by those forsaking his rest. Heb. 4:1, "*LET US THEREFORE FEAR, LEST A PROMISE BEING LEFT US OF ENTERING INTO HIS REST, ANY OF YOU SHOULD COME SHORT OF IT.*" We read in Isaiah

44:6, last sentence, "*I AM THE FIRST, AND THE LAST; AND BESIDE ME THERE IS NO **GOD***". Isaiah 44:8, last paragraph, "*IS THERE A **GOD** BESIDE ME? YEA, THERE IS NO **GOD**, I KNOW NOT ANY*" Isaiah 45:5 "*I AM THE **LORD**, AND THERE IS NONE ELSE, THERE IS NO **GOD** BESIDE ME*". We know GOD is now a Trinity of souls, being FATHER, his only begotten SON, and his HOLYSPIRIT. FATHER was and is the first, and all gifts of awareness and all creations, including all conscious and physical life itself are come forth from the soul and physical aspects of FATHER. All physical things are created by FATHER'S will upon the singularity points of his body's essence; and they consist in their complexity by his patterned forceful desires. We know FATHER can create, or perceive, all physical things; but not love. Which by GOD'S definition is our strongest devotion of desire towards another, and a choice made within the heart of each gifted free-will soul. FATHER cannot make others love him; but he can and did create a way for love to be born and endure. By FATHER being loving, he designed all things in the great complexity we see for the purpose of choice alone. To hopefully bring to him a loving holy righteous family, in whom he can trust and love, and with whom he can share all he is and has. A family that will always love and trust in him, and give all they are back to him. Always remember, that though it is nice to say and hear I love you; our love in its far greater essence, is not in what we say, it is in our faithful responsible way we live, show, and give our love. And GOD is never fooled in the least by anyone at any time, even if we fool ourselves. We can easily fool ourselves and each other about how much we each love GOD and others. But GOD sees by our every desire, word, and action, exactly what and who we love, and the very measure of our love for each. By this, GOD knows who searches for him and his ways like treasure, and who he trusts to live in his eternal Kingdom as one family. Let us look more to this great purpose of creation, and what the creator of all things has done to this point. From his first act of creation, of which we have his written chronology, FATHER has moved in one direction with one purpose, being to love and be loved back in an eternal, faithful, responsible, sustainable way. Scripture says we are created in the image of FATHER, but *our* highest potential can never compare to his. We know FATHER by his own desire, has taken an ancient wise loving spiritual and physical semblance. And the basic conscious and physical aspects of all eternal souls were gifted and fashioned in his basic image to share in his ways of love. FATHER has a face that shows he is pleased or not pleased;

eyes that shine in love, or peer in displeasure; a mouth to speak in loving tones, or stern warning. These with which we know he sees, hears, and speaks to us. FATHER perceived himself a body with arms hands and legs to hold his Children and creation within his life and his love, and to walk among us in pleasure, or leave us at our rebellion. Isaiah 45:12 "*I HAVE MADE THE EARTH, AND CREATED MAN UPON IT: I, EVEN MY HANDS HAVE STRETCHED OUT THE HEAVENS, AND ALL, THEIR HOST HAVE I COMMANDED*". Genesis 3:8 "*AND THEY HEARD THE VOICE OF THE* ***LORD GOD*** *WALKING IN THE GARDEN IN THE COOL OF THE DAY: AND ADAM AND HIS WIFE HID THEMSELVES FROM THE PRESENCE OF THE* ***LORD GOD*** *AMONGST THE TREES OF THE GARDEN*". JESUS said in John 14:9, "*IF YOU HAVE SEEN ME, YOU HAVE SEEN THE* ***FATHER***." To spiritually and physically see JESUS, is to see a soul and body of love and truth, born of and come forth out of FATHER. And being one in love, FATHER'S ancient loving appearance, is reflected as younger and smaller in JESUS. We know FATHER has all the loving emotions of compassion, hope, joy, jealousy, hurt, zeal, disappointment, righteous-anger, and wrath; these, and all the emotions found in giving, receiving, and protecting our love.

These loving conscious-physical aspects of FATHER, are also gifted to us in a far, far lesser smaller measure. Not enough thought is given by mankind to the fact that eternal love and family must be protected from every evil, and FATHER will do this above all things. We see sinners are destructive to GOD'S love and family. And seeing there will be eternal free-will in Heaven, do you think there will ever be free-will sin, or sinners, allowed into the new Heaven? There will not. GOD through the whole of the Bible says if one cannot quit sinning while here in their love for GOD; one cannot quit sinning anywhere in their same love for GOD. Lucifer and 1/3 of the angels sinned and fell *while in* Heaven, which is now mired in war. We know by simply looking at the example of the earth, that sin spreads and increase like a virus or a cancer. Would you allow sin or a sinner into your home, knowing it would spread to destroy your family? And beyond all doubt, it does, and it will. Why do souls sin in *this* world? Because it is their strong desire in this time of their choosing. This serves love's eternal purpose, in that it is free-will agency which separates the eternal love for GOD in the next world, from the fallen love of this current world. It is a fact that this greatest polarity of all abiding forces called eternal love, cannot come into existence without its antithesis. Is it only by the probabilities of free-will desire, and a true real choice between two great desires, that a more desired love can be truly known and shown. Only by our truest desire, can eternal love truly exist. The definition of evil is to desire and choose any words, actions, things, or thoughts, that do not perpetuate the perfect order of GOD'S creation in his loving ways and desires. If a desire, act, thing, or word, does not sustain, or increase the Kingdom of GOD'S love, it is evil. Everything is on one side or the other, there is *no* middle ground, *no* gray area, and there is no benign choice. Do not be defined as evil by GOD. Every *evil* word, act, thing, and desire, whether by knowing intent or not, all work against the Kingdom of GOD. Scripture says that without faith, we cannot please GOD; but what is this faith GOD seeks. Our faith is our desired hopeful belief, without absolute proof. But every-thing that is believed by everyone is accepted in faith. Every scientific understanding, declares no one has any absolute empirical proof of any one tiny single thing, because our perceptions of reality are all second-hand here-say biased creations of evidence. So, the faith GOD seeks, starts at this question; do we want GOD to be there more than anything else. At a point in the life of every soul, we ask and answer certain questions. One, do we believe this reality and life could all just happen; or was it designed and created by a knowing conscious forceful awareness. If we see it must have been designed and created, we then ask by whom. If one leans strongly towards love, and hears GOD'S call, and reads and studies Scripture, yet does not desire what it says, it will be rejected, or re-interpreted to suit that individual heart. Any soul can say a verse means something less or more than it does, and form a biased interpretation; this is hated by GOD. But if we see and love the sweetest brightest eternal beauty, and hears GOD'S loving voice; and it breaks our heart that GOD wants us back. We then hear the sounds of eternal life itself, and we truly long for the ways of GOD the world long ago rejected. At that moment, our faith, GOD seeks is born; but it is new, weak, and in peril. This new young faith of love is a tender bloom in need of great care and nurturing. If this great desire does not fade within, if our gaze is steadfast upon GOD, their nurturing is forever there beyond measure and without fail. Faith in, and love for GOD, are inseparable in all their children; and great love and faith in GOD cannot fail. I truly and deeply believe in GOD, but why? Above all the abundant evidence I see, it is because I so *love* what I have been shown. And though, as of yet, I have not physically seen or touched CHRIST, I truly know and love GOD, but why? It is because I so *believe* what I love. Faith holds love, and love holds faith, they are bound in the moment of their adjoined birth. Faith and love both grow, or die, together. Although we have not yet seen, we, out of desire for knowing and having a higher love, truth, and life, choose to devote ourselves to search for the one we worship and adore in all our hope. By maturing in HOLYSPIRIT'S nurturing, we learn better the lesson that true love and life, only come from a loving Creator. By continuing in this faith, we are led by love, and turned from the unbelieving world to seek GOD'S ways. Not seeing, touching, or knowing, his completeness, yet we wait on FATHER, not knowing *all* we wait for. There is this faith and love within our hearts on one side pulling us towards GOD. Yet echoes of temptations for things of this world, can pull us the other way. Temptations in the flesh can be strong; and sadly, they do overcome the faith and love of many. These have a love for GOD, but sadly their love and desire for GOD is not their greatest desire. Many asks, why are temptations for physical pleasure through mankind's fallen physical nature, set before us. Seeing mankind's strong desires for these are what brings all rebellion, that causes all pain, suffering, and death, causing many to fall from the faith. This is a tough obstacle for many to overcome and accept. Many ask why GOD did not create a family with a perfect nature and no temptation. Well, before we wonder further, we always accept GOD'S will, knowing all things in creation must have relevance one to another. And in knowing GOD'S thoughts and ways are infinitely higher than ours, we trust that if there were another way to eternal love, it would be so. Having said that, can FATHER do all things with his physical creation? Can he do all things he should desire, even with our free-will conscious soul? He could take control over the body and against our will make the body do and say anything he wants for eternity, but that would not be us. And GOD does not want to instill or force his desires upon us. Where in that, is found our love for FATHER he seeks and needs. This is the question, seeing real love requires absolute free-will, and our *true* choices born of our desires. We see GOD is under a multitude of self-imposed restrictions; and if these restrictions were removed one by one, each would reduce our free-will choice for love, as it reduces all free-will evil. We see *why* GOD will not do many of these things. GOD above all things, desires to find eternal love. Love cannot be engineered; it can only be allowed and nurtured. Can we see any other path to a great free-will, heart-breaking love for GOD, which must be born out of true hope and desire. If we had no true choice against GOD, or a robotic nature, we would be making only programmed choices, lacking any evidentiary desire of love. Our great wondering, and the mystery of GOD within our hearts would disappear. Our loving want, and desire to know GOD would be gone. Even our hope would disappear, for we hope for things in our not seeing or having them. In being born without knowledge or understanding, we see that it is truly our love, appearing through our learning and desires. By having a free-will of desire, we make our choices out of our rising wants within our soul. Since the fall of mankind, it is still FATHER'S plan and design to try and restore all who truly love, need, and fear him. We all see the great horror and destruction of imperfect choices made by the multitudes of mankind, stemming from their worldly physical desires. And these are their choices which come by not looking up in any spiritual desire for GOD, but by embracing their *individual* physical nature, and beliefs of their own design. But GOD'S joy comes to the fewer who do not embrace imperfection, and overcome the imperfect nature of the flesh by the forces found in their eternal love for GOD. This is attaining *our* perfection through emulating CHRIST. Every tiny choice we make in this life, is choosing between our love for our Creator, or our love for his Creation. Any influence or restrictions by GOD, causing us to make only perfect choices, would deny and destroy all choice for us. These people who wish GOD would take the choice away from them to sin, refuse to stop these sins on their own, because their sins are what they want and love. They need to define their own way to live, and deny there is any divisive aspect or nature in those ways. If you think all of mankind can live together in some form of *non-violent* coexistence, under all broad allowance of conflicting spiritual and physical ways to live; you are delusional. I promise you, there will be endless rising disagreements leading to constant violence. All conflicting spiritual physical ways and violence will bring a great downfall. If one closes their eyes to the *eternal* realities, one may think such ways can work in a short-lived *temporary* imperfect worldly sense, but never eternally. In seeing eternal love and peace *grow* out of our desire and learning of FATHER, please consider that if he could have gifted us souls with a pre-loaded omniscient understanding, we would again lose all learned, desired, and chosen love. We would lose our breaking heart, felt in GOD'S forgiveness of our past sins. If we were gifted to only make perfect choices, we could never see or know FATHER in the loving merciful way we do. Having received and felt GOD'S love and mercy, who could bear to lose such a gift and sign of true love from above. So, we see the essence and felt evidence of our love, must come at a cost. After mankind's fall, it was a redemptive part of GOD'S governing equation, to allow every soul to be led and find their costly way back to them. All cost, pain, and suffering, brought by our former imperfect will and imperfect knowledge are a consequence that just must be known and overcome, there being no other way. It makes all perfect sense, that if we *all* start together in a programmed omniscient, we all end up mechanical and designed, lacking love, desire, and hope, towards GOD and each other. But the wonder of wonders is, if we all start as a blank slate, we can learn, know, and overcome imperfection to find eternal love in emulating CHRIST. We can grow in love towards CHRIST, as Scripture commands, which is our resurrection and ascension into the bright light among the Holy Angels. But what about darkness? Look back and see the once certain perfection of creation, before there was any evil. GOD said to Lucifer "*THOU WAST PERFECT IN ALL THY WAYS, UNTIL INIQUITY WAS FOUND IN THEE.*" When Satan, because of his beauty and abundant perfection said; "*I WILL BE LIKE THE MOST-HIGH*" Does this not show that even a soul with a more perfect knowledge, can still succumb to an imperfect logic of self-desire, believing one's self equal to FATHER? Yes. So, we see receiving perfect understanding alone cannot protect us, it is our combined growing love, faith, and true knowledge of

GOD that bind and protect us. Rather than focusing on the individual rebellious events that bring pain and suffering, we should focus on the eternal outcome of GOD'S long endured work, so eternal love which gives life its eternal meaning, can exist in us. We know, and do not question why such painful things just must be. If we all freely followed GOD'S created ways in love, there would be no pain or suffering, and eternal love would be known and shared by all souls, bringing an eternal purposed reasoned conclusion. In all of this, we also try to illustrate the great point, that though FATHER can do so many things. We see he has chosen a path of his own painful suffering, self-impositions, and sacrifice, required to allow the existence of free-will love. We see *why* FATHER won't do all things; including for a while longer, not stopping all evil. We should still find great joy and take heart in that though we do not see FATHER stop all evil yet, he has promised he will soon stop all *new* evil and pain. And heal all things done to all his children who suffered unjustly for eternal love's sake. Revelation 21:4-5 "*AND* ***GOD*** *SHALL WIPE AWAY ALL TEARS FROM THEIR EYES; AND THERE SHALL BE NO MORE DEATH, NEITHER SORROW, NOR CRYING, NEITHER SHALL THERE BE ANY MORE PAIN: FOR THE FORMER THINGS ARE PASSED AWAY. AND HE THAT*

SAT UPON THE THRONE SAID, BEHOLD I MAKE ALL THINGS NEW. AND HE SAID UNTO ME, WRITE: FOR THESE WORDS ARE TRUE AND FAITHFUL". As for the world's innocent little children who suffer… Matthew 18:10, "*TAKE HEED THAT YE DESPISE NOT ONE OF THESE LITTLE ONES; FOR I SAY UNTO YOU, THAT IN HEAVEN THEIR ANGELS DO ALWAYS BEHOLD THE FACE OF MY* ***FATHER*** *WHICH IS IN HEAVEN*". And Matthew 18:14 "*EVEN SO IT IS NOT THE WILL OF YOUR* ***FATHER*** *WHICH IS IN HEAVEN, THAT ONE OF THESE LITTLE ONES SHOULD PERISH.*" Knowing all physical things in this world are now a temporary vapor, let us always look far beyond what our eyes can see. It is our great joy that FATHER promises at the fulfillment of this moment of our choosing, he will bring his Children into his Kingdom forever. Till then, FATHER says if we are attacked in this world for loving his SON, we suffer in the name of our LORD, we suffer for the cross, we suffer for righteousness sake, and Heaven's sake. And as CHRIST'S followers, we get to lay our suffering at his feet, being the good gifts of our love. Surely, we must consider in our hearts that we have not begun to see or feel the historic totality of suffering that has been endured for love to exist. Let us consider why GOD must suffer so much for love to exist.

GOD has so labored to find eternal love, and endured so much unfaithful irresponsible rejection, blasphemy, ridicule, hatred, mocking, and cursing accusations towards them; and even the lukewarm religious embrace of so many who still rebel against FATHER'S ways. FATHER has offered and given all he has in the greatest of all his gifts, his SON, yet that gift receives the greatest of all rejections. FATHER has felt and endured not just the pain of all mankind's evil against *him*, he has felt the pain that his SON and all his children have felt in their steadfast faith throughout all history. Who can know the sum of the pain that has over the moments of all creation, been felt by GOD. Such suffering inflicted upon them by the rebellious, that if we could achieve the impossible, and comprehend and hold the sum knowledge of all the pain felt by GOD in their hearts because of us all. If we could hold even the smallest part of this pain in our own heart; that pain, fully understood and felt by us, would bring in us a rise of guilt and sorrow of such a magnitude for what we have all done against GOD and life. We each would fall to the ground in pain of a never-ending broken heart we could not begin to bear. We could never again draw a breath, for the pain in our throat and chest welled up from the hurt of our shame. If it were possible for us each to know the pain and suffering GOD has endured by, not every soul combined, but by us each alone. We would cry unending, until the very last drop of moisture from our body is forced from our last dry tear duct, leaving us as dust among the winds. *And* before his gifted souls, and the physical creations were brought, FATHER was one soul, utterly alone in a *painful* loneliness no one can imagine. We should not ever ask why we, or why do any have to suffer, or witness suffering. It is all for love, it is because of sin and rebellion, it is because fallen angels and mankind can't desire, and refuse to love and live in FATHER'S created eternal ways. All of these are why GOD has suffered so much. And why does love itself suffer? Again, each sin is an act of irresponsible unfaithfulness towards GOD, in being unfaithful to the bonds of Holy Marriage. How can we love, yet cause hurt by being unfaithful to love? Does it not amaze us all, that in seeing the state of this world, that GOD'S mercy is still here at all? Being an extension of grace to many yet alive. Do we truly know that our physical lives are not our own, and truly strive to be one with the eternal purpose of GOD? Or do we truly live a physical life, like it is our own? By our beliefs and desires, do we only allow GOD into *our* life to the extent that *we* find the time and place which pleases us? Let's us each seek our perfection towards GOD, which is to live the eternal ways of GOD. If we strive to live by every command, and yet stumble in error or weakness, then shame must reign within us for it. We must humble ourselves seeking mercy, being broken and in a true repentance; thereby, having remained fully within the bounds of perfection as they are set towards us. For in all true fashion, CHRIST'S great sacrifice and provision of restoration, is of the *perfect* commands and love of FATHER. So, forgiveness being born in CHRIST'S *perfect* work of salvation, is set forth as a perfect work of FATHER to be obeyed; and we must obey this always. A certain *defined* line of forgiveness and restoration has become in part, with the perfect works of CHRIST towards mankind, and mankind towards CHRIST. Be ye perfect in it, for we are called to such things in Sacrificial Love. CHRIST shall not, and did not die in vain. Yet it shall always be beyond the understanding knowing of mankind, as to what are the true boundaries of GOD'S enduring mercy. The Scriptures express there are boundaries to GOD'S compassion; yet to precisely know these lines is beyond us. Be amazed by this compassion, never abuse it, or try its patience. Live in knowing that in our obedient love, is GOD'S promise which will never fail. Amen

How can we know the HOLY SON of GOD

Most souls have heard the story of the prince who put off his royal attire to hide his status, wealth, and crown. And travel to a faraway land appearing as a pauper, to find a bride who loved him for who he is, and not for what he had, and can give. And for great loving reasons, this is that original story. A greatest of all possible loves, was the sacrificial love of CHRIST, sent to span and mend all spiritual divides. The Holy Sacrifice of CHRIST is the greatest price for love ever paid in all reality; a price every combined good word, written and or spoken, fails to adequately describe. CHRIST'S life is also the highest price that *could ever* be paid for anything, and it was the only price that could bridge the great gulf opened up between the perfection of GOD and our lost souls; a gulf none can cross alone. What does a bridge do? It allows us to cross from where we are, to where we need to be. The SON came down from on high, coming down so far and so low, putting off his immortal body, and leave his Heavenly Throne. Why? So that JESUS'S soul, could be adjoined to the flesh of mankind, growing within the womb of Virgin Mary, using DNA from the loins of King David, put in her by HOLYSPIRIT. The SON of GOD came so close to us, to walk in our weakness as mortal flesh and know our physical experience. The HOLYSPIRIT was withdrawn and withheld from JESUS for about thirty years as he was tried and felt and withstood all worldly temptations. In this, CHRIST achieved experiential knowledge of our trials of existence. None can say to JESUS, but you don't know what it is like to be mortal in the flesh, and be tempted. CHRIST extends his Holy forgiving hand, to save all who would grab hold in true love, to rise up out of the depths of hell and death. To cross over his bridge to the higher perfect order, ways, and purpose of GOD. Through his love, JESUS, bearing a weight more than the whole of creation, paid a painful price that all the Holy Angels and all who ever drew breath combined, could not pay. All praise, love, honor, and glory, to YESHUA. Amen. JESUS is many faceted and beautifully described in Scripture as FATHER'S only begotten SON, second of the GODHEAD. JESUS is the picture of the perfect eternal love, both from and for FATHER. JESUS was the first eternal gifted soul from FATHER. So, CHRIST and then HOLYSPIRIT were at FATHERS side, for all his other gifts and creations. JESUS was asked, and obeyed, and came willingly into this fallen world because of *our* sin. He became as we now are, relying on the flesh to sense his reality. He became humbled, powerless, lonely man, walking without HOLYSPIRIT'S indwelling for about thirty years, having only his loving trust in FATHER. Walking sinless, subjected to all the hateful struggles and temptations of this world, and King David's past down proclivities. JESUS was tempted as a man by the flesh, the world, and Satan. He was also tempted by the Godlessness he saw in mankind, to turn and let us all die in our endless sins. Tempted beyond human endurance, yet he overcame every temptation from Satan, the world, and mankind. One of the greatest Chapters in the Bible, is JESUS speaking to all, of who he is, and his eternal purpose. Telling of his nature, his character, his responsibility, and the place of honor set upon him by FATHER, and more. This beautiful telling of JESUS coming into his own gifted existence, in his being divided out from, and born of FATHER'S soul. This occurred before any and all other creations. JESUS declares that for our every sin, our salvation rest in him alone. Read this short Chapter slowly; and remember that throughout Scripture, JESUS often calls FATHER, his GOD or LORD. JESUS speaks of himself, to Solomon in Proverbs chapter 8 *"DOTH NOT WISDOM CRY? AND UNDERSTANDING PUT FORTH HER VOICE? SHE STANDETH IN THE TOP OF THE HIGH PLACES, BY THE WAY IN THE PLACES OF THE PATHS. SHE CRIETH AT THE GATES, AT THE ENTRY OF THE CITY, AT THE COMING IN AT THE DOORS. UNTO YOU, O MEN, I CALL; AND MY VOICE IS TO THE SONS OF MAN. O YE SIMPLE, UNDERSTAND WISDOM: AND YE FOOLS, BE YE OF AN UNDERSTANDING HEART. HEAR; FOR I WILL SPEAK OF*

EXCELLENT THINGS; AND THE OPENING OF MY LIPS SHALL BE RIGHT THINGS. FOR MY MOUTH SHALL SPEAK TRUTH; AND WICKEDNESS IS AN ABOMINATION TO MY LIPS. ALL THE WORDS OF MY MOUTH ARE IN RIGHTEOUSNESS; THERE IS NOTHING FROWARD OR PERVERSE IN THEM. THEY ARE ALL PLAIN TO HIM THAT UNDERSTANDETH, AND RIGHT TO THEM THAT FIND KNOWLEDGE. FOR WISDOM IS BETTER THAN RUBIES; AND ALL THINGS THAT MAY BE DESIRED ARE NOT TO BE COMPARED TO IT. I WISDOM DWELL WITH PRUDENCE, AND FIND OUT KNOWLEDGE OF WITTY INVENTIONS. THE FEAR OF THE ***LORD*** *IS TO HATE EVIL: PRIDE, AND ARROGANCY, AND THE EVIL WAY, AND THE FROWARD MOUTH, DO I HATE. COUNSEL IS MINE, AND SOUND WISDOM: I AM UNDERSTANDING; I HAVE STRENGTH. BY ME KINGS REIGN, AND PRINCES DECREE JUSTICE. BY ME PRINCES RULE, AND NOBLES, EVEN ALL THE JUDGES OF THE EARTH. I LOVE THEM THAT LOVE ME; AND THOSE THAT SEEK ME EARLY SHALL FIND ME. RICHES AND HONOR ARE WITH ME; YEA, DURABLE RICHES AND RIGHTEOUSNESS. MY FRUIT IS BETTER THAN GOLD, YEA, THAN FINE GOLD; AND MY REVENUE THAN CHOICE SILVER. I LEAD IN*

THE WAY OF RIGHTEOUSNESS, IN THE MIDST OF THE PATHS OF JUDGMENT: THAT I MAY CAUSE THOSE THAT LOVE ME TO INHERIT SUBSTANCE; AND I WILL FILL THEIR TREASURES. THE ***LORD*** *POSSESSED ME IN THE BEGINNING OF HIS WAY, BEFORE HIS WORKS OF OLD. I WAS SET UP FROM EVERLASTING, FROM THE BEGINNING, OR EVER THE EARTH WAS. WHEN THERE WERE NO DEPTHS, I WAS BROUGHT FORTH; WHEN THERE WERE NO FOUNTAINS ABOUNDING WITH WATER. BEFORE THE MOUNTAINS WERE SETTLED, BEFORE THE HILLS WAS I BROUGHT FORTH; WHILE AS YET HE HAD NOT MADE THE EARTH, NOR THE FIELDS, NOR THE HIGHEST PART OF THE DUST OF THE WORLD. WHEN HE PREPARED THE HEAVENS, I WAS THERE: WHEN HE SET A COMPASS UPON THE FACE OF THE DEEP: WHEN HE GAVE TO THE SEA HIS DECREE, THAT THE WATERS SHOULD NOT PASS HIS COMMANDMENTS: WHEN HE APPOINTED THE FOUNDATION OF THE EARTH: THEN I WAS BY HIM: AND I WAS DAILY HIS DELIGHT, REJOICING ALWAYS BEFORE HIM; REJOICING IN THE HABITABLE PART OF HIS EARTH; AND MY DELIGHTS WERE WITH THE SONS OF MEN. NOW THEREFOR HEARKEN UNTO ME, O YE*

CHILDREN: FOR BLESSED ARE THEY THAT KEEP MY WAYS. HEAR INSTRUCTION, AND BE WISE, AND REFUSE IT NOT. BLESSED IS THE MAN THAT HEARETH ME, WATCHING DAILY AT MY GATES, WAITING AT THE POST OF MY DOORS. FOR WHOSO FINDETH ME, FINDETH LIFE, AND SHALL OBTAIN FAVOUR OF THE ***LORD***. *BUT HE THAT SINNETH AGAINST ME WRONGETH HIS OWN SOUL: ALL THEY THAT HATE ME LOVE DEATH.*" We hear JESUS sharing his life of love, joy, and wonder with FATHER as he Created all things. We surely see how leaving behind FATHER, his eternal body, and his Throne in Heaven, to come walk here among all the evil filth, vile, hell, and death this world had become, was a suffering no one on earth has ever felt. JESUS beyond that daily suffering, felt and tasted the harshest life this world had to offer in that moment. He lived homeless, in poverty, and oppressed, and at his last reason and purpose, he endured a far more painful physical torture and physical death than any soul ever, before or since. He suffered in his pure understanding, that FATHER and all the Holy Angels in Heaven, had to turn their eyes away from him, as our sin was then finally placed upon him. He then descended into the lowest conscious pits for three days, where death and hell abound without measure, to reassure the saved souls of the dead. JESUS, received *all* love from FATHER, and as love itself commands, he poured out that same love to those in this world who would receive it. MESSIAH taught us that all who receive GOD'S love, must give all, as we receive all. JESUS is our perfect example of Godly loving obedience to FATHER, and responsible charity to others. JESUS came to us, and died as he lived, being for others and to please FATHER. We have heard the price of sin is death. That is because FATHER only creates perfect order to last for eternity, and sin is every imperfect interjection that destroys created perfection. And as it is a contagion, sin cannot be allowed. We see a second death, removes all sinners and their destructive forces towards the perfection of GOD'S next New Kingdom. JESUS entered into this fallen creation, like us, born separated from FATHER. We all need CHRIST as our Savior, as our salvation requires one of such unknowable value, to take upon himself, not just our shame of irresponsible unfaithfulness to life eternal, but to bare our actual pain of punishment. JESUS is the sum of righteousness as he came to make manifest and fulfill all the commandments, laws, ways, and Prophecies of GOD in the Old Testament. Do you want to accept his wondrous gift? If so, understand that JESUS must suffer real pain and bleed for *your* sins. JESUS suffered pain beyond our imaginations ability to comprehend, as his body was literally torn in every spot. Scripture says he was "*marred more than any man.*" So, we know that no man *before or since*, ever felt and endured this level of pain and punishment before finding physical death. We need to understand, that before his greatest final suffering, JESUS endured an earthly life of only sadness and grief. Our LORD found no joy in this sinful world. JESUS loved, but this world never once brought him comfort. In Scripture, our KING was about FATHER'S work, even as a child. The full knowing mind of CHRIST was born into the flesh of a babe, and as that babe grew, JESUS lived a life of pure sorrow and grief, surrounded by a world filled with sins against FATHER. Each day seeing only rebellion, apostasy, deception, and the confusion of hell in a world of flesh, being a cauldron of death. And each day drawing nearer to the final reason he came. Do we understand what it must have been like, for the Holy Righteous SON, who had never sinned, to for the first time have sin put upon his soul: our sin. What a horror to endure, that CHRIST was then and there, in all ways, accountable to FATHER, who he forever obeyed and loved above life itself. By Scripture, we see and know the exact second that our forgiven sins were placed upon our KING. It was at the very second when JESUS died. We know this, because the price of sin is death. Our KING having never sinned, could never die until he had sin placed upon him. He could not die at any point in his life until that second. And the sins of many could not be placed upon him, *until* the full price of absolution for those horrors of imperfection was first paid in blood. This is why and how he was marred more than any other man, and endured more physical punishment than is humanly possible. Seeing the price of sin is death, no matter how much, or how long he was beaten cut and bruised, JESUS could not die without sin. FATHER could not require of him, the separation of his soul without sin. So, by the power of his perfect eternal life, JESUS continued to live and pay for, and feel and suffer the pain and death of many. The men who beat him, surely wondered by what power does this man live on and endure, asking when will death finally find this man. One pagan centurion who witnessed most, if not all of his suffering from the first strike in Gethsemane, to his giving up the ghost. His having seen many men tortured to death, he had no earthly explanation for a man to have such an enduring force of life that continued long and far beyond his belief. This Centurian was the one, who after witnessing all these super natural facts, said in Matt 27:54 "*TRULY THIS WAS THE* ***SON*** *OF* ***GOD***." Can we try to comprehend what it was like for JESUS, to for the first time, feel shame and need in that exact moment our sin was placed upon his soul, for all in Heaven to turn away in fear, and for JESUS to cry out in fear to FATHER for mercy. JESUS did not die painless in his sleep for our sins; he suffered unimaginably; but why? Such a great suffering, was required to cover the cost of each and every sin that will be truly repented by the fewer among mankind. We see each and every sin forgiven, was paid for in real suffering. And each of us, who accepts MESSIAH'S suffering for our past sins, shall we see the SON beaten more for us. Can we continue to sin and repent, knowing he must suffer for every forgiven sin. It is that personal if we claim CHRIST'S blood and suffering for our *past* sins, seeing CHRIST does not forgive us all future sins, that being a license to sin. Romans 3:25 says, *"WHOM **GOD** HATH SET FORTH TO BE A PROPITIATION THROUGH FAITH IN HIS BLOOD, TO DECLARE HIS RIGHTEOUSNESS FOR THE REMISSION OF SINS THAT ARE PAST, THROUGH THE FORBEARANCE OF **GOD**;"* Please note and never forget the word *past,* in this Scripture. FATHER did not say that by mistake, but for great purpose. Is there no strength in seeing CHRIST'S pain, to stop sinning; will we add to his suffering. In our Baptism of HOLYSPIRIT, is there not great strength in that we are called to be a Holy

Temple for HOLYSPIRIT. What strength do we derive from these things to stand and be Holy? GOD and their words will judge. FATHER gave us his great gift of our soul, and guiding love. And after imperfection entered the world, FATHER'S Angels still called all souls back to his ways, his law, and his love from a distance. Yet imperfection grew, and so FATHER gave mankind the written law and the animal sacrifice for our stronger sense and understanding of repentance and cleansing. But mankind grew immune to the meaning in the loss of the life force of an animal, given in their place to pay for their past imperfection. FATHER then went further to pour out his all in his most precious SON. FATHER gave us JESUS to be all things: a Teacher, Priest, King, and a Holy Sacrifice and price paid in love's blood for others. FATHER then added to this greatest of gifts, the indwelling HOLYSPIRIT. If this is not enough to keep us from violating GOD'S commands, nothing is, and we are *truly unworthy*. If we cannot learn and grow to obey the laws of perfection here, because of our free-will desires; what will we do in the millennial reign, or in Heaven? Scripture says sin entered the heart of Satan *in Heaven*, and sin was spread by him to one third of the Angels, who also chose to leave their first estate. There will always be free-will in all souls. Can JESUS forgive and knowingly allow a proven lustful soul who does not have the desire to cease from sin into the next perfect eternal home of his family. Will empty prayers and hollow hearts come into Heaven to sin, and bring again an eventual destruction to many of GOD'S children? Where is the eternal purpose in that? Why did that painful price our LORD paid, have to be so great to satisfy FATHER'S wrath against us? How great is the sum of imperfection to be forgiven? CHRIST'S Sacrifice "*pleased*" meaning *satisfied* FATHER. In Isaiah 52 and 53 we see a perfect picture of the beauty of love. Yet Scripture says CHRIST'S earthly semblance was uncomely, minus physical beauty, so none would accept or desire him for his outward appearance, for that is worldly physical lust. If we cannot come to understand the magnitude of what JESUS did for us, how can we come to him at all. And how can JESUS ever present us to FATHER. Once we see *we* contributed to the amount and severity of pain JESUS suffered, in payment for *our* every imperfection, we see the weight of *our* forgiven imperfections on one side of the scale, and the weight of the suffering of our SAVIOR for us on the other side as payment in full. And can or will we add to that final price. How many more sins, and their painful price, can *we* each be allowed to put on that scale. Be careful about saying JESUS already paid for every sin that was ever committed. If that were true, then all souls are saved and can sin in abandon. If you believe this, you hold a false doctrine. We are told in Scripture that all are not saved, but scarcely the few. JESUS suffered and died, only for those who want to come back to love and live in the eternal perfection of GOD. It is a personal debt owed by each, and there was actual pain felt, and blood lost because of every sin on that scale. Can we require another stripe on his back, are we allowed to use and abuse this gift. FATHER commands we worship in a personal way, and not just corporately. We must in all aspects, accept and do this. We each must see and accept our part in JESUS'S pain and hurt, and ultimate death for our *past* forgiven, and yes, if we should rarely stumble in shame, our future sins. By understanding this, we gain love and strength to overcome and stop sinning. When JESUS said on the cross it is finished, that meant the price he *could* pay, was truly paid, for what *could* be truly forgiven. Yet GOD'S pain of hearing the un-forgiven, curse, blaspheme, and mock him; and seeing so many falsely claim his loving gift without turning from their evil ways, goes on. Do any think FATHER'S heart does not ache, when the Holy names of GOD are taken in vain, or when the SON and HOLYSPIRIT are mocked. FATHER in his greatest love and mercy, gave us CHRIST'S ultimate sacrifice that held ultimate power and strength. So, to disrespect or dishonor such a gift as this, being more valuable than the whole of creation; I dare not think of what FATHER must do to each soul who does not give all eternal reverence. Who instead use and abuse, or give dishonor or disrespect to his SON. When a soul hears of the things of GOD, and is drawn, they are wise to quickly see that at their calling and drawing, still being in the flesh, there is no real strength or knowledge of what is the full eternal love of GOD. And in their flesh and soul, still abides an abundance of false confused beliefs and love for this world. Any desire for this world and all it holds, has no communion with, and is not worthy of being called eternal love. GOD calls these desires, lust, and by this knowledge we seek a new spiritual love, and reject physical lust. If we try to come to GOD holding to lust, believing they are in any way a part of eternal love, our soul is burdened, weak, and cannot grow towards GOD. We will always be as a babe, needing and wanting the constant feeding of milk, being unknowing, and in constant peril. Before the ultimate sacrifice of CHRIST, we were to give ten percent of all we held. After the Trinity gave, and offered all they are and have, CHRIST said we must give all, in return. To whomever much is given, much is now required. GOD'S words will judge. Remember we need HOLYSPIRIT in the fullness of love and truth, and not in part. The truth and love in part, is GOD in part. Who wants and waits to see a great miracle and wonder and a glorious eternal Throne? A Throne like none before, and there shall never be another. Who waits to see the *one* for whom this coming New Jerusalem is made, the *one* who overcame the breadth of creation and all moments. The *one* who endured all things in a responsible faithfulness that cannot be compared to any before, since, or forever more. The *one* who has given every measure, and every thought without fault. The *one* whose love for FATHER is a purest form of radiant, fervent, consuming, eternal desire. The *one* whose desire has never moved from the needs of FATHER'S loving truth. The *one* whose gaze is forever steeled, immovable, immutable, and impenetrable, and whose shoulders have stood forth from his footing, forever victorious against every mighty evil wind. Who waits to see this longed for moment, when the entirety of the greatest love for FATHER, sits and reigns. The *one* who is the pure joy in the heart of FATHER. FATHER asks, *who*, where are the souls, wanting JESUS beyond and high above all things that are good and to be desired?

How can we know the HOLYSPIRIT of GOD

HOLYSPIRIT is, and translates and dispenses and delivers, all the mysterious unifying forceful desires of GOD'S loving forceful truth, strength, and comfort. HOLYSPIRIT is he who searches out, and spiritually adjoins and unifies FATHER with all his loving eternal Children. HOLYSPIRIT is described in Scripture, as quite ethereal, and moving throughout the aether and creation, it all being his physical platform, or body if you will. So, he has no observable localized semblance of a body. As with JESUS, at HOLYSPIRIT'S division from FATHER, he became the third soul of GOD'S Trinity. But HOLYSPIRIT does not sit upon a physical throne in any physical form as both FATHER and JESUS. The entire aether and creation have become his Throne. HOLYSPIRIT seeks out GOD'S children as a teacher, bringing communal strength and comfort to all who love GOD above all else. He comes to us for the great Holy cause of teaching us FATHER'S truth and love. He comes to strengthen us against temptation, and lead all who yield in true desire, into a highly preferred understanding and force of life. So, we see ONE GOD in their three incarnations. Scripture says for us to be quite aware of how our tiniest sin grieves HOLYSPIRIT; and Scripture says, "grieve not the HOLYSPIRIT". Forgive us HOLYSPIRIT, for we surely have grieved you in the past, but in our love for you, we want to never grieve you again. Scripture says, at JESUS'S Baptism, HOLYSPIRIT returned to our LORD in a semblance of an ethereal dove like winged form, descending upon him as he rose from the water. And after JESUS was crucified and had ascended, HOLYSPIRIT came into the sinless temples of CHRIST'S disciples in the sounds of a rushing whirlwind on the day of Pentecost, and evidenced himself above each in a cloven tongue of fire, as a sign they are now one with GOD. Many others received this gift to lesser degrees in the laying on of hands by a soul having a strong measure. Which is how, I first received this true wonder. Scripture tells us to honor HOLYSPIRIT'S wondrous mysterious presence within us, so HOLYSPIRIT may dwell with us in peace, as we are called to present ourselves a Holy Temple. Do not take this lightly, or half-heartedly, and exhibit the pure evident ways and conduct associated with GOD. Do we reverence this responsible honor and command toward us, to maintain the body as the body of CHRIST, and a Temple of the Holy Righteousness of GOD. We seek to grow in HOLYSPIRIT'S strength, to overcome all covetous desires of the flesh. In this, GOD will not be driven from us. If we sin, HOLYSPIRIT is grieved and surely turns his gaze. If we backslide into sin and weakness, and remain there, at some point we can and will fall away and lose all hope of reunion with HOLYSPIRIT for eternity. GOD and their words will Judge us. What wisdom, strength, and comfort, have we each received from HOLYSPIRIT. If you do not feel the truth of GOD'S will being continually written within you, if you don't feel you are being given an ever-growing strength over sin; you have a very fearful question to ask yourself. HOLYSPIRIT is true, so know every contrary man is a liar. If we live in the commands of sacrifice and Godly service and ways of love, we are promised HOLYSPIRIT will come to us. We must be worthy of the gifts of HOLYSPIRIT, promised in Scripture. These gifts enable us to serve our true purpose of increasing and preserving GOD'S Kingdom, and our ongoing edification of the faithful. All these gifts and attributes, are promised to *all* who seek GOD with the fullness of their heart. Do we see these gifts within, or around us; are they evident in our Church and ourself. If we don't see these gifts, our next question must be why not? Can a person be self-deceived into claiming they have an indwelling of HOLYSPIRIT for self-comfort, but in fact have nothing? Can a lukewarm community Church, be completely void of HOLYSPIRIT'S forces from the pastor on down? Or do you believe as many say, that these gifts are not for this time. These gifts are not abundant at this time, because we live in a time of immense apostasy. Today's worldly Churches, can't receive any gifts of HOLYSPIRIT. This is why the many needing the comfort of false assurance, claim everyday things of chanceful reality, as their miracles and evidence of grace. We believe every word of GOD, which says great signs follow those who truly believe and love. Scripture says those who truly believe, are those who obey every word of GOD. Are these signs following you, do you have these gifts, do you live by every word? Have you given this prayerful thought? Do I? I strive and pray that I do. HOLYSPIRIT, help us write these truths upon our hearts and minds in the blood and strength of JESUS. For all *professed* Christians, still in the flesh, the Revelation of CHRIST'S Prophecy is the least understood book of the Bible, just as HOLYSPIRIT is the least understood of the TRINITY. The FATHER is head of the Holy Trinity, and the SON is the Second of GOD and head of the Church, who came out from and obeys FATHER. We have our experiential reference as to what a physically embodied dad and son are; so, many form a false physical understanding of a kindred relationship to these first two souls of GOD. By doing this, HOLYSPIRIT becomes less relatable to many. We must seek a spiritual relationship and understanding with the entire Trinity, and not a physical representation. On one hand, JESUS said blasphemy against him, *while* he was in the flesh, being part man, can be forgiven; but blasphemy of HOLYSPIRIT, being not part man, being purely GOD, this can never be forgiven. Do you accept this direct command from JESUS; or in rebellion, do you believe it *is* forgivable? Where does one's destructive power to create individual personal beliefs comes from? It comes from free-will and the desired confusion of hell. Scripture tells us that all power and honor are given to the SON. Yet JESUS said he can do nothing with that power but what he sees FATHER do, and FATHER taught him all things, and he only has power through his emulation of FATHER. JESUS, from *his* gifted beginning, was taught as the first born of every free-will creature, brought forth out from FATHER'S Soul and body; thus, being the only begotten SON. Who later upon FATHER'S request, put off his gifted immortal body and his connection to HOLYSPIRIT, to be born in the earth and fallen flesh of a man. Why? To suffer and know the temptation and pain of fallen mankind, yet never succumb to temptation. In this, learning his holy righteous referenced understanding of Judgment over *our* condition and choices. We read JESUS'S connection to HOLYSPIRIT was returned at around thirty years of age, to complete his ministry and painful purpose. We see clearly in the Gospel of John, that our LORD without fail, is an eternal loving obedient SON. We see verse after verse in which JESUS says, he always does only those things, that please FATHER, and only those things

FATHER has shown him, and commanded him to do. And *every* experience between FATHER and JESUS, pass through HOLYSPIRIT between them. So, we see JESUS did miracles with HOLYSPIRIT and FATHER as one; they are connected and know, as one in love, belief, and purpose. HOLYSPIRIT also connects and adjoins all the *Holy* Angels and Saints, to GOD. This connection, is forever lost to the Angels who fell. This gift is offered to fallen mankind, that we might be added as Holy Sons and Daughters of GOD. GOD is three separate incarnations, or souls, yet one GOD of one pure shared belief understanding and love. The three souls of GOD are above all. Then we see Seven and four most powerful Angels just below them in the Holy of Holies. Then the lower angels of the lesser Heaven, like the Cherubim. We are all gifted souls in ordered size, reach, and force. Yet we all need to be united through HOLYSPIRIT as one in loving true belief and eternal purpose. The saddest fact about HOLYSPIRIT, is that in most Churches and professed Christians, he is never truly known or felt in any eternal way or purpose. HOLYSPIRIT is neglected the most in worship, prayers, and biblical study. He is often considered a Christian benefit or perk, like an angel there to serve us, or a force field of favors and granted wishes. And not like the core desires and awareness of GOD, needing to be known, loved, communed with, *and* obeyed. Any lesser misapplication of communion towards HOLYSPIRIT, is an offensive afront to GOD'S heart and purpose. Why does HOLYSPIRIT receive such treatment by most professed believers. It comes down to still being a creature of the flesh. To the religious minded still in the flesh, their Soul and Spirit are things they don't quite comprehend as they play along. It is a very different way of life to be a spiritual, and not a fleshly Creature. Can spiritual qualia be explained to, or understood by someone in the flesh? How can we encourage souls to see the difference? How do we describe something spiritually ethereal, that has no worldly physical reference? How do we each see ourselves, and why do most believers still identify as, and cling to the flesh? Is it because they see and feel the body to much, and see the body in the mirror too much; seeing a deceitful reflection, and not seeing their soul. Again, the immature towards GOD, being in the flesh, have a more kindred feeling towards FATHER and SON, holding a fleshly visual of GOD. JESUS said the hour comes that we shall not worship in the flesh or the mountain, or in Jerusalem, but we shall worship in Spirit and in truth. And FATHER wants all to worship him in this way. We hear JESUS *tell* a man "to let the dead, bury the dead" who asked leave to attend to his dad's funeral. Signifying that all that would be there were dead temporary physical aspects; and all who walk around, in and of this physical world, are spiritually dead, who profit nothing. Once a soul loses all Godly desires, there is nothing left for GOD to love, or help. We must see ourselves, as souls dwelling with HOLYSPIRIT, adjoined in love and truth within a temporary fleshly Temple. We must dwell and stay in oneness with HOLYSPIRIT, not living as a fleshly body with a soul tucked inside that we don't feel or understand. We must give much prayer love and study to all three souls of GOD'S Trinity. GOD and their words will judge all. We know by Scripture and experience, HOLYSPIRIT is on the front line, for GOD'S Children in all Spiritual battles. For this reason and cause, HOLYSPIRIT has no singular firm body through which he can be grounded, or tempted; but HOLYSPIRIT can be greatly grieved and hurt spiritually as he is the third aware soul of GOD. We see how the semblance and attributes of FATHER are necessary for our desiring and finding eternal life and love. We see how the semblance and attributes of JESUS, are necessary for our finding eternal life and love. But we must see how the same Spiritual attributes, minus a semblance in HOLYSPIRIT, is equally necessary for our finding eternal life and love. Three ruling souls, differing in needed ways, yet one in every spiritual knowing way of awareness being necessary together. A greater and greater perfect design is revealed as we go further into GOD. We know when GOD sees, touches, or communes, with our soul in this world, it is through HOLYSPIRIT. Scripture says, FATHER is too Holy to look upon sin; that it is painful and grieves him, so grieve not the HOLYSPIRIT who is GOD. If we study all Scriptures, for each soul of GOD above, we see HOLYSPIRIT comes in his love to a most painful task. Adore and worship HOLYSPIRIT, do *not* treat him like he is living in us, but treat the body like his temple, and like we live there with him as obedient students. Do not offend GOD, who is the master of every eternal house. We would give you here the best, most useful definition of iniquity. Simply being all works, not being pure and of GOD. Now most all mankind, physically give out of their abundance, and out of hope for self-preservation. There is deep within the flesh, a physical desire to share a little for the reason there is safety and benefits in being part of a group. This is not a physical evolutionary trait in mankind, it is a devolutionary fallen trait, in becoming more like an animal. But Godly souls hold no thoughts of physical weakness or need. When the man who had followed every physical law his whole life, asked JESUS how can I be perfect. JESUS told him to sell all that he had, and give it away. The man went away sad, possessed by and dependent on his many physical possessions. Most all physical charity is not of GOD, and it is important to learn to distinguish spiritual giving from giving in the flesh if our giving is to be seen and accepted by GOD. Able gave his best physical gift in a loving spiritual way; but Cane gave his best physical gift in a physical way, and it was rejected. So, Cane become confused, envious, and angry, and killed his brother. Growth in Spirit, requires we ask and answer many questions. If we find any spiritual question too hard, the fact we find it hard has in part answered another question for us. If we are called to choose between the things of this world the flesh covets, and a possession-less servitude to GOD, and this choice is hard. This tells us, where our hearts strong desires *currently* are. As our LORD said, at some point it is about counting the cost. Can we all pay the physical and spiritual cost of eternal life, or will many not be able to pay the price? Physical sacrifice is required for spiritual growth. Spiritual growth is *only* found in turning from the flesh. We never see a soul get closer to GOD, as their flesh remains strong. And we know that HOLYSPIRIT is not found in a continual changing truth, which is not the truth. Changing truth is the spirit of confusion. HOLYSPIRIT, have patience and mercy towards us. Continue to teach, strengthen, rebuke, and chastise us towards goodness and all the ways, truth, and love of GOD. All our praise and love and desire towards you HOLYSPIRIT. Amen

How do we know the Truth.

The truth of the ever-so-many things in GOD'S great creation are all truly wondrous, and many truths are stranger than fiction. But mankind's truth has become truly subjective from their small slow physical perspectives. To know the one pure truth, one must truly and intimately know FATHER. Who created his founding truth in the very beginning that governs and brings eternal consistency to all creation. GOD'S truth is not strange or changing, it is immutable, pure, and undeniable. Yet in both its simplicity and complexity, it is not known to mankind, as they under and over-complicate reality in endless false beliefs. All false beliefs spring from curious desires, and temporary godless perspectives. It is amazing how GOD'S truth, strength, and purity, are allusive to professed believers. Not by being hidden, or too hard to understand. But because it is not the subjective truth they agree with, or desire to see. Most souls complicate and make the truth about a multitude of things it is not; re-making the truth into endless ambiguous, incomplete, self-pleasing beliefs. Souls with hardened hearts, have altered and hidden GOD'S Truth from their eyes. Most pollute GOD'S Truth with worldly needs and illusions, born of physical desires for themselves and others they hold dear. There is more than one way to think about this great wondrous aspect of reality called truth. There is something called telling the truth; there is something called knowing the truth; and there is the Holy aspect called doing or living the truth of GOD. And one can literally know and tell GOD'S truth, every-second of every-day, and yet never live it. John 3:16-21, "*FOR* ***GOD*** *SO LOVED THE WORLD, THAT HE GAVE HIS ONLY BEGOTTEN SON, THAT WHOSOEVER BELEIVETH IN HIM SHOULD NOT PERISH, BUT HAVE EVERLASTING LIFE. FOR GOD SENT NOT HIS SON INTO THE WORLD TO CONDEMN THE WORLD; BUT THAT THE WORLD THROUGH HIM* <u>*MIGHT*</u> *BE SAVED. HE THAT BELEIVETH ON HIM IS NOT CONDEMNED: BUT HE THAT BELEIVETH NOT IS CONDEMNED ALREADY, BECAUSE HE HATH NOT BELEIVED IN THE NAME OF THE ONLY BEGOTTEN* ***SON*** *OF* ***GOD****. AND* <u>*THIS IS THE CONDEMNATION*</u>*, THAT THE LIGHT IS COME INTO THE WORLD, AND MEN* <u>*LOVED DARKNESS RATHER THAN THE LIGHT*</u>*, BECAUSE* <u>*THEIR DEEDS WERE EVIL*</u>*. FOR EVERY ONE THAT* <u>*DOETH TRUTH*</u> *COMETH TO THE LIGHT, THAT HIS DEEDS MAY BE MANIFEST, THAT*

*THEY ARE WROUGHT IN **GOD**.*" We hear our LORD say, that in our living out the full pure truth of GOD, there is eternal life. But just knowing and telling GOD'S full truth will not set us free. We can only be free from condemnation when we *do*, and *live* the truth through desired one-ness with GOD. But if one denies or refuses to accept what the full truth of GOD is, how can they live it? Scripture says many believe they live in GOD'S truth, yet live in something less of their own subjective design, being false, impure, and temporary. When we give our accurate account of events we witness, this is our relaying *our* truth of experience; but that's not the truth we seek. At its very foundation, GOD created the one truth, which he made to be eternally lived in his love. If we would desire to know the greatest and first truth upon which eternal life is derived and understood, it is just this simple. If GOD creates an extremely complex functioning model (all of creation) for the one reason of being a perfect environment, to allow an eternal Godly love between FATHER and many souls (to serve and please him) and it functions accordingly; this accomplishes GOD'S truth set forth. The best definition of creations truth is its truth of purpose. Or, reality functions according to its *true* purpose. Or, it functions as it was meant to, and made to by FATHER. If we live as GOD created us to live, we live the truth, his truth.

This is as foundational and basic as the truth can get. And to fully and truly speak of it, as I do here, is only to tell the truth. I can know, write, and speak, every perfect word of truth, yet have nothing if I fail to spiritually live it. To know it, tell it, and live it, is the trinity of the truth. We see many souls in this vast complex creation, having free-will, say to themselves; I believe *I* can alter part of my purpose, to seek and fulfill my individual desires, and not live my *full* intended purpose for GOD. This is then a lie of life and purpose; and so, a lie of eternal life itself because it is not as FATHER created and intended it to be. All rebellion to life's eternal purpose, is called individualism, bringing divisions into GOD'S intended perfect order of creation. It brings conscious and physical disorder, called entropy, atrophy, confusion, fission, hell, which manifest every spiritual physical confusion of illness and death. We see billions of people living what they are fond to call *their* own individual physical life. Living by *their* own physical truth of purpose, governed by *their* own physical desires and beliefs. And resulting in billions of lies of purpose of created life in each moment of reality. So, we now see and separate truth from belief. A soul's belief can be *their* accepted truth, yet not be GOD'S truth or belief. To seek only the truth and belief of FATHER, we must know it can only exist in its purity and fullness, and cannot come from us. The creation does not get to tell its creator, or define for anyone, what is the slightest part of the original created truth. The one defining truth, comes only from its origin and author. The first step to know and live the truth of GOD'S intended *eternal* purpose, is to strive in love to remove and deny our individuality. Many say if we do that, we will no longer be our selves at all. And in a way, regarding all the evil one now is, this is true. We will indeed then have been lost to all the evil that we were and have committed, and become new and pure to only all that is good and of GOD. Yet we will still be us. And what else will be different? We are all gifted out from, and once being part of FATHER, so the saving difference is that the true followers of CHRIST know, love, desire, and emulate his ways, above all things. Those who rebel, say they do not want these ways in which FATHER has defined what we are created to be and do. A rebel soul, deep within believes in autonomy and that the body is theirs. Even most believers give to GOD, the amount of control they find physically comfortable, and keep the rest for their self. GOD and their words will judge us. Please seek and love everything about GOD. We are warned over and over that GOD can't abide rebellious pride, or lust, which bring all imperfect aspects called sin. Let us have no love or desire for this sinful destructive world, and reject lust for the temporary things within it. Seek GOD in a search for all eternal spiritual things, and find the *truth* of all things, it is nowhere else to be found. The author who made the truth, is eternally its source, and it cannot be effectively rewritten or reinterpreted by any other. Knowing this foundational first truth is the beginning of all true Scriptural interpretation. Most professed Christians do not begin or focus their search upon FATHER'S created truth. That truth being GOD'S will, plans, needs, desires, warnings, and commands as to how we are to live, worship, and seek them, and why. If to form our belief, we focus or start with Scriptures about what GOD has and will do for us, and what we should do for each other. We build our temple of belief without a foundation of the first truth. The simple truth is we are part of GOD'S planned creation, and it *all* but our gifted soul, belongs to GOD. We have no right to a single physical breath or beat, no matter how often we think we do. Our gifted souls are ours, but the body is GOD'S. We are adjoined to a body in the hope it would be our platform to love and emulate JESUS. And we should not say that *we* live "*our*" life for GOD; but only that we find life in oneness with GOD. We should know with our whole heart that GOD created all things to work in a perfect eternal way. This is the definition of *the*

Truth. And I have now told you the Truth, being as it is meant to be by GOD. Must we each live the whole and perfect truth? Or are we only to live it in the parts we like and accept? If we know this pure truth, and never deny it, desiring all GOD'S perfect eternal ways, not yielding to souls in the flesh who run to death. Then we can never be deceived, for in this we dwell in the impenetrable strength of GOD'S mind. 1st John 3:9, "*WHOSOEVER IS BORN OF* ***GOD*** *DOTH NOT COMMIT SIN; FOR HIS SEED REMAINETH IN HIM: AND HE CANNOT SIN, BECAUSE HE IS BORN OF* ***GOD***." 1st Cor. 2:15-16, "*BUT HE THAT IS SPIRITUAL JUDGETH ALL THINGS, YET HE HIMSELF IS JUDGED OF NO MAN. FOR WHO HATH KNOWN THE MIND OF THE* ***LORD****, THAT HE MAY INSTRUCT HIM? BUT WE HAVE THE MIND OF* ***CHRIST***." If we have the same belief and desires as GOD, we are then of one mind and truth with JESUS. Remember we were all created in GOD'S image, so, our spiritual needs, do reflect GOD'S needs on a minute scale. We have the ability to love, so we need and want to be loved back. GOD needs and desires to be loved back also. Love has to be given freely of desire, so again, free-will, chance, and a true choice are all required. Which allows for rebellion in the desires for the evil ways and things of this world. Many choose to focus their love on themselves, their earthly family, and this physical creation, more than the one Creator of it. Many are not searching for GOD'S Truth, but for their own truth of GOD. Jeremiah 17:9-10, "*THE HEART IS DECEITFUL ABOVE ALL THINGS, AND DESPERATELY WICKED: WHO CAN KNOW IT? I THE **LORD** SEARCH THE HEART, I TRY THE REINS, EVEN TO GIVE EVERY MAN ACCORDING TO HIS WAYS, AND ACCORDING TO THE FRUIT OF HIS DOINGS.*" To live professing a belief in GOD, while asking and wanting for worldly things from GOD, is an impure direction. GOD sees the measure of our desire to receive and return his love in his ways. FATHER created nothing beyond his reach of knowing and understanding; and if our soul and love rise to his personal attention, he searches and tries our heart to see exactly how much do we love him; or do we love the creation more. On our side, it is beyond the ability for any created being to ever perfectly know the last and final outcome of all these things. Can one see their self being created to know when and how it was done? On our own, we can't find any complete knowledge of anything, and must be taught all things by GOD. Physical sciences say they know our physical presence is reality; then contrarily admit they don't know in any absolute way, where we are, why we're here, or even if we are here. It is all just their own beliefs, born through desired second-hand hearsay perceptions. Yes, our perceptions of all that is around us, are created as the body senses and relays the physical information of GOD'S creation to our souls; which we use to create our experiential qualia of self-awareness. But we must aspire to see far more than can be seen in sensing a fallen six percent sliver of the physical reality alone. We perceive all things through the greatest ethereal mystery in creation, being our own soul. When you think of yourself, do you only see yourself in a mirror? Why would you see the body as yourself? Look into a mirror, and see if you can know in your heart beyond doubt, that you are not looking at yourself? The flesh's worldly connections feel very powerful. Do you feel or think the body is more than a temporary vehicle or platform? If so, your feelings and thoughts are wrong. Standing in the truth of GOD, in every mirror I see a body-temple for HOLYSPIRIT and my soul to share. This truth brings us to an uncomfortable reality, that at first most don't like to know. Being that *what we are*, if stripped of the physical body, provided by GOD. Essentially, we are each a defined dimensional size, reach, understanding, and belief, of a conscious soul, inherently capable of forcefully generating patterned projections of *creative* desire. And through our soul's desires, we interpret all sensed things, to forcefully project, create, and perceive our experience of qualia. We then make choices, based on built-up beliefs, desired emotions, and memorialized experiences. This is who, and *all* that we are, and no more. How confused can many become, and how hard is it for many to know the truth and accept their true state with GOD, once they become captive to the desires and beliefs of the flesh. We recommend all souls take care, and consider the feelings Adam and Eve must have felt, after sinning for the very first time when GOD asked, *what is this thou hast done*? Was it shattering? Imagine never seeing or hearing lightning strike before, then have the loudest strike in all eternity right at your feet, shaking you to your core as your heart pounds so hard you feel it in every cell of your body with sheer fear. Was it on that scale? Or do you think they stood there calmly like it was something that just happened, and not a big deal? What did it feel like to instantly fall so far from the heights of eternal life? We were born fallen, and so aspire to be redeemed, and returned on high, but Adam and Eve knew a pure clean state of heart and mind we have never known. How does a soul knowing only the perfection of love and eternal care of GOD, deal with the knowledge they have lost it all, and interjected destruction, hell, and death, into GOD'S vast perfect work. Knowing the once perfect body they inhabit, will corrupt and decay back into the dust of the earth; and having brought the same upon all children after them. They had never before felt any shame or sorrow, these were the very first such feelings ever felt within any soul of mankind, and they came on a scale of emotional upheaval of sorrow and shame, no one else has felt since. On the subject of truth, let's consider 2nd Cor. 5:9-10, *"WHEREFOR WE LABOR, THAT, WHETHER PRESENT OR ABSENT, WE MAY BE ACCEPTED OF HIM. FOR WE MUST ALL APPEAR BEFORE THE JUDGMENT SEAT OF* ***CHRIST****; THAT EVERY ONE MAY RECEIVE THE THINGS DONE IN HIS BODY, ACCORDING TO THAT HE HATH DONE, WHETHER IT BE GOOD OR BAD."* 2nd Cor. 13:5, *"EXAMINE YOURSELVES, WHETHER YE BE IN THE FAITH; PROVE YOUR OWN SELVES. KNOW YE NOT YOUR OWN SELVES, HOW THAT* ***JESUS CHRIST*** *IS IN YOU, EXCEPT YE BE REPROBATES?"* 2nd Cor. 12:21, *"AND LEST, WHEN I COME AGAIN, MY* ***GOD*** *WILL HUMBLE ME AMONG YOU, AND THAT I SHALL BEWAIL MANY WHICH HAVE SINNED ALREADY, AND HAVE NOT REPENTED OF THE UNCLEANNESS AND FORNICATION AND LACIVIOUSNESS WHICH THEY HAVE COMMITTED."* 2nd Cor. 11:11-15 *"WHEREFOR? BECAUSE*

*I LOVE YOU NOT? **GOD** KNOWETH. BUT WHAT I DO, THAT I WILL DO, THAT I MAY CUT OFF OCCASION FROM THEM WHICH DESIRE OCCASION; THAT WHEREIN THEY GLORY, THEY MAY BE FOUND EVEN AS WE. FOR SUCH ARE FALSE APOSTLES, DECEITFUL WORKERS, TRANSFORMING THEMSELVES INTO THE APOSTLES OF **CHRIST.** AND NO MARVEL; FOR SATAN HIMSELF IS TRANSFORMED INTO AN ANGEL OF LIGHT. THEREFOR IT IS NO GREAT THING IF HIS MINISTERS ALSO BE TRANSFORMED AS THE MINISTERS OF RIGHTEOUSNESS; WHOSE END SHALL BE ACCORDING TO THEIR WORKS.*" How true and fearful a warning.

How do we know GOD'S WORD

GOD'S WORD, as with all qualities in reality, is impossible to know if we cannot overcome our fleshly desires and worldly beliefs. We must deny any and all primacy of the flesh, in how we see and hear all things spiritual. A spirit of the flesh, and a spirit of GOD, can never see eye to eye on one single thing. The flesh will always interpret GOD'S words to *its* benefit. Those who believe in GOD, have two choices; they can align their beliefs to GOD'S, or they must in self-deceit, interpret GOD'S words to align with their desires. But then a strange thing happens, because in both ways, one can *believe* they are in GOD'S grace. But the fearful thing is that only in one of those two ways, does JESUS see us in his grace. And that alone, is what is true. I see souls, forgo personal relationship with GOD, for relationships with personalized gods. What does this say about the many newer weaker Bible translations. In days long ago, to consider translating the Scripture of GOD into another language, was feared by those who love GOD, sensing great trepidation in offending GOD'S word in the tiniest way. GOD allows mankind to misinterpret his words, and even allowed those to be printed and labeled as being his truth. But we also know FATHER has kept the strength of his truth in this world in its original form for his Children. So, choosing the true Bible in these last days requires prayer and guidance. For the few earliest English translations, this work was done in great fear and need, so people could study GOD'S commands for themselves. Stay with the first translations. The Authorized King James Bible, was a work of HOLYSPIRIT. A monumental effort involving the input of over 1000 Godly leaders and language experts, highly on guard of each other's ability and motive. All fearful for its holy integrity and purity, and led by the HOLYSPIRIT. They watched and checked each other over and over for every coma and period, striving for the perfect original word for word of GOD, and not any subjective context or intent. These many servants knew and feared that GOD oversaw that translation. It took a long time, and was *not* done as a matter of so-called improvement or denominational convenience like modern versions. It did not contain any footnotes, or interpretive guidance; that being the domain of HOLYSPIRIT. Does anyone find it fearful that in such a time of apostasy as today, the Holy Scriptures are rewritten for all weak ungodly reasons. Often overseen by single individuals who sell their version for profit. You can write a weak Bible that people will prefer, and it will sell and be followed by many. As Scripture warns, there will be false Doctrines. All new versions, are in many parts, not true translations, but weak reinterpretations. Each so-called new improved edition of GOD'S truth, grows weaker in the fearful admonition of GOD. Now when we read the Old Testament, we see a long history of how over the centuries, there is a perpetual sliding back by mankind into weaker states away from the full strength of GOD'S truth. As weak believers are continually divided into weak sects and denominations by their own desires. Scripture describes this as all having gone astray. Why can't everyone see this is happening in our day on an exponential scale, when it not only is, but our LORD plainly said it would happen. Who sees the endless Christian denominations, and these many new

Biblical interpretations, and does not see this is the spirit of antichrist, confusion, and division crept in. Blindness in any part to GOD'S truth, can be nothing short of weakness. Understand there is money to be made from people looking for hope, and they most often prefer a false worldly hope. How are many gone after these false Doctrines, do these weak false doctrines weaken GOD'S command to not appease the flesh? Yes. Ask yourself, how can people read the exact same words, yet hear different meanings? Do you hear the warning in Scripture to try every Spirit to see if they are of GOD; which is required in finding the truth? Do not think we simply ask for truth, and whatever comes to our mind is sent to us by GOD, that is not Scriptural. Hearing GOD'S truth, requires a fervent desirous effort and labor of eternal love, and we each are responsible for choosing our belief in GOD. Why do most souls hold different beliefs? It is because they desire different things and ways. It is in FATHER'S design, that only those who passionately love him, have the desire to fully search him out? GOD and their words will judge us. It is not a perfect analogy, but useful to see that since the fall, GOD has moved off, and hid themselves from us; knowing those who love them, will search. And the stronger that love, the more desperate their search. Isaiah 45:15, "*VERILY THOU ART A*

GOD THAT HIDEST THYSELF, O **GOD** *OF ISRAEL THE SAVIOUR.*" Proverbs 8:17-21 "*I LOVE THEM THAT LOVE ME; AND THOSE THAT SEEK ME EARLY SHALL FIND ME. RICHES AND HONOR ARE WITH ME; YE, DURABLE RICHES AND RIGHTEOUSNESS. MY FRUIT IS BETTER THAN GOLD, YE THAN FINE GOLD; AND MY REVENUE THAN CHOICE SILVER. I LEAD IN THE WAY OF RIGHTEOUSNESS, IN THE MIDST OF THE PATHS OF JUDGMENT THAT I MAY CAUSE THOSE THAT LOVE ME TO INHERIT SUBSTANCE, AND I WILL FILL THEIR TREASURE.*" These words speak of eternal spiritual treasure, not lucre. Our love for GOD, needs to be above all else. Many say how do we know our love for GOD is true? We can ask ourself why do we love GOD? If all one can come up with is what GOD has done for us and our loved ones; and what we believe GOD will do for us and our loved ones in the future. This does not sound like the fullness of love, but self-serving, and not knowing or loving GOD intimately. We must be driven by a strong loving desire to learn and understand all the attributes of GOD, to find and love and understand the purpose of GOD and their designed ways. And we must want to live in all ways that please them, by living their every perfect word. Do you desire to *not* hurt FATHER or his creation with your forceful imperfections of destruction. Then yes, you have at least a little *responsible* true love for GOD, and the closer you get to GOD, the more your passion will strengthen. If along the way, you come to any Scriptural knowledge of GOD'S will, plan, and desire, that you find hard to accept, you can weaken and stumble if you can't accept every truth in faith and obedience. You must see *yourself* in err for lack of knowledge and wisdom, and yield in faith alone; or you are then in the same danger as was Eve and then Adam. Call out to FATHER, JESUS, and HOLYSPIRIT, call out to Holy Angels for help in prayer for good Spiritual wisdom. The truth will never fail, if we seek and wait for it in GOD'S pure purpose. And when by a pure love for GOD, we find a good measure of truth is given to us, it is assured and can withstand the storms to come. Prepare yourself, and share all strength you have with all who can discern. Warn all to try all words and Spirit's to see if they are of GOD. The result of any compromising, or negotiating between the ways of GOD and this world, bring only hell and death. Any belief, not being the full and pure truth of GOD, is a lie. Part truth, means part lie. All paths and Doctrines are not true and right, only one is true and right. Truth mixed with any lies, blinds us to the evils that seek to destroy our souls. Do not be satisfied with any lesser measure of wisdom.

We know from Scripture, that differing orders of angels were created to serve in different capacities. We know the semblance of the angel's great ethereal bodies, are mostly like FATHER, JESUS, and us. And as with us, their souls, were divided out of FATHER'S soul in differing measures, as an eternal gift. JESUS said if *we* are brought into Heaven, we will all be as the Angels are. So, if you wonder what life will feel like in heaven, study the lives of the Holy Angels. Angels are still very much a mystery; yet we know some are translated here among us in the earth. We know they are sent back and forth from Heaven to here, as GOD instructs and physically sends them. Though there are ascending orders of Angels, we know that when ethereal angel's bodies are translated down to earth, they are condensed, firm, and genetically compatible with the semblance of mankind. Yet they have far larger more forceful souls than mankind. We know angels have free-will and can rebel as mankind can, and a third of them have. We *do not* know if there are mechanisms of forgiveness for any small rare willful sins by angels. We do see in Revelation 2:1-5, that GOD tells the Angel of one of the Seven Churches to repent. But this may refer to the earth born leader of that Church, seeing the words Angels, Sons, and Children of GOD, interchange, concerning responsible leadership within GOD'S Kingdom. We do see from Prophetic Scripture, there is no more hope for the father of lies and his Angels. Matthew 25:41, "*THEN SHALL HE SAY ALSO UNTO THEM ON THE LEFT HAND, DEPART FROM ME YE CURSED, INTO EVERLASTING FIRE, PREPARED FOR THE DEVIL AND HIS ANGELS.*" The *Holy* Angels, hold an eternal perspective on all things, unlike the lost angels and lost mankind. The Holy Angels are not omnipresent, omnipotent, or 100% omniscient; but they are, and live perfect. We know as with mankind's perfection; the perfection of the Holy Angels is less than the sum of GOD'S perfection. We know that angels fight great battles having great power; battles we could never enter. Matthew 11:12, "*AND FROM THE DAYS OF JOHN THE BAPTIST UNTIL NOW THE KINGDOM OF HEAVEN SUFFERETH VIOLENCE, AND THE VIOLENT TAKE IT BY FORCE.*" Luke 10:18, "*I BEHELD SATAN AS LIGHTENING FALL FROM HEAVEN.*" Revelation 12:7-8, "*AND THERE WAS WAR IN HEAVEN: MICHAEL AND HIS ANGELS FOUGHT AGAINST THE DRAGON; AND THE DRAGON FOUGHT AND HIS ANGELS, AND PREVAILED NOT; NEITHER WAS THEIR PLACE FOUND ANY MORE IN HEAVEN. AND THE GREAT DRAGON WAS CAST OUT,*

THAT OLD SERPENT, CALLED THE DEVIL, AND SATAN, WHICH DECEIVETH THE WHOLE WORLD: HE WAS CAST OUT INTO THE EARTH, AND <u>HIS</u> ANGELS WERE CAST OUT WITH HIM." Now Satan, being a powerful angel, will during the Millennial Reign, be stripped of all ability and power, and be bound. Revelation 20:1-3, "*AND I SAW AN ANGEL COME DOWN FROM HEAVEN, HAVING THE KEY TO THE BOTTOMLESS PIT AND A GREAT CHAIN IN HIS HAND. AND HE LAID HOLD ON THE DRAGON, THAT OLD SERPENT, WHICH IS THE DEVIL, AND SATAN, AND BOUND HIM A THOUSAND YEARS, AND CAST HIM INTO THE BOTTOMLESS PIT, AND SHUT HIM UP, AND SET A SEAL UPON HIM, THAT HE SHOULD DECEIVE THE NATIONS NO MORE, TILL THE THOUSAND YEARS SHOULD BE FULFILLED: AND AFTER THAT HE MUST BE LOOSED A LITTLE SEASON.*" At the final Judgment of Satan. Isaiah 14:9-16, "*HELL FROM BENEATH IS MOVED FOR THEE TO MEET THEE AT THY COMING: IT STIRRETH UP THE DEAD FOR THEE, EVEN ALL THE CHEIF ONES OF THE EARTH; IT HATH RAISED UP FROM THEIR THRONES ALL THE KINGS OF THE NATIONS. ALL THEY SHALL SPEAK AND SAY UNTO THEE, ART THOU ALSO BECOME WEAK AS WE? ART*

THOU BECOME LIKE UNTO US? THY POMP IS BROUGHT DOWN TO THE GRAVE, AND THE NOISE OF THY VIOLS: THE WORM IS SPREAD UNDER THEE. AND THE WORMS COVER THEE. HOW ART THOU FALLEN FROM HEAVEN O LUCIFER, SON OF THE MORNING! HOW ART THOU CUT DOWN TO THE GROUND, WHICH DIDST WEAKEN THE NATIONS! FOR THOU HAST SAID IN THINE HEART, I WILL ASCEND INTO HEAVEN, I WILL EXALT MY THRONE ABOVE THE STARS OF ***GOD****: I WILL SIT ALSO UPON THE MOUNT OF THE CONGREGATION, IN THE SIDES OF THE NORTH: I WILL ASCEND ABOVE THE HEIGHTS OF THE CLOUDS; I WILL BE LIKE THE MOST HIGH. YET THOU SHALT BE BROUGHT DOWN TO HELL, TO THE SIDES OF THE PIT. THEY THAT SEE THEE SHALL NARROWLY LOOK UPON THEE, AND CONSIDER THEE SAYING, IS THIS THE MAN THAT MADE THE EARTH TO TREMBLE, THAT DID SHAKE KINGDOMS.*" Satan chose his final love, and now we choose ours. Heavens Holy Angels, as far as dealing with GOD'S lesser Children in the earth are by nature, loving, ministering Spirits. But they are also messengers of warning, death, and destruction, if it be GOD'S will. We know their lives teach us as Gifted Souls, a most fearful lesson we must write on our hearts; being that temptation and self-deception are powerful enough to bring down angels if they do not watch and abide in GOD'S truth of love and life. Remember, GOD did not make all things, and *then* add righteousness; GOD made reality to work eternally with righteousness as its core. Can we pray to, or for, one or all Holy Angels? Why not, we can communicate in love and hope, praying *for* them as fellow Children of GOD; we can ask the Holy Angels to watch and pray for us and ours, this seems good and needful. And pray for them as they battle great evil forces and temptations. But we must be careful if we hear an angel, as there are fallen angels looking to take advantage of us. If GOD sends an angel to us to communicate directly, we will also see them. GOD in all Righteousness has set his chain of command in reality to include each of us. Let us find good wisdom in looking to CHRIST, and to the Holy Angels made in their ways above us. Let us not want more than what we are created to be. In FATHER'S wisdom, he blessed mankind far above all the lower creatures in the Earth. Let us look up and know we are meant to bare the same faithful responsibilities as the Holy Angels. There will always be souls FATHER has gifted above us in greater responsibility, and those below us for which we are greatly responsible. Be you man, woman, or child, old or young, of any degree of gift or wisdom given by

GOD. Be satisfied with what you might be, and strive, for FATHER has set the perfect order of all things. In the many responsibilities we each must bear, we have all said wishfully, what I would give to go back and be a child, and not bear these weights. Bearing all we are created to bear is not grievous, and it is our calling from GOD. Amen

How do we know a Blessing

We can rarely know a blessing or curse in this world, at the time of it being received. We first have to see the final results brought to our Spirit, caused by anything given, happening to, or taken from us. It matters not, if at first it is perceived as good, bad, or unknown. Can a temporary physical abundance be a blessing if our selfish indulgence flourishes and we do not need and drift from GOD? That could not be a blessing. What if great physical loss struck our life; is this always a spiritual curse? What if we learned and gained great spiritual strength by that loss, we could receive no other way? Yes, a great personal loss, can in a certain way, become a spiritual blessing. When CHRIST died, there was great loss felt in that day by many; but it became the world's greatest blessing. Let's agree that the definition of a spiritual blessing is anything that in the end, brings us spiritually closer to GOD; and anything taking us further from GOD in the end, is a true curse. In this world, we can know at first, if something *appears* to be spiritually or physically rewarding or threatening, but that is all we can know at first. Remember we live in a stochastic world, filled with free-will choice and chance occurrence; we should not think certain things come our way by divine intervention; but we must know all random things that befall us, can affect our Spirit. So, in the end, every experience is a Spiritual curse or blessing, whether by divine intervention or not. Who among us who returns GOD'S love, can always know if GOD touches our life through what appears to be a chance occurrence? Who can know the limits of GOD'S love and wisdom towards us? How do we see GOD'S hand in our life, when for love, GOD hides his hand so well? We must know, being given a non-coveted stewardship over the provision of many souls is one thing, yet abundant personal possession is quite another. I have never understood those who claim JESUS, yet say all physical things in abundance are their personal physical blessings from GOD. Many believe in their heart, that GOD blesses them with physical comfort, security, and possessions. Why would GOD bestow worldly temptations and comforts of the flesh upon us, after warning us to resist these things? The flesh is covetous and deceitful; and physical gratification is our soul's greatest enemy. Many say to us, if you die and there is nothing after, you have wasted your life. I know life is eternal, but I will play along as if it is not. If all things vanish and cease to exist, life holds no real eternal meaning, purpose, or value, for me or you in anyway. How can one in any intelligent way, say pursuing a life of temporary worldly desires, will at, or beyond death, bring them anything. If you have something that is physically desirable to all flesh, and you let me have it for one month, then take it back, what can that be worth? What can a temporary experience of possession, mean to anyone in the eternal? We spend our life, seeking things that make us happy. And for those like me who desire the knowledge of, and the pursuit of the one true GOD we have come to know; who has revealed himself to us in varied measures at his desired times. This has brought us more joy and peace, than if we owned, controlled, and had complete adoration, of the entire world. And what does this do for us at death; we see unimaginable love and life for all the eternal eons. We see that the worship and possession of this world, could never bring us happiness. To live for the flesh, would be but for a blink in eternity. Esau sold his birthright for a meal to satisfy his flesh in a moment. And he cried the rest of his life to be restored; but it was gone and too late. We seek only *eternal* love, and true lasting joy. Nothing in this temporary world compares to our joy that comes in knowing the one true loving GOD has named us, and prepares an eternal place for all who love him. Now one *can* also lust in the flesh for GOD through false doctrine if we are not careful. One can believe they are building a lifelong relationship with GOD, but if it is based on what they hear GOD will do for them, it is false. We must overcome this, as Scripture says a selfish love is lust. Our love towards GOD for who they are, in all glory, is the reason and purpose for all creation. And all who spiritually love GOD above all, will be saved; and that too is a blessing. Those who just want GOD'S physical provision, will not be saved. So, a *physical* desire for GOD, is a curse. Does GOD seek our highest eternal love, or just anything temporary that anyone describes as their love? If we say we love GOD, yet we offend and hurt GOD. How is this love? Love does not exist in forever hurting, and saying you're sorry. The one thing a loving FATHER cannot create, is returned free-will love. FATHER did create the complex conditions required, to allow all choices between perfection or imperfection that will distinguish and separate all eternal love for GOD, from all lust. Eternal love breaks open our heart and plants itself, and being a young tender plant, we seek to learn more about GOD'S good eternal ways to nourish and grow our love. We learn of GOD'S ways in wanting to be with those above, who loved us first. Early on in returning to GOD, still being weak, we see and are joyed by the milk of what GOD promises us out of his love. Yet as we grow past needing milk, we find the meat, sinew, and bones of Scripture that reveal the pain GOD endures for each of us, teaching us GOD'S need to be loved in a perfect understanding. We feel this loving need cut our heart with shame each time we sin; for each sin is our irresponsible unfaithfulness to GOD. What does our heart feel when we hear FATHER say how he would have gathered all together with him, but they would not? Have we not felt the pain of trying to help our child, or another child out of love, but they would not be helped? If so, we can barely begin to comprehend the pain felt by GOD for billions of lost souls. Each sin towards GOD, is like the unfaithfulness of an earthly spouse towards their beloved. And if we know this, we hold a true love for GOD. Would a loving parent hurt their newborn baby, or deny them what they need? Would they give them things, not good for them? Of course not, because they love them. To be in the flow of faithful responsible love, not just from, but back to FATHER, is to be a true child. JESUS said we must become as a *little child.* One part of the meaning in this, is in the way we love. If we see the differences between an adult and a child's love, it is a blessing. A powerful difference, is adults have matured in the flesh, and succumbed to knowing it far more than children. Adults take freedom and money to buy the desires of their flesh, and influence over others. And when we first experience that intimate physical relationship, this changes our soul far more than most realize. If the whole world did not see it as a natural biological function, science would declare sex to be the strongest, most addictive habit in the world, consuming nearly all of mankind, across all walks of life. Look at advertisers, who's number one commandment is sex sells. But sex is meant by GOD to be a physical necessity, only for married procreation, and not for recreation. I am trying to show how all fallen souls experience a great change in their spiritual to physical balance after their first experience, as a strong reversal of control occurs between the soul and the flesh in most. The understanding of spiritual things become blurred by strong physical desires; and most are turned to sustain the desires of the flesh. So, they are definitely not child like any more. We see a good *little child* has only a loving *subservient* familial understanding and perspective. What does it take for an adult to return to this understanding and perspective of a child towards FATHER, and is it possible. Observe the last supper. Scripture says John, in our eyes a grown man, was like a child, as he laid his head on the chest of JESUS, whom he loved and trusted like a child would. Why, well he had come to the great knowledge that JESUS was the SON of GOD, having no physical desires; being so high above who John saw himself to be. So, he felt childlike in his presence. We see a grown man, doing something, a five-year-old little brother would do in simple pure trust, humbly leaning on the *much* older stronger brother he looked to in all trust, adoration, and reliance, resting in a protection of familial love. And in that purely spiritual moment, there were zero thoughts, cares, or communication of the flesh, between John and CHRIST. When we enter into FATHER'S rest, on *his* Sabbath, we see ourself as FATHER'S child, climbing up and sitting safely in his lap all day. In our true familial love, seeing FATHER far above us, without effort we become as little children looking to him for guidance and care. With nothing more to offer or give in return, but childlike obedient loving trust. Our wanting to see FATHER smile at us, like a small child seeking the loving approval from a dad. This is our created place and purpose. What distinguishes an adult from a child? Is it size, age, understanding, knowledge, truth, or just years. Do you want FATHER to view you as his child; or as a mature adult, being more equal? Remember, *love* for GOD can never be measured in our profession that GOD consciously and physically exist; but only in our loving desire to please GOD and know their love for us. If our faithful love grows, this is our greatest of blessings. Do we have enough faith? Our faith is our desire to be and live perfect before all untouched glory, through the power of CHRIST and HOLYSPIRIT. Are we still running up the road of emulation towards GOD, who is our blessing? Or are we now walking? Do we only take a casual stroll closer now and then? Or are we camped on the side of the road, thinking we have surpassed others, so we see no need to get closer? Do we have a stagnant faith, or a continual growing strength of faith? Have we come to a firm conclusion through our loving hopes and desires to please GOD without end? Do we have a firm belief by the evidence and knowledge we have acquired? Do we have faith enough to lay down our life in all accordance, given by the word of GOD? Faith is the root word of faithful; meaning to stay, to love, to care, to protect and serve that which you are faithful to. Are we faithful to GOD almighty first, and then to his Children? And what is faith as a mustard seed? In its great blessing, it is so small, but it is real; being actual, not almost, not sometimes, not maybe, not I think so, not I will try. It is tiny, yet firm, assured, and without doubt. Do we have this tiny firm faith? Is this firmness of faith gained through undeniable contact come down from Heaven? Is it taught by GOD to only certain souls for a special purpose, such as Prophets or the first Apostles? Is it something we all can attain through growth in love and understanding? Or, is it that when we each grow ourselves in faith and love by practice and obedience, and we come to the precipice of absolute faith; that then, JESUS comes down in that reality to reveal himself and give the touch of his pure loving seal? CHRIST being both the beginning and finisher of our faith. We each are called to achieve this closest purest firm relationship with GOD while still in this world. What would we each do or give up for that seal? Ask yourself sincerely, *would* you really and truly want that? Or the world you have now? Be careful about saying yes, because you can surely have that at its cost. Run now, never look back, run from this world right into the arms of GOD. Give no great thought for anything you leave behind. GOD has promised to be there with arms out waiting. What could possibly hold us back, but doubt, temptation, and covetous desires of the flesh? We have asked in this chapter what blessings are? They are the strong lessons of a good Spirit found in faith. But in this physical world, a blessing is only made manifest (true and evident) through our true works of GOD. Faith founded in love is a greater thing than works; but the greater cannot exist without the lesser. Judge for yourself which is our greater blessing; that GOD loves us, or that we love GOD? They both must be pure and eternal. We must desire as much of each as we can hold. One can do a great multitude of physical works in Churches without a true faith in CHRIST. But we cannot have a true faith in CHRIST, without true works. These being first, our loving obedience to FATHER in all his eternal ways. Then after that, what is our calling? There dwells the unknowable lengths and breadth of the vast mercy of the Triune GOD, who are able to save all they choose, even to and from the ends of all darkness, despair, and pain. Amen.

How do we know Salvation and Eternal Life

The WORD says to a lost world, GOD can restore you to perfection by your true repentant profession of obedient belief, born of true eternal love. In knowing YESHUA is the only begotten SON, born first, and out of FATHER alone. And in praying and turning from all our sins, asking and believing CHRIST can, and may forgive us our *past* sins by his great sacrifice. MESSIAH Judges our hearts and knows if we are sincere; and if so, he forgives us not at, or by our asking for it, but only in his desire to give it. This seemingly simple moment of granted mercy, moves us a great step out of time, hell, and death, and into grace, love, and eternity. One vast step, that moves us a spiritual distance, greater than the breadth of all creation, from where we were, to where we will be. But what about our next steps; we can't stay and stand on that first step for eternity, though many do try. One can also turn and go back into the lost; and many do. But if we would continue in GOD, Scripture says many things about our abstinence from all sin, and our continued fervent study of the Scripture. It says we are lesser Gods, made to now see the physical world, as fallen, lost, and evil. We are of another Kingdom, no longer of this world, crucifying all fleshly desires for the things of this world, crucifying our individuality daily. It is our Spiritually turning away from this world, and yielding in loving desire to every teaching of HOLYSPIRIT. And by steadfastly growing in truth and keeping the body a Temple for HOLYSPIRIT, we will be saved to GOD. Each of these steps needs to be given serious study as to what they entail. Luke 14:27-30, "*AND WHOSOEVER DOTH NOT BEAR HIS CROSS, AND COME AFTER ME, CANNOT BE MY DECIPLE. FOR WHICH OF YOU, INTENDING TO BUILD A TOWER, SITTETH NOT DOWN FIRST, AND COUNTETH THE COST, WETHER HE HAVE SUFFICIENT TO FINISH IT? LEST HAPLY, AFTER HE HATH LAID THE FOUNDATION, AND IS NOT ABLE TO FINISH IT, ALL THAT BEHOLD IT BEGIN TO MOCK*

HIM, SAYING, THIS MAN BEGAN TO BUILD, AND WAS NOT ABLE TO FINISH." Some souls start this transformative work of CHRIST in turning from a creature of the flesh and this world, to become a creature of GOD and a New Kingdom, yet are unable to finish. It is a work of love to be perfect before GOD. In understanding Scripture, repentance is not saying we are sorry. Saying the words "I am sorry" has little to do with repenting, repentance is not found in words. Repentance has everything to do with our broken heart and shame, stemming from the knowledge we have caused hurt and destruction to love, truth, light, life, and all that is good. There is our painful acceptance in our hearts that *we* have been *unfaithful* to the one who has always been faithful. It comes by our knowing that through our every little sinful action, we contributed to the destruction of GOD'S created and desired spiritual physical world around us. That we added to the level of pain and misery that permeates all we see. Our every tiny sin, does not hurt us alone, and does not offend GOD alone. There are real, evil, and continuing consequences to our every violation of the true way things were designed and created to be and work by FATHER. All the horror and suffering we see and hear about in this world, are the affects brought by the accumulation of constant ever-increasing individual sins. How can we continue to sin, knowing we then help destroy all that is. We pierce the heart of GOD every time we reject the eternal perfect ways. In each sin, we reject GOD'S love and authority by our inward desire to be the god of our own life. It is all and every sin that destroys love, life, and creation, in bringing hell and death. When GOD says he will punish those who hurt the earth, GOD is speaking of sinners. They are not pointing first, at any who physically pollute the earth, or abuse its resources. Can GOD truly love *us* forever if we continue to wantonly sin? Do *we* truly love GOD above all if we willfully continue to sin against them? What is our visible proof we see in our life, that we love GOD? Do you sin against your parents, spouse, siblings, children, and friends; where is your love for them in doing that? Do you live seeking and always wanting to know more about GOD? Do you want more and more of GOD in your life, never feeling satisfied with what you have and know of them? In studying the WORD, do you get revelations about GOD, through HOLYSPIRIT, that break your heart for all they each have endured? Do you still want to trade more of this world for GOD'S; more than you have already? Do you see this world stands between you and GOD; and something to be overcome, that pulls and tempts and tries to confuse and lure you from GOD? Will you chase GOD'S face if he hides from you? Will you search the depths of the ocean, the heights of the mountains, and the breadth of the earth and stars for more of GOD and their love? Do you crave ever-more knowledge of GOD'S ways and creations? Will you truly die for GOD? Are you dying from your past and this world for GOD now on a daily basis? Do you live to emulate JESUS? Is this your strength of love you have and give to GOD? Is this not the love they deserve? Do you obey all commandments and sayings? Are you filled with love, or filled only with shame for the weakness of your love for GOD in your heart? As with family and friends, the more time we spend with them, giving of ourself spiritually, as we care for their needs, the more they know we love them. GOD'S love has been given in full from the very beginning; and the commandment of love itself is for us to return that love in full. JESUS came into this sinful world to reveal, illuminate, and purchase, our only way out. CHRIST did not come to surrender, or stay and dwell among us in continuous sin. We are asked on one side, to give and to obey. But on the other side we are gifted by GOD with the strong power of love and truth to live by those commands; *if*, we want GOD more than anything else. We have all been asked from the Garden of Eden until now to do one thing; to out of love and understanding, obey every perfect word of GOD. If any say on that day, I tried LORD but I *could* not. GOD will say you lie to yourself and to GOD, for it was by the desires of your own heart that you *would* not. FATHER will say I gave you my soul and commandments; I gave you my SON to suffer in front of you; I offered you HOLYSPIRIT for strength and understanding. JESUS and HOLYSPIRIT will say we offered you all the loving forces that exist, to do and overcome all things. If we fall short, FATHER will say I told you to fear me for your final outcome. But after all these things, your head turned back to obey your own desires. What will GOD say on the day of Judgment to backsliders, and those who have fallen away? Do all understand the differences between backsliding, and falling away? These are two very different degrees and distances of moving away from GOD and towards the sins of this world. Backsliding can happen to any soul after their initial first true repentance and forgiveness. But falling away, can only happen to a child who has attained the Baptism of HOLYSPIRIT. Many think they are forgiven children, but are not. Many who have been forgiven at least once, believe they have the Baptism of HOLYGHOST, but do not. Coming to GOD is a growth process, not a sudden on or off switch. GOD and their words will Judge us. Backsliding is a Child losing ground, stumbling back into a sin they had overcome to a certain degree; similar to an addict's relapse back to one or more certain drug. From this stumbled confused state, though it is hard, backsliders whose love and faith in GOD has not been lost, may again recover their strength to repent, and again receive grace and try to please GOD. But to fall away, in *all* strongest sense, is to fall too far, after being so very close to GOD, with no possible return. This is something far deeper and it is an irreparable change. It is a renewed fracture and break from true love and faith occurring at the very core of the soul. Such a soul is then an irreparable broken vessel, such as Satan. This can happen for all sinful reasons, such as being greatly deceived; to become so confused again, one can no longer know the true GOD from a false god. A great example being believers who turn to take the coming mark of the Beast, in believing Satan is GOD. Or in reverting to a conclusion, that though one believes in a god, they no longer believe the Bible is GOD'S truth. Another way is to believe the Bible, but find unacceptable disagreements with GOD'S ways; or putting our love for any soul above our love for GOD. By these, our love for GOD is weakened and we no longer love eternally. There are other examples, coming by all confusions and deceptions. Read Hebrews chapter 6, for it speaks of those for which there is no further remedy. Fear every word of GOD, they are true and will never change. Strengthen your love, and find hope in the mercies of GOD.

How can we know ourselves

By what standard and measure can we ever try to know ourselves. Do we use GOD'S standards, ours, or a consensus of mankind. So, a great eternal question is, who are we, and *what* are we. If we try to define ourselves, using only physical existential parameters and having the eyes of this world, we can never see or know our spiritual nature. But if we see and listen to ourselves and the creation through GOD'S eternal perspective, with no other bias; we can see ourselves in ways that guide us in aspiring to reach for Heaven. If we are true believers, we know our only strength and compass, is our love for FATHER, JESUS, and HOLYSPIRIT. And in knowing the Holy Bible is the strong foundation of GOD'S truth towards us. *Then* we see ourselves through the knowledge that first, we are minute conscious souls, set deep within FATHER'S vast Creation. We know the body we indwell is part of FATHER'S perceived creation, which is sustained by him. And we see we are unable to provide for ourselves a physical platform for life or truth. So, we are bound to, and reliant on our Creator for all things. All bodies are created and sustained by FATHER'S forceful desires that perceive the same. But *our* soul is ours, an eternal gift of a minute yet adequate portion of FATHER'S own Soul. So, all bodies in all strict definition, belong in every way to their Creator. After we accept these eternal facts, we are then divided into two main groups. Group one are those who seek GOD, knowing and accepting all obligations that come in submitting to the truth of FATHER'S purpose. Group two, are a rebellious proud people, who believe that all or part of their physical lives, belongs under the rule of themselves; believing it is theirs, to do with as they see fit in living a lie of their created purpose. In two additional sub groups; mankind is ascribed as existing in *two* spiritual and physical levels, purposes, and appearance, being made male and female under FATHER'S created intention. But the growing confusions of hell, blind many to GOD'S created immutable distinctions. After that created distinction of purpose, we are each a soul, gifted in orders of differing size reach and responsibilities by GOD. These perfect ways and orders, are set and delineated in FATHER'S wisdom and purpose. This is a basic definition of who we all are at our most basic foundation. Scripture declares we all exist through dualism of two distinct components; and for our continuity of experiential awareness, we require both. A gifted soul, and created body, are defined in science as soul Dualism. Do you believe in these two distinct components?

Are you a pure Soul Dualist? GOD then commands us to know and feel each component in quite separate ways; to know, see, and keep them separate and never see them as one. As long as you see or feel yourself as one with the body, not seeing a true separation, you can never overcome the flesh. Since mankind fell, and surrendered their control to fleshly desires, the flesh and Soul have switched in their created controlling order, now born of evil desires and contrary purposes. We must switch them back, now! There is nothing more important to the flesh than the current moment, and the things that bring physical experience. Being lost and trapped in hell, death, and time, we should seek nothing more than the ways that perpetuate eternal love. There is, since the fall, a raging war for control between the two. The flesh is now who most strongly want to be, and are; while the Soul is who we need, and should desire to be in Holiness. We cannot be both. Rev 3:15-16, "*I KNOW THY WORKS, THAT THOU ART NEITHER COLD NOR HOT: I WOULD THOU WERT COLD OR HOT. SO THEN BECAUSE THOU ART LUKEWARM, AND NEITHER COLD NOR HOT, I WILL SPUE THEE OUT OF MY MOUTH.*" From our beginning, we soon find big questions. And our core desires towards these questions, define and differentiate us all in many ways. We see to one extreme side, many who do not want to waste a second in this world wondering why, or who, or where they are. They just want physical pleasure from the ride until it is over. On the other extreme side, are souls who spend their days and dreams searching for the *eternal* truth, reason, and purpose of life. And then, there are all the souls holding differing levels of desire and concern for both knowing, and not knowing, wanting and not wanting, across a vast spectrum in between. We each are found at one of these extremes, or in the countless degrees along that spectrum of couldn't care less, to consumed by caring. And in listening to believers across this spectrum, especially near the lukewarm middle. We hear them say, that those on the weaker more physical side of them, are in danger of judgment, but those on the stronger desiring spiritual side of them, are fanatical and legalistic. JESUS, our example to *emulate*, was and is found to be most fanatical and legalistic. We should not think we are ever complete, or in our strong perfect place. Or ever at a place to stop learning to perfect ourselves. Does this answer the question for you, as to which people GOD seeks across this spectrum? This question has an obvious answer. Scripture in its fullness says GOD searches for those who fervently and lovingly searching for them. Scripture teaches us repeatedly that the harder we look to GOD in love, the more they look upon us. And the less we look to who they are, the less they see us. If you need any convincing of this, it is all there for your careful reading throughout the Bible. Scripture commands us to love GOD with all our heart, mind, and soul. This is obviously describing the person to the extreme side of seeking truth and answers. Those seeking GOD in their personal longing, consumed in their belief and knowledge that our loving Creator wants the same thing we do; to be eternally loved on a most personal level. FATHER seeks a true Child relationship with us. Do you seek something less or more? What is it you righteously seek with your children? There is no way one can say by any real authority, that FATHER wants us to show our love for him by staying at a distance, to get lost in physical pleasures of abusing all he has made. It is impossible to truly know ourselves, if we can't stop and truly examine ourselves. It can be just as hard to know our-self deep-down, as it is to know someone else deep down. We often see and know only our surface, and their surface. How can fallen mankind ever know the full truth of each other, when souls led by their flesh, instinctively hide and minimize their shame and weakness to each other, and to their selves. And to some extent, most exaggerate their attributes. Consider that within all souls, most are *temporarily* making what should be *eternal* decisions. We make all our decisions based on two *of* three basic aspects. Two being, do we hold an eternal love and desire for GOD, that controls us? Or do we hold a love for the world and the flesh that controls us? And three being the chanceful trials and temptations that come upon us each in our lives. This simple equation, times the sum of all inter-soul experiences in all creation, allow for a near infinite number of desired outcomes in each life. Each soul grows different levels of basic want towards GOD, and or, the things in this world. So, by these aspects bringing desires and chance, we make our choices in all situations. So, again, how can we deeply know ourselves in true detail? All alone, we can only get lost in ourself; and only through our knowing obedience to GOD, can we ever know ourselves. Can we try to know ourselves or someone from the minimal amount of detail we sense and process in our awareness of the physical? How can we ever know another from such limited information, not seeing the influences and lessons of another's experiences, preferences, and dislikes? If we have had a long, good, easy physical life, can we begin to know a person who has had a long hard and bad physical life? Or visa-versa? What does a short hard physical life, spiritually do to, or for a soul? And what does a long and easy physical life, spiritually do to, or for a soul?

Does it depend on the soul, or the life? It is the belief, love, and desire for GOD that rises or falls within each soul that is the great consideration. And in not one thing being static or stable in this fallen creation, except for the truth and love of GOD; the influence of all of these considerations upon each soul, must then change with a multitude of other contributing factors. By the laws of love, temptation, chance, and endless combinations of the same, all souls without the true mooring guidance of GOD, are set adrift in the confused variations of spiritual and physical perspectives called hell. Surely good souls can arise out of a hard-physical life; and bad souls can arise out of an easy physical life, and visa-versa. So, as to the physical life sensed by us, our spiritual outcome can only be known and judged at its conclusion. Should we not conclude that as Scripture says, it is a great benefit to all who remove and flee temptation in their lives. We must separate from this world's iniquity and influence upon our choices. Don't blame your decisions concerning GOD, on your physical life; or your decisions concerning physical life, on GOD. What if someone seemed to receive not just an unusual amount of physical bad in life, but it seemed to come at always the worst of times. Or if one had an easy physical life all the way through, experiencing abundant rich fortune and pleasure from birth.

Would these take different degrees of spiritual strength to come to GOD, leaving these different ungodly lives behind? And which would be harder? Scripture and reality teach us there is great purpose and an equalizer to all things; being our belief and love for GOD. In that CHRIST put off his eternal body, and came down so low to be in the flesh of mankind, to sinlessly endure a harsh impoverished life, and the harshest death. And at *our* Judgment, by this fact, can we ever declare physical injustice or unfair advantage as an excuse for our sin. The only surety given us, is that if we try to reach GOD with all we are, GOD will always meet us in our efforts with their spiritual strength and comfort to overcome any and all things. What we are trying to show, is that we can never judge ourselves, or another from the perspective of one's physical life's experience. We must first judge the *Spirit* (good or bad attributes) of each soul, *only and always* in the moment at hand, through wisdom, love, hope, and discernment. Or, we can find ourself faulting and denying someone who is in grace, and guiltless through the blood of CHRIST; maybe even attacking a work of GOD. Let's not find ourselves working against a servant of the GOD'S Kingdom, in that through pride or ignorance we cannot set one's past aside, including our own. What does one's past mean; ten years ago, one year ago, yesterday, or one second ago. Do we think if someone commits a crime one minute ago, but then *truly* repents and *is* forgiven, they must still go to jail? Aren't they now altogether a new creature? Does not GOD say they are now altogether clean? Does this often occur? No. Now we cannot know if they are truly forgiven, and must in faith try their spirit to see if they are of GOD, or if they are still evil. Have you ever done anything illegal long ago, for which you were not caught and paid your societal penance. Do you feel obligated to confess and pay a fine, or serve time in jail for everything GOD has forgiven you for. In Scripture, when the law said the woman taken in the act of adultery, should get the death penalty; did JESUS say to her you are forgiven, but you still have to pay a lawful price for your crime? No. Why not? Because when she truly and shamefully bowed and vowed and desired to never do it again for all eternity. JESUS paid the price of death for her. So, those sinful rulers were cut within, in their demand for her punishment. If our LORD paid the price for a sin, and we say we accept that; but we say they should still go to jail and pay again. We are denying the power and worth of what our LORD did. I do not say all who repent with their lips should be believed. If found unrepentant, we must banish souls until they are. But if we discern that they have repented, we must relent in belief. We could for decades trust and love someone who in their past, committed a crime they repented of, and lived a life for GOD, never being caught. Not knowing this, all see them in a certain loving light; but if one day we learn of their long past repented and forgiven deed, would you suddenly see them differently? Would you see someone you don't know, who should pay a price? We *are* commanded to know and judge sin, to see and know whether something is good or evil, and to separate ourselves from all ongoing evil. But GOD'S true Children are never to execute any *eternal* judgment upon another for sin. All punishment of death, was taken from mankind and given to CHRIST alone, as we can never know or deliver a righteous death penalty. Yes, parents and Governments have a need to banish or incarcerate those found violently beyond societal moral and lawful practice. Communities and families, who lived before CHRIST'S sacrifice, strove to live righteously, and were commanded to remove rebellion whenever it arose, to preserve the sanctity of the people. But the people are now in mass, *all* gone astray. So, FATHER reserved that final sentence and power of death to CHRIST alone. This is the fullness of what JESUS meant when he said, let any who are without sin, cast the first stone. Did not Paul imprison and demand the stoning to death of the

Godly such as Stephen. Should Paul have been sent to prison or executed for false imprisonment and murder, for which he confessed guilt in Scripture. When the Scriptures say "judge" here, and "judge not" there. We see it delineates these statements many times in saying we should always see, judge, and know sin for sin, and evil for evil; all according to GOD'S commands; and separate ourselves from the unrepentant. Thereby exhibiting and living within the commanded wisdom of GOD. But to condemn and demand a *judgment* of death upon another, this is now a Righteous Judgment that our LORD, having died for sin, said is now his purchased right alone. Even most of the professed Christian Society has differentiated, and looked down on those who have been caught and convicted of crimes, thinking them more-evil than those who have never been caught. Most cops and judges have committed at least a few petty crimes in their lives; yet were never caught. And if they had been caught and convicted, they may not have the position they now hold. They know this, but hold a position over others anyway. If a Judge or a Policeman commits a petty crime today, do they turn themselves in with much to lose? I doubt this occurs often. We further can say without hesitation, that in every single case upon the face of this earth, where mankind has judged and sentenced another, there has never been *perfect* justice. As GOD'S Children, we dwell in his higher standards, which surely extend to a thing as valuable as justice. We seek a *perfect* justice that mankind is horribly incapable of rendering. In every case in history, the judgment handed down by mankind, was and is now, by great varying degrees, too severe, or not severe enough. In many cases the guilty are found innocent, and in some cases the innocent are found guilty. But not once ever by mankind, has there been handed down a sentence, in which a judge knows and considers every possible aspect without bias, to render a perfect penance. Should the Children of GOD be party to any imperfect injustice? We are commanded to come out of, and not touch any unclean thing. Consider the other systems of justice around the world. Many governments are a terror upon the people, oppressing, abusing, using, and even murdering them under the terms of a national or state system of justice. And what of this so-called forgiving nation called the U.S.A, which must many times break apart families, taking mothers and fathers from fearful crying screaming children over lesser matters which should not be. More and more people break laws in abandon, it is an impossible situation. We see souls who try hard, can stumble and get caught up in the machinery of this world to be harshly punished, while violent criminals are coddled and encouraged. We are not condemning law and order in the least; we are pointing out mankind's weakness and imperfection. Now of course we know most places would be worse off without their imperfect systems of justice; and a few godless countries may be better off without theirs. But GOD'S Children hold no part or hope with this world's justice. We hold with the next, where a perfect Justice will sit. Scripture says that whatever a person has truly repented of in their past in GOD'S forgiveness, is set aside as if it never happened. Live in this truth, or boldly live a lie before GOD who will Judge you with perfect Judgment. I've always found the tough stand of the Christian belief in the death penalty is wrong. Just as wrong as the Christian left who support abortion, unholy marriage, taking prayer out of school, and all other forms of immorality. And in Government debt, we see the current, steal money from future generations without permission, to address their current desires. All are wrong on these and many issues. Many compromise the words of Scripture towards a certain worldly consensus of comfort and power. GOD and their words will judge us. This does not in any way, mean to bring any and all into our camp who profess the name of JESUS. We are called to protect our children, to try all Spirits to see if they are truly of GOD. We are commanded to give what we have freely, where we see a *Godly* need. We are commanded just as often to protect and strengthen all GOD'S Children, lest we fall to temptation. We can do both, and are responsible for our choices and their consequences; some requiring repentance. Who can see that this life and what happens here, is only important as to what we choose to give back to GOD in our belief. The sum of the reasons for the fallen flesh to resist the ways of GOD, is unknowable and growing, because they are all individual lies which perpetuate more after their kind. The false reasons for a proud soul to resist the instruction of GOD are many; but GOD has set before us, all pure truthful reasons for our soul to hear and follow. And when we share these truths in this world, we are most always disappointed at the lack of reception. We see people trapped inside the error of false doctrine, or a religion of their own design. It is hard to impossible for most to see CHRIST is Spiritually supreme, and he brings to them all great truth. For many it is like trying to comprehend another is much stronger or more intelligent than they. Faith is tested by experience; but Faith is never founded in our experience, it is a leap beyond our experience. How can one know certain qualia, in not experiencing it for their self. How can a weak person know what it feels like to be able to bend thick steel. How can a person with an I.Q. of

90, understand what it means to have an I.Q. of 180. The souls trapped in false doctrines, know only what they want to see and feel. They dismiss truth in their own way, as they contort all they do not understand or want. The Scribes and Pharisees rejected MESSIAH in this same way. It is up to each to find their way by following HOLYSPIRIT, by hearing, searching, and praying. GOD has always suffered to make us a way to come back. Mankind has gone so far away, that GOD has had to give *all* they are and have. And so the end must now come. We soon come to the fullness of GOD'S work for love, and his purposeful creation. What shall we love; I love all ways and commands of GOD, and all who believe and truly love GOD in obedience to his every word. I also love in hope and caution, all my salvageable enemies. But the fallen things and ways of this world, I avoid and I hate. What do we see as the big picture? GOD for a short time, still holds their hopeful love for the fewer in this world. GOD gives all souls free-will choice, and surrounds us all with choice and chance. FATHER gave all who would receive them, his words and gifts of love. Yet most all receive FATHER'S words and love differently, revealing their true core desires. We are known to JESUS by our love for him, being our obedient countenance. Our love to each other is second, and we must surely remember this. Our love for GOD is shown in our loving obedience to every commandment of love. And that our love for each other is only possible from the source and the well from which we draw, being GOD. Above all this, who can know the whole mercy of GOD and the ends of his love and forgiveness? Amen

How do we know Godly Religion and Governance

Why does this evil world say the Bible is filled with bigotry and contradictions. By their contrary evil beliefs, they hear and see a confused temporary physical reality. Evil interpretations over and under emphasize, under and over-value, and deny GOD'S true intent, allowing endless contradictions to arise in *their* individual interpretation. The original continuity of any meaningful writing, fills with error and conflictions when misinterpreted. The Bible is written in a most special Godly way. And to be properly received, it has to be read in the exact same way. If many souls read a verse, and all hear different meanings. At most, only one can be right; or they can all be wrong. Scripture was given through HOLYSPIRIT, and it can only be interpreted through the same. HOLYSPIRIT indwells us, mainly to teach us GOD'S eternal truth and love. Have you been taught the meaning of Biblical Scripture by mankind, or by yourself, and not by HOLYSPIRIT. GOD'S words must be received in all proper entirety, and not in part. Consider a soup analogy, where we are careful to always taste all the elements at once, to get its true over-all flavor. But if you have favorite aspects in the soup, and pick around certain parts that don't please as much, this will mislead you. We can never minimize verses that come against our worldly desires. This is the root cause for different Doctrines. A Pastor trying to please people instead of GOD, *must* preach a misinterpretation. And who follows the words of a Pastor who casually strolls along with no appearance of alarm or concern. And as liberalism works to eliminate all Godly effort and understanding in all school, work, play, and life, deceitfully saying we must all accept and respect the non-believer; and the lukewarm Church follows suit in political correctness. We must strive and run to GOD with all our heart, speed, and strength, as commanded, not lingering as a herd of cattle to be slaughtered. So many, by reason of a weak desire for GOD, and strong desires for the world, fall behind and drop out of this race. Most souls cannot be saved from their own hearts; this is Scriptural wisdom. Many destroy themselves and others who lag behind with them. If we weaken the Church of GOD'S truth and purpose, trying to fit everyone into Heaven, *all* will perish, and the great perfect loving work of GOD will have been sacrificed for nothing. If GOD allowed everyone into Heaven, it would look *just* like this world. Satan brought war to Heaven, and his children keep weakening the truth to allow more sin here. Scripture says that after JESUS spoke certain things, many went away and followed him no more. JESUS did not call them back and try to weaken the truth for them. We must never look to a weakened worldly religion, or earthly government for spiritual leadership. Look to JESUS and HOLYSPIRIT, upon whose shoulders the eternal governance of FATHER'S Kingdom rests. Worldly Religion and Governments of every type fail, due to corruption, power, privilege, and possessions, that fill them. The true purpose of governance, is to promote and enforce eternal values; but in all mankind's governance, we see debating, compromising, re-defining, and the constant re-writing of truth and values as strife builds, bringing the end of that country. What form of governance in this fallen world, is least destructive. Taking Spiritual Biblical direction, we find that in all honesty, a true strong *Godly* Priestly Monarchy is the best form of governance. But only if you have GOD loving and fearing souls as Priest, King, and subjects. The lesser this quality exist in a Priest, King, or people, the worse and less affective that government will be. No Godly man seeks or has a desire to rule over other men; but evil men have these needs and desires. A Priestly King must be selected by GOD for their qualities, and be anointed and appointed in Righteousness. All forms of earthly government in these days, are wholly contrary to the laws of GOD. Yes, governance is essential to peace and stability, yet the fall of every society's peace, was the result of its people being divided from within by a divided government. Every war is a product of governments building fragile tangled webs of agreements and treaties within this world; building financial alliances, breeding mistrust among un-natural allies. And ultimately, small offences birth a hunger for vengeance. All powers and orders of the flesh, contradict Scripture; and a desire for vengeance is underestimated. Upon a careful study of the human condition, we find souls are not great believers that people can change completely, or permanently. And most struggle with the question of, are all cultures equal, or are do some exhibit more loving Godly qualities than others, being set in their traditions of life by their ancestors. And to what extent do we blame evil environment that brings temptation. Are we all born in a fragile unpredictable condition, which by our core desires builds our strengths and weaknesses, as our internal desires make our every choice? What is the worst thing you have ever done? Can you say if circumstances were different, or you had been stronger, you would not have done it? Could you have also done worse, being weaker, and under worse conditions? Without a great love for GOD, and given the worst possible scenario and weakness for each of us, to what level of evil are we each capable; raising the old proverb, there but for the grace of GOD, go I. Now the opposing question is, what good are we each capable of under the right strengths and conditions? Could we each really lay down our life for another? The Biblical term “the book of” means the story of a person’s life: like the book of Moses, or Isaiah. What if there was a book of you in the Bible, what would its pages show you to be? A Prophet, a Preacher, a Disciple? A simple soul needing healing? A Samaritan, or a Scribe or Pharisee? A sinner or publican, hot cold or lukewarm? A Gentile, a worker of iniquity that GOD does not know? Or a child of Israel and the most-high? Are you of the flesh, or of GOD? Have you ever felt really close to GOD, then lost ground in a spiritual battle and felt you were not as close to GOD as before? We feel loss, because backsliding is exactly what has happened; we also feel closeness, with any gain. Our every choice, leads us one way or the other. Think of the closest you have ever felt to GOD’s presence. Do you think that was as close as you can ever be? Do you think FATHER wants you or I to settle in at a certain distance from him? Can we grow closer every day to GOD? Yes. Does Scripture teach us that GOD wants us to grow closer to him every day? Yes. What is this force that can slow our run towards GOD, or stop our progress and pulls us back from GOD? It is always our lustful desires and worldly entanglements. There are only two opposite pulling forces of desire at work: eternal love, or temporary physical lust. One force is projected from and back towards our Creator, and one towards his Creation. What is lust for GOD'S Creation? It is a line in the sand we should never cross. It is doubtful most know or feel when they have crossed it, and most *professed* Christians spend most of their time on the wrong side. By curiosity, desensitization, and growing desires, most yield to and become the things they see, hear, and feel daily. Should we make sure we and our children do not see, hear, or feel, anything that is not of GOD'S ways? We can feel joy while existing in GOD'S creation among our brothers and sisters in CHRIST, as long as we obey GOD'S every word; and always see the Creation on the other side of GOD, and never between us. We must see reality, looking through GOD. We never want to do anything in his Creation, or with his Creation that is not according to FATHER'S designed purpose. If the Creation ever brings us more joy and pleasure, than the Creator of it, we are in lustful idolatry and great danger. I find something common in most newly professed Christians, including my own early start. Most new Christians focus first, on removing the things from their life that are evil and destructive, not just in the eyes of GOD, but mankind as well. This can be hard, yet rewarding, because we feel we are doing things pleasing to GOD, and yes, friends, family, and our self. Most on earth support souls quitting smoking, drinking, drugs, gambling, fornication, using foul language, and other such things. So, we receive strength and encouragement from GOD and Mankind. But once we get past that common list, and come to the things GOD says to remove from, or add to our life, that our friends and family do not agree with. That encouragement from GOD continues, but from mankind, it stops for the most part. So, at this point for many, their growth towards GOD slows or stops. And if some push through this level of difficulty; what about continued growth into the most Righteous and Holy things GOD asks of us? Things that actually bring ridicule, mocking, and rejection, from our friends and family. If we obey all of GOD'S words without fail, we will be seen as fanatical, even by most *professed* Christians. In his work here, most saw CHRIST as a fanatical usurper. GOD'S brighter and brightest righteous ways, are not wanted by most, and CHRIST waits for us on the other side of them all. In going all the way to GOD, the help we once received from mankind, turns to ignoring, and then to attacking. Coming up against this, *and* one's remaining desires for the world, is where and why most professed Christians stop growing. They either fall away, or settle in for the rest of their lives at a certain distance from GOD that is acceptable to their family, friends, and their flesh. So, we see at its true source, all true Godly governance resides within ourself. If we yield to the forces in this world, we are lukewarm and compromised. GOD and their words will Judge. FATHER said; I would that you were hot or cold (pure good or evil) but because you are lukewarm (part evil) I will spew you out of my mouth (vomit you out as an offensive vile aspect). How will such souls who bow to peers, stand against the coming great pressures to bow to the embodied Antichrist. The honest answer is, if they can't withstand this world, they won't withstand him. If this describes any*one's* current state, whether one knows and admits it or not, they already bow to the beast. The Beast in all ungodly practice, is a world-wide order of direction against GOD, as the world en-masse, goes along to get along. And as the world goes that way, we go the uncompromised way. Please know that GOD never negotiates or compromises with evil, and GOD does not change. *Our* righteousness is manifest in living the loving truth authored and ordained by GOD, which does not change. If GOD says to us friend, we will help you climb to the top of GOD'S highest mountain, where is found your perfection in us. And you say, GOD, with your help, I will climb with you to the top. But short of the summit you say, though you walk with me each step, the physical pressures, hardship, and discomfort of the journey are too hard, and you stop short of the top. Your word and commitment to GOD have become untrue. Scripture says those who endure unto the last, will be given a robe and a crown. First count the cost, to see if you truly want to build your entire house. Do not trust in following partial religious doctrines for salvation. We must break past all levels of desire for the flesh in our walk with GOD; but this is not what every soul wants. There is no doubt, that the level of true love and desire we have for GOD'S ways, determines how far we each climb toward GOD. Our desire *is* our drive, need, and strength. For a last example, consider a soul who says they love GOD, yet they cannot stop all sinning. When they say they love GOD, do not doubt them; but their love for the sin they cannot stop, is still their greater love. They don't love GOD *enough* to stop all things that are destructive to GOD'S creation. We are each commanded to love the LORD our GOD with *all* our heart, mind, and soul. Where does the whole word of GOD place a sinner's salvation? It places it in dire peril. Let's say just for arguments sake, we could qualify and quantify the type and amount of love for GOD, and our lust for the world, we each hold. Then maybe we could say a sinner's love for GOD was weaker, say 35% love for GOD, and 65% lust for this world. This is definitely not all their heart mind and soul. If *their* salvation is in danger? Where then are we? We should always be worried and fearful, I know I am. JESUS said fear not man, but fear GOD who can eternally imprison your soul. What spiritual to physical equation qualifies us as being in GOD'S loving obedience and grace. Must it be 100% for GOD and 0% want for the things in the world? No. For all of Scripture, it is this; we are to above all, desire to be always perfect before the eyes of GOD. So, 70% love for GOD and 30% love for those souls in the world seems a minimum balance. We must desire and seek a closer understanding walk with GOD, fleeing temptations, refraining from all lusts as we learn to recognize them. We must confront and overcome our faults as we find them, for *this* is our love for GOD. We must strive to emulate JESUS, denying the flesh in our love for him, FATHER, and HOLYSPIRIT. Although we can know with some strength, that in our certain moments, we are in grace; yet seeing brothers and sisters fall away, we are never to assume we too, are not vulnerable to fall from grace. It is a great-evidence to GOD of our love, that we remove temptation from our life, not wanting to sin. We won't live life surrounded by temptation, unless we want to be tempted. We cannot then cry to GOD that our sin nature should bear the blame, that is deceitful. Even if we fool our self, GOD is not fooled. Who among us who loves GOD, will endure to the end my friend? Why is it that those in the Spirit, and those in the flesh, see all things so differently, given identical circumstances? This is a strange reality. Now what of political groups, should a Child of GOD hold any earthly alliance or allegiance. We can hold to issues important to GOD: such as blessings towards Israel, protecting religious liberty, the sanctity of Holy Marriage, outlawing *murderous* abortions; and we can vote for those who live by such standards. What does Scripture say about abortion? Many things. Jeremiah 2:33-34, "*WHY TRIMMEST THOU THY WAY TO SEEK LOVE? THEREFORE HAST THOU ALSO TAUGHT THE WICKED ONES THY WAYS. ALSO IN THY SKIRTS IS FOUND THE BLOOD OF THE SOULS OF THE POOR INNOCENTS: I HAVE NOT FOUND IT BY SECRET SEARCH, BUT UPON ALL THESE.*" Abortion is murder. And GOD hates the ungodly attack on his Ancient Holy Institution of Marriage, created and sanctified thousands of years ago by FATHER; *not* government. These are orders and observances to live, as GOD commands. If we vote issues and not party or person, and avoid the liars and close choices of the lesser of two evils, we can be forgiven if we err in our due diligence. Is it a coincidence that GOD'S issues are all under attack by liberal politicians and groups, or is it by Satan's orders? Who is governing this world, who is at the very top, leading the direction of this world order? Luke 4:5-8, "*AND THE DEVIL, TAKING HIM UP INTO AN HIGH MOUNTAIN, SHEWED UNTO HIM ALL THE KINGDOMS OF THE WORLD IN A MOMENT OF TIME. AND THE DEVIL SAID UNTO HIM, ALL THIS POWER WILL I GIVE THEE, AND THE GLORY OF THEM: FOR THAT IS DELIVERED UNTO ME; AND TO WHOMSOEVER I WILL I GIVE IT. IF THOU THEREFORE WILT WORSHIP ME, ALL SHALL BE THINE. AND JESUS ANSWERED AND SAID UNTO HIM, GET THEE BEHIND ME, SATAN: FOR IT IS WRITTEN, THOU SHALT WORSHIP THE LORD THY GOD, AND HIM ONLY SHALT THOU SERVE.*" The Dominion of this world was surrendered by Adam, to Satan when Adam fell; this world for now is Satan's Kingdom. What can and should we do according to Scripture; be a blind Republican or Democrat? GOD surely despises all weakness of compromise. Most professed Christians, who align as Republicans, are lukewarm; yet the far fewer professed Christians aligned as Democrats, are much farther from GOD. There is no eternal hope in earthly politics, we must only look up, and not to the right or left where there is no eternal hope. Most people on both sides are weak, confused, and compromised. They borrow and sell their soul and future for the present pleasures, comforts, and securities in their lives. All around, we see souls deep in their own desires, striving over things of this world and the flesh, not caring for the eternal things. To accept any lesser evil cannot be pure or perfect, and can only be accepted by mankind. I watch and do vote only for leaders who advance GOD'S Commands and Prophecy. Being anti-abortion, Israel's one state solution with its Capital of Jerusalem, and a free Christian nation. I vote for a vanguard of constitutional guarantees that enforce our Christian liberty. Let's discuss the great divide and contention in internal Christian governance, being the *law* of GOD. Why and how has Satan made the *law,* the greatest debate and division in the Church? Because the law defines sin, and Satan knows it cuts and divides the flesh the most. The flesh hates and rebels against GOD'S law more than all things. Satan knows that the love for GOD'S law, guides and strengthens the goodness of our soul to love and please FATHER. GOD'S law is a natural enemy of Satan.

Even in the perfect beginning, there was still one law. Why? To show love for GOD is to trust and obey GOD. Genesis 2:16-17, "*AND THE* ***LORD GOD*** *COMMANDED THE MAN, SAYING, OF EVERY TREE OF THE GARDEN THOU MAYEST FREELY EAT: BUT OF THE TREE OF THE KNOWLEDGE OF GOOD AND EVIL, THOU SHALT NOT EAT OF IT: FOR IN THE DAY THAT THOU EATEST THEREOF THOU SHALT SURELY DIE.*" Breaking this law, shows they have turned to a greater love for themselves and things of this world; and would by *experience,* then know and understand the difference between being good and evil. And as Satan did, they became like GOD in that they *then* knew the difference between the good, bringing life with GOD, and evil bringing death's separation from GOD. As soon as mankind chose the Created thing over its Creator, the struggle for GOD became trying to save as many as would turn back, and again love GOD above all creation. GOD all throughout the Old Testament, tried to repair a broken mankind. From the beginning, GOD has never changed once from the first word in genesis to the last word in revelation. You can search the Scriptures the rest of your life, and not find GOD say anything other than if we love them, we will obey every commandment and saying, for in them is eternal life. It is very significant, that the closing words in the Old Testament, as GOD went into mournful preparation to send his SON to suffer and die, were these; in Malachi 4:4-5, "*REMEMBER YE THE LAW OF MOSES MY SERVANT, WHICH I COMMANDED UNTO HIM IN HOREB FOR ALL ISRAEL, WITH THE STATUTES AND THE JUDGMENTS.*" And as the New Testament dawns, we hear JESUS in his obedient ministry, deliver not a different Covenant, but a continuity to fulfill FATHER'S Prophesied perfecting of his original covenant. CHRIST reaffirms our required continued complete obedience to the law of GOD. Matthew 3:15, "*SUFFER IT TO BE SO NOW: FOR THUS IT BECOMETH US TO FULFILL ALL RIGHTEOUSNESS.*" JESUS in Matthew 4:4 "*IT IS WRITTEN, MAN SHALL NOT LIVE BY BREAD ALONE, BUT EVERY WORD THAT PROCEEDETH OUT OF THE MOUTH OF* ***GOD***." Matthew 4:7, "*IT IS WRITTEN AGAIN, THOU SHALT NOT TEMPT THE* ***LORD*** *THY* ***GOD***" Matthew 5:17-19, "*THINK NOT THAT I AM COME TO DESTROY THE LAW, OR THE PROPHETS: I AM NOT COME TO DESTROY BUT FULFILL. FOR VERILY I SAY UNTO YOU, TILL HEAVEN AND EARTH PASS, ONE JOT OR ONE TITTLE SHALL IN NO WISE PASS FROM THE LAW, TILL ALL BE FULFILLED. WHOSOEVER THEREFORE SHALL BREAK*

ONE OF THESE LEAST COMMANDMENTS, AND SHALL TEACH MEN SO, HE SHALL BE CALLED THE LEAST IN THE KINGDOM OF HEAVEN: BUT WHOSOEVER SHALL DO AND TEACH THEM, THE SAME SHALL BE CALLED GREAT IN THE KINGDOM OF HEAVEN." The term, used here by CHRIST, *to fulfill,* in this context means not only to live by, but to make manifest and complete in his obedience to death. The leaders and people had strayed from GOD'S law, and JESUS rebuked everyone often for their misinterpretation of the law, for working traditions into the law to usurp and lessen the commandments and statutes of GOD. The leaders of that time were surely not responsible for all changes, those changes had occurred over generations. But they still received and accepted them in contradiction to Scripture, as they too desired and perpetuated the same. JESUS would surely say the same today, for most have gone after their worldly traditions, disregarding GOD'S commandments. When JESUS told the Pharisees they did not judge righteously for his healing people on the Sabbath; giving the example that where laws overlap, *they* righteously still performed circumcisions on the Sabbath. They could not understand that every act or work of GOD'S love and righteousness are lawful on the Sabbath, like teaching the Scripture. They could not discern the difference between all lawful spiritual works of GOD, and their daily physical worldly work of gathering sustenance and abundance. It is unlawful on the Sabbath to work to perpetuate exchange and increase from the creation. The flesh has a consuming urge to gather, prepare, and protect the things it physically needs, and desires. This is the entire existence and purpose of all flesh, having no eternal promise or communion with GOD. We can surely see that on the Sabbath we cannot earn money, or buy, sell, bank, or pay bills. We cannot improve or maintain things in this world that is not a true-life emergency, or make plans or preparations to do so. We cannot cut grass, or wash clothes or cars. We may affect our basic requirements of bathing, dressing, and feeding the body, in it being the Temple of HOLYSPIRIT. We may physically aid another if it is truly urgent, as not doing so will cause suffering or death. We are commanded to rest from our work and life of this world, and keep the day solely for our sinless personal and corporate loving acts and worship of GOD. And what do most think of this beloved fourth commandment, which begins by telling us to *remember it*, the one most professed Christians have forgot. Have you traded GOD'S Commands for tradition of man. Do you honor the Seventh Day, have you allowed it be changed to the first, or the sixth day? What does GOD ask of us on this day? We read in Exodus 31:14-17, *"YE SHALL KEEP THE SABBATH THERFORE; FOR IT IS HOLY UNTO YOU: EVERY ONE THAT DEFILETH IT SHALL SHURELY BE PUT TO DEATH: FOR WHOSOEVER DOETH ANY WORK THEREIN, THAT SOUL SHALL BE CUT OFF FROM AMONG HIS PEOPLE. SIX DAYS MAY WORK BE DONE; BUT IN THE SEVENTH IS THE SABBATH OF REST, HOLY TO THE LORD: WHOSOEVER DOETH ANY WORK IN THE SABBATH DAY, HE SHALL SHURELY BE PUT TO DEATH. WHEREFORE THE CHILDREN OF ISRAEL SHALL KEEP THE SABBATH, TO OBSERVE THE SABBATH THROUGHOUT THEIR GENERATIONS, FOR A PERPETUAL COVENANT. IT IS A SIGN BETWEEN ME AND THE CHILDREN OF ISRAEL FOR EVER."* Isaiah 58:13-14, *"IF THOU TURN AWAY THY FOOT FROM THE SABBATH, FROM DOING THY PLEASURE ON MY HOLY DAY; AND CALL THE SABBATH A DELIGHT, THE HOLY OF THE* ***LORD****, HONOURABLE; AND SHALT HONOR HIM, NOT DOING THINE OWN WAYS, NOR FINDING THINE OWN PLEASURE, NOR SPEAKING THINE OWN WORDS: THEN SHALT THY DELIGHT THYSELF IN THE* ***LORD****, AND I WILL CAUSE THEE TO RIDE UPON THE HIGH PLACES OF THE EARTH, AND FEED THEE WITH*

*THE HERITAGE OF JACOB THY FATHER: FOR THE MOUTH OF THE **LORD** HATH SPOKEN IT*." For the exact same New Testament understanding of the Sabbath. The first thing to remember, is JESUS followed it as commanded and exactly as is written, even in his death and resurrection. Study Paul's New Testament writings in Hebrews Chapters 3 and 4., which are about obedience to the Sabbath. Pray and study Heb, 4:8, and HOLYSPIRIT will teach you that GOD'S-day has always been the seventh day, and still is. Isaiah 66:22-24 says it will always be so forever, even in Heaven. GOD'S words will judge us. Back to the whole governing law of GOD in general. We ask, what does the law do? It tells us how not to hurt GOD, others, or his Creation. The law reveals all aspects of sin, which introduce disorder and bring destruction to all FATHER'S Creations, thereby destroying eternal peace, joy, and love. FATHER has told mankind from the beginning, and CHRIST repeated it when he said if you love GOD with all your heart mind and soul, and love your neighbor as yourself, you will fulfill all the law and the prophets. After the fall, as now, by their love for the flesh rising above their love for GOD, people lost the meaning of the 10 Commandments, this to allow for their sins and ways. So, GOD gave a more detailed defined outline in the Levitical Law of what a people loving

GOD and each other should do. And if one should offend, GOD stated corrective measures, saying this is how love itself will best make the people whole again. The flesh must be controlled in how we worship and show our love for GOD, as a great eternal love for GOD is unnatural to the fallen flesh. The law contains severe penalties, because sin is the opposite of love. Love heals and helps, but sin spreads and kills, so, it can *never* be permitted. The *laws* of love can never go away, not without love also going away. If you think a couple little sins, or even one sin is no big deal to GOD, then think hard about the entire world, and all that has ever been lost. Every child born, was born separated and on the side of death for what most say was one small sin. The contagion and infection from that first sin, started, led to, and brought all the evil and destruction, pain, suffering, and death, from then to now. That is why GOD said in the Garden, "*WHAT IS THIS THAT THOU HAST DONE*?". Adam and Eve could not see what they had done, just as most do not see the damage, pain, and death brought to the world by their small to great sins. There is not, and can never be a small acceptable sin. In eternal love, we should not in rebellion see any over-bearing or unnecessary sternness in GOD. But in our understanding, learn the so great lesson of why. See how destructive sin must truly be upon reality and souls. GOD in all their great love, wisdom, and sorrow, knew sin could spread and destroy the whole of creation. GOD has no desire that any should be lost, and it brings him no pleasure that any should be cut off. I see this with all my heart. We see beyond all doubt that sin cannot be allowed, then, now, or ever, for eternal loves sake. So, it seems a bit of a mystery, that by GOD allowing all evil for a short moment. That evil's contrast, shows us and helps us see and feel eternal love. Evil reveals to us how eternal love, is ever so beautiful. Yet any *eternal* allowance of sin, in the end destroys all things. There must and will come an end to all evil and sin; so, trust in GOD'S every word. The need for the beautiful suffering of CHRIST, was not in the law being wrong or evil, but in mankind's weakness in having the law alone; it being insufficient to redeem to GOD all who could know a true love and deep desire for GOD'S ways. FATHER surely did not want his SON to suffer, and tried everything else first; but the need remained and was great, so more might be saved. So, FATHER obeyed his own laws of love and asked his only begotten SON to suffer and die for us. Yet even this greatest gift of love cannot break the cold hearts of stone in most souls. Paul wrote much on the law in Scripture; some of which is too hard for those in the flesh, or weak in the Spirit to understand.

2nd Peter 3:15-16, "*EVEN AS OUR BELOVED BROTHER PAUL ALSO ACCORDING TO THE WISDOM GIVEN UNTO HIM HATH WRITTEN UNTO YOU; AS ALSO IN ALL HIS EPISTLES, SPEAKING IN THEM OF THESE THINGS, IN WHICH ARE SOME THINGS HARD TO BE UNDERSTOOD, WHICH THEY THAT ARE UNLEARNED AND UNSTABLE WREST, AS THEY DO ALSO THE OTHER SCRIPTURES, UNTO THEIR OWN DESTRUCTION.*" We hear in this, that many go to their destruction, studying Scripture in weakness, or the flesh. Scripture says they turn truth into lasciviousness. Which means to interpret Scripture in fleshly way to allow sin. This is no game; souls cannot trade Scriptures back and forth in misinterpretations against each other like a card game, to see who plays the last best card. We must at the first, through HOLYSPIRIT, seek strong definitive Scripture which cannot be misinterpreted, to build our foundation. 1st John 3:1-10, "*BEHOLD WHAT MANNER OF LOVE THE **FATHER** HATH BESTOWED UPON US, THAT WE SHOULD BE CALLED THE SONS OF **GOD**: THEREFORE THE WORLD KNOWETH US NOT, BECAUSE IT KNEW HIM NOT. BELOVED, NOW ARE WE THE SONS OF **GOD**, AND IT DOTH NOT YET APPEAR WHAT WE SHALL BE: BUT WE KNOW THAT, WHEN HE SHALL APPEAR, WE SHALL BE*

LIKE HIM; FOR WE SHALL SEE HIM AS HE IS. AND EVERY MAN THAT HATH THIS HOPE IN HIM PURIFIETH HIMSELF, EVEN AS HE IS PURE. WHOSOEVER COMMITETH SIN TRANSGRESSETH ALSO THE LAW: FOR SIN IS THE TRANSGRESSION OF THE LAW. AND YE KNOW THAT HE WAS MANIFESTED TO TAKE AWAY OUR SINS; AND IN HIM IS NO SIN. WHOSOEVER ABIDETH IN HIM SINNETH NOT: WHOSOEVER SINNETH HATH NOT SEEN HIM, NEITHER KNOWN HIM. LITTLE CHILDREN, LET NO MAN DECEIVE YOU: HE THAT DOETH RIGHTEOUSNESS IS RIGHTEOUS, EVEN AS HE IS RIGHTEOUS. HE THAT COMMITETH SIN IS OF THE DEVIL; FOR THE DEVIL SINNETH FROM THE BEGINNING. FOR THIS PURPOSE THE ***SON*** *OF* ***GOD*** *WAS MANIFESTED, THAT HE MIGHT DESTROY THE WORKS OF THE DEVIL. WHOSOEVER IS BORN OF* ***GOD*** *DOTH NOT COMMIT SIN; FOR HIS SEED REMAINETH IN HIM: AND HE CANNOT SIN, BECAUSE HE IS BORN OF* ***GOD****. IN THIS THE CHILDREN OF* ***GOD*** *ARE MANIFEST, AND THE CHILDREN OF THE DEVIL: WHOSOEVER DOETH NOT RIGHTEOUSNESS IS NOT OF* ***GOD****, NEITHER HE THAT LOVETH NOT HIS BROTHER.”* Romans, 3:31, *“DO WE THEN MAKE VOID THE LAW*

THROUGH FAITH? ***GOD*** *FORBID: YEA, WE ESTABLISH THE LAW.*" The law has always been, and always will be our minimum standard of love, righteousness, and responsible behavior. But the law alone, was and is weak in conveying the love and strength of GOD needed by many to overcome the flesh. One should never say we are not under GOD'S law, for this is improper. Hebrews 8:7-10, "*FOR IF THE FIRST COVENANT HAD BEEN FAULTLESS, THEN SHOULD NO PLACE HAVE BEEN SOUGHT FOR THE SECOND. FOR FINDING FAULT WITH THEM, HE SAITH, BEHOLD THE DAYS COME, SAITH THE* ***LORD****, WHEN I WILL MAKE A NEW COVENANT WITH THE HOUSE OF ISRAEL AND WITH THE HOUSE OF JUDAH: NOT ACCORDING TO THE COVENANT THAT I MADE WITH THEIR FATHERS IN THE DAY WHEN I TOOK THEM BY THE HAND TO LEAD THEM OUT OF THE LAND OF EGYPT; BECAUSE THEY CONTINUED NOT IN MY COVENANT, AND I REGARDED THEM NOT, SAITH THE* ***LORD****. FOR THIS IS THE COVENANT THAT I WILL MAKE WITH THE HOUSE OF ISRAEL AFTER THOSE DAYS, SAITH THE LORD; I WILL PUT MY LAWS INTO THEIR MIND, AND WRITE THEM IN THEIR HEARTS: AND I WILL BE TO THEM A* ***GOD****, AND THEY SHALL BE TO ME A PEOPLE:*" We should say clearly that we are to live by all the Oracles of GOD in love, without fail. But we are currently, not under GOD'S law and statutes given to Moses for the administration of death as punishment, or the remission of our sins by animal sacrifice; or for our faith and strength alone to deliver our eternal salvation. These are now found in our love for, our belief in, our obedience to, and our emulation of CHRIST, after he perfected FATHER'S original Covenant. We should in joy, write the commandments and every word of GOD on our hearts and mind as commanded. We should see the law, love the law, obey the law. The law is the love for GOD'S ways and the brethren. By this inviting HOLYSPIRIT, for it is GOD'S truth. We should never speak evil of the law, or speak against the law, or deny the law. And we should never speak of the glory and beauty of GOD'S law, without then in all amazement, praise, and wonder, speak joyfully of JESUS, the far greater glory, beauty, and love of GOD'S law. Which is GOD'S purest covenant, now written in not just stone, but in the love of his Holy Blood. CHRIST is the Crown upon FATHER'S Eternal Covenant and Promise, ruling from atop, and highly preferred. This is also our fulfilling and perfecting the law through JESUS in us. The following Scriptures touch the heart of this debate and our discussion. Romans 8:1-14, "*THERE IS THEREFORE NOW*

NO CONDEMNATION TO THEM WHICH ARE IN **CHRIST JESUS**, WHO WALK NOT AFTER THE FLESH, BUT AFTER THE **SPIRIT**. FOR THE LAW OF THE **SPIRIT** OF LIFE IN **CHRIST JESUS** HATH MADE ME FREE FROM THE LAW OF SIN AND DEATH. FOR WHAT THE LAW COULD NOT DO, IN THAT IT WAS WEAK THROUGH THE FLESH, **GOD** SENDING HIS OWN **SON** IN THE LIKENESS OF SINFUL FLESH, AND FOR SIN, CONDEMNED SIN IN THE FLESH: THAT THE RIGHTEOUSNESS OF THE LAW MIGHT BE FULLFILLED IN US, WHO WALK NOT AFTER THE FLESH, BUT AFTER_THE **SPIRIT**. FOR THEY THAT ARE AFTER THE FLESH DO MIND THE THINGS OF THE FLESH; BUT THEY THAT ARE AFTER THE **SPIRIT** THE THINGS OF THE **SPIRIT**. FOR TO BE CARNALLY MINDED IS DEATH; BUT TO BE SPIRITUALLY MINDED IS LIFE AND PEACE. BECAUSE THE CARNAL MIND IS ENMITY AGAINST **GOD**: FOR IT IS NOT SUBJECT TO THE LAW OF **GOD**, NEITHER INDEED CAN BE. SO THEN THEY THAT ARE IN THE FLESH CANNOT PLEASE **GOD**. BUT YE ARE NOT IN THE FLESH, BUT IN THE **SPIRIT**, IF SO BE THAT THE SPIRIT OF **GOD** DWELL IN YOU. NOW IF ANY MAN HAVE NOT THE SPIRIT OF **CHRIST**, HE IS NONE OF HIS. AND IF **CHRIST** BE IN YOU, THE BODY IS

DEAD BECAUSE OF SIN; BUT THE ***SPIRIT*** *IS LIFE BECAUSE OF RIGHTEOUSNESS. BUT IF THE* ***SPIRIT*** *OF HIM THAT RAISED* ***JESUS*** *FROM THE DEAD DWELL IN YOU, HE THAT RAISED UP CHRIST FROM THE DEAD SHALL ALSO QUICKEN YOUR MORTAL BODIES BY HIS SPIRIT THAT DWELLETH IN YOU. THEREFORE, BRETHEREN, WE ARE DEBTORS, NOT TO THE FLESH, TO LIVE AFTER THE FLESH. FOR <u>IF YE LIVE AFTER THE FLESH, YE SHALL DIE</u>: BUT IF YE THROUGH THE* ***SPIRIT*** *DO MORTIFY THE DEEDS OF THE BODY, YE SHALL LIVE. FOR AS MANY AS ARE LED BY THE* ***SPIRIT*** *OF* ***GOD****, THEY ARE THE SONS OF* ***GOD****.*" These verses of Scripture center everything on the facts, that by the power of CHRIST, we must kill the desires of the flesh, and be filled by HOLYSPIRIT, raising a new spiritual creature; and a body now being a Temple of GOD. Walking in and always obeying the laws of love, we are truly in and part of FATHER, being his Children. We need to each examine our lives honestly and continually, making two lists. One being the deeds, time, thought, and effort, given daily to worldly fleshly needs like: seeking prosperity, attending to physical health, comfort, beauty, and all the entertainment and pleasures of this world. And on a second list, write how much love, time, and effort, we bring for GOD, to deny and resist the things on our first list. In all this, it brings us to one answer and one question. Why the answer first, because that is the type of question it is. The answer is yes, the suffering sacrifice and love of the Trinity of GOD, is sufficient to erase all forgivable sin. And the question is; knowing this loving sacrifice of love, from FATHER, SON, and HOLYGHOST towards us, do we now love GOD and life enough, to deny this horrid world and the flesh, and receive GOD'S truth and power to stop sinning. To lovingly and joyfully hold to all the Holy Oracles as our profession of faith and love. GOD'S words will judge us all. On the law, the flesh pulls hard one way, and our love for GOD the other. Let your Doctrine be received through your desire to please GOD, and not ever through what will please you or others. Truly, the law and physical works alone cannot save us; but show me how we are saved by displacing the law with sin. We have two choices in all and every single thing we do or say. That is to sin, or obey GOD'S laws of creation. How are we not under the law by our love. Nothing in life is benign; everything we do, think, or say, is lawful by FATHER'S commands, or it is a sin. New-Testament, Romans 2:12, "*FOR AS MANY AS HAVE SINNED WITHOUT LAW SHALL ALSO PERISH WITHOUT LAW, AND AS MANY AS HAVE SINNED*

IN THE LAW SHALL BE JUDGED BY THE LAW:" The law remains. We find that no matter your doctrine, no matter your belief. Scripture says by sin we die, and by obeying all laws of love, we resist and deny sin to live in GOD'S righteousness and salvation. Love GOD, obeying their every word. Know that eternal love and life are our most precious gift from such a love, and such a one, that to truly declare him, there exist no words, thoughts, or imaginings, for FATHER is beyond all things. Praise him.

How do we know Evil

Exactly what is evil? Evil is any and every lie of purpose. It is everyone and everything that in any way, works against or alters FATHER'S eternal perfect plan and purpose. FATHER designed all things to fulfill his and our omniscient perfect eternal desires; and if we find ourselves desiring anything, contrary to that perfect plan, we hold an evil desire, wanting a divisive thing. And by wrongful desires, we are tempted to act out and become evil, like Eve and then Adam. GOD knows, and we see how evil spreads. And like a virus, evil needs a host to spread. Evil introduces disorder and imperfection into the eternal perfect information of everything and every soul it possibly can; thereby making whatever in infects, temporary. Thankfully, GOD will not abide evil *eternally*. GOD allows us all our expanding moment of choice to learn of, turn from, and refrain from all of evil's destruction. GOD has done, and given everything to teach us that if we passionately love the good eternal things and ways of GOD, we can remove, and keep all evil from our life. If we desire GOD'S perfect goodness above all things, they can restore us as before Eve and Adam ate of that tree. Mankind for the most part in ignorance and lust, are so far gone, they can no longer want or even consider the truth. Most do not see how the forceful effects of their choices, reach and eternally extend beyond their subjective understanding and temporary perspective. And they surely don't equate their evil as being in league with Satan. Scripture says quite plainly that each time we sin, we are in that moment, of our father the devil. Most professed Christians don't see Satan as having any effective rule over any part of reality, or them. But when Satan showed JESUS the Kingdoms of this world, and said worship *me* and I will give you all you see. Could Satan really give JESUS all he showed him at that time? Yes. There are multiple Scriptures that show us this world is currently under Satan's dominion. Adam surrendered his dominion over this world to Satan. Satan's power in no way compares to GOD'S, yet his power of persuasion and allure, in this world, are equal to each soul's desire to be a god over their own life, and so, his power is underestimated. Professed Christians can't imagine Satan's current level of control over this world. For JESUS, Satan created a physical image or vision of all the kingdoms of the world. We must know, that required a greater conscious reach and force than our souls currently experience. And read what Satan did to Job's life by consciously forcefully creating real tornados and fire. Scripture says the powerful Holy Angels of GOD, do not contend against Satan, unless commanded to by FATHER. We must know Satan has been about his own insane purpose since his fall. This is Satan's own desire; nothing was forced on him as some Satanists contend. Satan's desire is for an impossible result; yet he studies, plans, and works to fulfill his desires. This world, thousands of years ago was far more religious, with nearly all mankind believing in GOD, or false gods. Now mankind's desires are for the comforts, pleasures, and life improving and extending technologies discovered by modern science; even to the point of worship. Their spell bound gaze at everything technology creates, are like children, waiting to see what new thing will be next to take and keep their needful minds off GOD. This world is full of distractions and endless temptations. Do we flee all these temptations as commanded for Spiritual safety? Or do we run to them, trying to fall under the next new spell before our family and friends do? So many technological controls of dependence are now in place. Like modern farming creating the inability in most to farm for themselves. Satan builds this world's dependance on convenience, entertainment, pleasure, security, comfort, and lies; using things like electricity, automobiles, airplanes, nice homes, furnishings, computers, supermarkets, hospitals, cell phones, cable and satellite entertainment, with government-controlled medical care science, and religion. All helping to create a new world order of a blind dependent under-class. If those are the icing on the cake, here are some sprinkles on top. If you have not followed the statistics, you would not believe the percentage of children and adults who are on prescribed mind-altering drugs, and let's add in illegal drug dependance. Satan's wants to increase the world's technological and chemical dependency, to control and redirect all thoughts and emotions away from GOD. Souls in Churches everywhere are taking pills to cope, to wake up, and to go to sleep. In most modern Church services, I hear a weak watered-down sermon, having a few of GOD'S words, out of proper context, mixed with psychology and modern motivational self-help measures, with maybe some current events and the latest diet advice. Of course, the jokes are also popular. All followed up by video, book, and bake sales in the lobby. Has it become barely about

GOD, and mostly about religious entertainment, comfort, and distraction. All modern religious doctrines are physically and emotionally addictive, yet spiritually destructive. And if you subscribe to a worldly life style, even a religious one, when looking into a mirror you will see an addict. If you live like this, yet don't believe you are addicted, then stop, go off the grid, and avoid it all for a week. You will learn what emotional and physical withdrawal feels like. All these facts will make it easy for Satan at the end, to coercively unite the world under his control. We know the Third Beast, or embodied Antichrist, will appear on Government controlled TV, soon after a third world war, and formation of a ten-region world-government. He will say he is here to unite the world under God. He will say he is here to keep everyone's desires (needs) fulfilled on a regular basis. Don't be surprised if everyone you know goes along in great joy: your parents, siblings, children, and friends, maybe your entire Church and you. This Prophesied *great falling away* from the truth to an abomination, *will* happen; just as it always has. But the great majority who fall, will not have a clue, or care to know they are being *spiritually* deceived, because they were already deceived. Remember falling away is not like backsliding, in which one can possibly recover. Falling away is an eternal thing one can never recover from. The Antichrist and False Prophet, don't appear in power until the Fifth Trumpet of the Tribulation sounds, as they then prepare the way for a false savior of the world, being Satan. The modern world has a secular wealthy population that is easy for Satan to control and use with wealth, fame, and allure. Satan has studied mankind's psychological and physiological weaknesses throughout the ages. Satan knows more about us, than we know about ourselves; only GOD can know our hearts better than Satan. This makes us, third on the list of knowing our true weakness and desires. If this scares you, good. For Spiritual comfort we offer this wisdom; if GOD but *willed* it, right now. Satan's body and entire extended physical reality would dissolve as if they never were. We know Satan, just like us, is a gifted soul adjoined to a body created and loaned by FATHER. Satan holds no eternal physical power or control like GOD. So, we see Satan and all his Angels were actually defeated the moment they fell; yet GOD allows his search for love to play out. We read how Satan's anger grows as the end of his delusional hope draws near. Mankind is the weakest of those who possess an eternal soul, and we can see and know evil, only to the extent we can see and know GOD'S truth in its contrast that defines all evil. Otherwise, one can falsely define good and evil for themselves. For clarity in visualizing anything, using all opposing contrast is a tool in observation. So, if one is far from seeing the contrast of GOD'S truth, one cannot see real evil. Such souls will also not be under attack from Satan, as he has no need to do so. But if you are on the true path of GOD, growing in their words, love, and strength; Satan and his Angels have reason and cause to come against you in persistence. What entity or army at war, wastes battle resources on an ally, or weak self-defeated enemy; and ignores a strong enemy. Satan wants every soul to abandon GOD, and to destroy their hopes for an eternal family. If we choose and seek GOD, we must prepare ourselves for internal spiritual and physical battles. Our battle is over the essence, reason, and eternal purpose of life on earth and in Heaven. To allow GOD to win our battle for us; holding nothing back, we must love GOD with all our heart, mind, and soul. Find, live, and stand strong in GOD'S truthful ways, and not our own. Seek out FATHER, JESUS, and HOLYSPIRIT in love and deed till the end, and rest in peaceful joy and be assured. The fast-coming last evil days, and global system of Antichrist, is gradually being accepted by much of the world and even the church at this time; and all for the same reasons it is accepted by non-believers. It is simply spiritual-weakness and strong physical desires. It is by not wanting to offend sinners with GOD'S truth that the Church has been muzzled and being desensitized to evil, and into gradual acceptance. The Antichrist will not appear as a shocking world-shattering revelation, to overthrow and alter a man-made system of governance, or force a new evil system upon mankind. It is a stronger and stronger system of global societal control that is being put in place in its steps and pieces right now. The Antichrist will walk right into a pre-prepared global system, being built at this very moment; and it has been in this process for a long time. Its final stages, if not for divine intervention of the Four Horseman, culminating in a third world war, it would be a somewhat peaceful progression. Most professed Christians will not sense it has happened. This final one-world system, is foretold to come *only* after much disease, famine, and a third world war, that combined, kill a fourth of all people. For Satan, it is about turning souls away from the created good things beliefs and ways of GOD, and to all individual evil things, beliefs, and ways. It is no small matter to convince billions of free-willed people to let go of guiding beliefs, and accept new ones on a global scale. Satan has made huge progress using the evil media, the secular population, lukewarm religious doctrines, and all angry physical perversions. And it is all growing at a blinding speed. Manufactured goods are made cheaper and faster, needing fewer workers, while at the same time, more people need jobs, creating more dependence on government, as it all becomes less and less sufficient. This all gives to Satan a great planned advantage. The world does not see this evil, because the world does not know what is evil. Weak souls, fearing loss of food, shelter, comfort, and healthcare, are easy to control. If a government controls these things, it controls most people. Beware of people wanting government control over these things, they are communist spirits of antichrist. This is the danger of false once saved always saved, predestination, inherited perfection, and the prosperity, doctrines. Also, any pre, or mid-tribulation rapture. Many souls have fallen under combinations of these false doctrines, and are not prepared to see, know, or stand. So many choose to believe they will not be here for the Tribulation, or the Mark of the Beast. Who can comprehend the confusion and fear that will enter the heart of these souls at finding their beliefs in a pre, or mid-tribulation rapture and other doctrines were false. Coming simultaneously with fear of losing their lives and standard of living, to which they are addicted. It is painful to know that most of these souls have already, or will fall away; yet there is hope for some. Now Satan uses every natural and created crisis, disasters, and conflict to accomplish his bigger and faster changes among mankind. So, through all the coming changes of the end times, and the coming loss of all peace on earth, Satan's total control will be implemented; which he has worked towards since his fall. To change people more than they desire, Satan uses natural and engineered inflation, shortages, propaganda, wars, terrorism, and disease. These raise fear, being the greatest emotion for controlling mankind. Satan will keep people focused on certain things, as he changes so many other things behind their back. How many crises will it take to bring about another world war. We are told in Scripture that the final crisis will include the exchange of weapons of mass destruction. This evil world for some time now, has been experiencing the Prophecies of Revelation 6: 1-8, known as the Four-Horseman of the Apocalypse. These four strong spirits work to change the world; 3 towards evil, and one towards good. On the near horizon we see all looming crisis, being world-wide energy shortages, hunger, disease, religious conflict, terrorism, anarchy, mass psychoses, global insanity, and chemical, biological, and nuclear war. We have the ever-accelerating global financial crisis of each country's debt. There is the great *false* portrayal of global climate crisis. And at the center of love and the world and all that is Holy, we have the Middle East and GOD'S beloved Jerusalem and Israel. How can this world accept a one-world government? A global financial collapse alone, won't be enough to get the strong countries of the world to give up their independence. A global energy crisis alone will not be enough. Would an energy and financial crisis combined be enough? That is still doubtful. But what if these crises are all started by Islamic religious fervor, causing the Middle East to erupt in a way that places Israel and the west in dire imminent risk as Prophesied. Such a war, would create a devastating energy crisis, triggering economic collapse on a global scale never seen. This world is now already at war, before the first official shot is fired. We hope you see the first dominoes fell long ago, and the inevitable evil comes soon. Why will this war go biological, chemical, and nuclear: well, because the world has grown quite insane, tensions are too high, and no country has the money, fuel, time, or will, to fight a world-wide conventional war. So, as we see global insanity increase, what could be the only need that *could* unify most souls under a world governance; nothing more or less than the survival of the species and planet will work. Putting the entire world under one new government will require a single global currency. A Global World Order must tell people that different currencies created division, unpredictability, instability, unfair practices, corruption, and are part of the reason for the world being at war. A Global Order must say different Constitutions cause the same problems. *But* an insurmountable impasse, is trying to unite many religions. This will be the new Global Government's greatest problem. Negotiating a one-world religion is impossible. So, as Prophesied in Scripture, we see this Global Treaty contains a seven-year religious protection of peaceful worship for all major religions. This allows countries to surrender their sovereignty, and unite under one covenant of government. For peace, there must be one global policing force, with no advanced militaries. Globalization makes governmental and economic sense, once you step outside GOD'S truth. Logically it could eliminate all large wars, international corruption and competition, and bring an appearance of peace and prosperity to the world. But the great paradox is, Globalization is the evil system of the Prophesied Beast. And we must know that regardless of our accepting a physical mark or a tattoo of a number or anything like that; our acceptance on a personal level for any global system, works against GOD'S Family. It is the evil that makes war on the Saints. Any acceptance in any way, will be a mark of evil belief in our forehead. Any daily work in that system, will be an evil mark of future blood on our hands. Any acceptance of globalization in our lives, is a form of worshipping mankind.

If we participate, those stains will be the blood of the saints upon us. Evil now moves with ever-increasing momentum. It is impossible to say when, but we approach an exponential point where things will happen so fast, that most will not be able to grasp and keep up with it all. By the time most figure out the first thing, Satan is many moves ahead. At a certain future early point of a Global Government, Satan will have all the elements of his plan in place, having real control over more than ever. So, FATHER does something merciful; he brings unimaginable plagues and super-natural events upon the entire world. Why is this merciful? This is GOD crying out I am here, in his wrath to the world with one event after another, growing more miraculous and fearful with each one. This is a two-fold outpouring in one element: it is GOD in his wrath, against those who will never hear his truth; yet on the other side, it is GOD'S loudest cries ever to mankind out of love, showing us he is truly there for those who might see GOD is bringing an end to evil. If these otherwise unexplainable last miraculous plagues cannot turn hearts back to the love and fear for GOD, those hearts can never be turned. There will come a certain point according to Scripture, that deception within all hearts will be so strong, GOD will sadly see that not one more soul can, or will be saved. JESUS says, in Matthew 24:22, that then, the destructive elements known as time, have to end for the elect's sake. Scripture shows us, our merciful FATHER goes to the last possible moment, before he has his Angel declare that mankind's moment of grace and choice is over. Satan eternally lost his love for GOD, and the same is true for all who have fallen away, as described in Hebrews chapter 6, which was not written for Satan, but for mankind. It is a most sobering fearful chapter in Scripture. I would again touch on the power of death that Satan holds. Satan cannot touch a Child of GOD without certain permission; but he has the power to tempt and lure us to choose hell and death with his lies, using fear and deception. Never fear Satan more than you fear GOD, and you will be fine. Satan is not the only one with this power to lure souls to their destruction, he is just the father of it. All fallen mankind has this power. If anyone lies and deceives someone away from the truth of GOD, they have done the evil work of their father Satan. Scripture says the truth is singular, beautiful, pure, narrow, and of GOD alone. Evil is all and everything imagined, not being from or of GOD. Each variableness, partiality, or compromising desire for any single thing beyond GOD'S Truth, is the evil gate that is as wide and as broad as the imagination of Satan, his Angels, and the lost among mankind. The greatest evils have never been hard to see, but wanting to see *every* lesser evil is difficult. Mankind wants in their own eyes to define the lesser evil as good; and this is too hard for most to resist. Yet this is still mankind's separation from GOD. The Spirit of Antichrist within the pulpits, allows much evil and falsehood, including, changing the definitions of two evil sister desires. One being, to covet, whose definition has been altered to be more about things taken or stolen. The biblical definition of covetous is wanting a physical thing not needed to physically live; being anything above simple food and raiment. The other desire is lust, which is now seen as a sexual reference. The biblical definition of lust, is the fleshly desires for any and all pleasures from all the physical things of this world, sex being one. The lust for food or power or sex are the same. These two terms are similar in meaning and nature, and are both sins against GOD. We are called to enter by the straight and narrow Gate, being the narrow Godly interpretation of the truth, which says that evil is defined by GOD alone.

The Scientific Creations of GOD

I have truly struggled to adequately condense a scientific understanding of FATHER'S created reality to fit as a chapter in this book, and prayed, asking if I should leave it out. All I received was the truth is only seen in strength by its beginning, middle, end, and all its lesser and greater parts and qualities in between. If you have not read "The Theory of all Conscious Created Perception" listed first among our books on pg. 2. This is a condensed sense of that book, in our hope of raising your interest in an understandable account of reality. It is a detailed spiritual and scientific journey into the foundations of GOD'S Creation. If you have read that book, we ask you still read this chapter. We ask why is there no room in the physical sciences for our soul of awareness? For what cause and reason is it shunned, denied, and even ridiculed? This world is full of hollow adaptations of belief that say physical complexities are the likely candidate for our otherwise enigmatic emergence of self-awareness. Being a belief built, minus any firm reasoning. They have zero evidence, so, it is their profound religious hope and faith. They have nothing beyond the correlations of neural activity, being formatted and sent or received by our soul. Science admonishes any lay person who conflates correlations with cause. But if asked to show a physical cause that explains our awareness, they offer only those correlations as physical-evidence of a non-physical phenomenon. You can *only* see the science of FATHER'S Creation through pure Dualism; being all our forceful creative souls, adjoined to a shared reality. Together, we all create our representations of reality. We must know that our forcefully created perceptions are formed from quantum particles and waves. We must know that creating our perceptions is conscious work; and that all such work requires projected force. Force has never been observed or defined by mankind, only its affects. Force attracts or repels, it pushes or pulls, this is the affective nature of force. Our every desired forceful patterned *intent* is projected to raise up ours, or tear at another's perceived beliefs within our soul's spherical reach of influence. Or to attract or repel some*one* within that same sphere. We forcefully push and pull as we form the substantive aspects of our intended created perceptions. Our intentions are observed in our every projected forceful desire to exist. These are the dual aspects of FATHER, who projected his forceful patterned beliefs to create the physical existence of *his* desire, called the universe. To this day, there remains the same ancient physical questions about the creative force, underlying all our conscious awareness. Being what is the essence and nature of our conscious awareness of qualia? And where do our soul's creative structured patterns of forceful projected information originate? Forceful patterns are required to build 3D particle-wave-based perceptions, so, what unseen particle-wave-based complexities are all our perceptions made of? How do we instantly, effortlessly, and forcefully create them? Why do elegant patterned fields of generated projected force, entangle and surround every created complexity? And what do these forceful occurrences found in our created particle-wave-based perceptions indicate or substantiate? The conscious question, has been studied in a physical way for too long, producing not one good physical answer. The forceful answers to conscious questions, are our key to understanding reality. What is true science, and who gets to define it? Is all reality confined to the things a human body can physically sense? No. A body cannot sense or know the essence of force, only its affects. The body senses patterns and levels of force that science calls physical information; but we deal with all information consciously. So, consciousness is the foundation of science, not the physical. How can not believing in GOD, give one a right to define science. Did disbelievers call dibs. So, science is only about things they physically sense, test, and accept. If so, why do they proffer so many things of faith, desire, and belief, that can never be physically proven. Things they conjur in desperation. They have painted themselves into a corner with no physical proof or direction. Why have their views failed. Ecclesiastes 3:9-11, "*WHAT PROFIT HATH HE THAT WORKETH IN THAT WHEREIN HE LABOURETH? I HAVE SEEN THE TRAVAIL, WHICH GOD HATH GIVEN TO THE SONS OF*

MEN TO BE EXERCISED IN IT. HE HATH MADE EVERY THING BEAUTIFUL IN HIS TIME: ALSO HE HATH SET THE WORLD IN THEIR HEART, SO THAT NO MAN CAN FIND OUT THE WORK THAT GOD MAKETH FROM THE BEGINNING TO THE END." By constantly violating their own firm definition of science, they must surrender their claim to define science. Who is the true scientist, and the true science denier? As with all things in such a divided confused world, filled with endless individual beliefs. A personal answer to this is made and accepted by each soul. You see it all comes down to what we each desire science to prove or disprove about GOD; it is no more complicated than that. Do we want GOD to be there, or not? We all trust a science that best substantiates our desire. Now in metaphysical nomenclature, pure Dualism, refers to a soul and body. Some call it, Soul Dualism. Dualists clearly see the interconnected existential aspects and reasons for how and why all Consciousness and all Physical Reality are entirely distinct separate things. Now there are other forms of dualism that try to blur these distinctions; but the science of *pure* Soul Dualism is central to the truth of reality. When Materialist talk about the foundations of reality, they speak in between, above, below, and around the subject. But they never want or need a full perfect answer for anything. Materialists define everything that exists in three physical dimensions, as being part of a purely physical and energetic reality, but what then is force. Is force not entirely different? Is force not appear and work as three-dimensional? Does force not occupy the same space as all created realities. Mankind can't begin to understand or define the very essence or quality of force, they can only measure its lawful effects. But in knowing this, why do they endlessly imply and infer things that cause and allow the world to believe science knows all about force, when they don't. Ask a Materialist, what physical laws of reality say desirous other-worldly *forceful* souls cannot have three-dimensional aspects, and occupy space? Materialists dwell in profound contradictions, as on one-hand saying all things are energy or mass. Then on the other-hand they say our ethereal holographic perceptions of imagery, sound, emotion, and touch that we all *consciously* create; for these they deny there is any physical structure or energetic substance to their nature. We see, hear, and in all ways, physically sense our created perceptions. What then is a perception's ethereal composition that we physically sense. What forms the complex structure of these qualia we consciously create, indwell, and experience within our own conscious estate? Do our created perceptions of sight and sound not physically exist? Are they physically non-existent, being formed out of nothing? What is nothing, and how can we create, or physically sense things made from nothing? Surely for metaphysicians *and* physicists alike, this can't be true. Our created perceptions are ethereal and ghostly in the physical sense, but we see a quantum realm filled with ethereal and ghostly occurrences. All things quantum, are ethereal, yet perceived to be barely physical and real. From GOD, down to us, we know the external physical reality, is only evidenced to us by our created and perceived conscious experiences. And we know our ethereal perceptions must be formed by forcefully arranging patterns of unseen unknown particles and waves. But how? From where do our projected creative desires originate? The *only* possible answer is that each soul's generated projected 3D *patterns* of desirous force, align and form our ghostly streaming elementary particle-wave-based perceptions. Secularist must reject particle-wave-based-perceptions. Because once they accept that we create 3D particle-wave-based quantum perceptions, that takes them down the rabbit-hole where they must see the Primacy of the Conscious Observer Effect. Secularist have tried to deny, twist, and erase this known observation for over a century, but have failed. We know that to accomplish any work, requires a generation of force, *and* a founding substrate or aether to work with or upon. This aether we all use, is comprised of the most basic form of reality, spoken into existence by FATHER as a sacrificial gift of creation for all lesser souls. Consider how FATHER does not breathe air to live; so, he does not use air to speak. When FATHER speaks to create, a certain amount of forceful patterned resonance and the pure physical point singularities that constitute his body, must be spoken forth or breathed out, *and* aligned according to the desirous creative patterned conscious force that expelled them. These are the two qualities, forming the breath or words FATHER speaks. Psalms 33:6, "*BY THE WORD OF THE LORD WERE THE HEAVENS MADE; AND ALL THE HOST OF THEM BY THE BREATH OF HIS MOUTH.*" When FATHER speaks forth to create, he projects his substantive perceptions to exist outside his person. And holds his universe between his outstretched hands to maintain his continuity of control. GOD'S spoken creative words, forcefully and *physically* exist. And this same generating, projecting, forceful creative desire that is the soul of GOD, exists in a minute amount as us, as our gifted soul. We are an eternal gifted space, or portion, or soul, eternally divided out from FATHER'S soul. We consciously project and forcefully create our particle-wave-based perceptions of desired reality, just as FATHER does. This while existing deep within FATHER'S great physically projected perception that surrounds us all. For their rebellious beliefs, GOD said to his people, in Psalms 82:6-7, "*I HAVE SAID, YE ARE GODS; AND ALL OF YOU ARE CHILDREN OF THE MOST HIGH. BUT YE SHALL DIE LIKE MEN, AND FALL LIKE ONE OF THE PRINCES.*" Children come from within their parents, to exist in their parent's likeness, qualities, and attributes. We are gifted and created in FATHER'S image, which is bicameral. To include our forceful creative soul and a sensing interfacing physical body. All children are born as lesser facsimiles of their parent, or parents. John 10:34-36, "*JESUS ANSWERED THEM, IS IT NOT WRITTEN IN YOUR LAW, I SAID, YE ARE GODS? IF HE CALLED THEM GODS, UNTO WHOM THE WORD OF GOD CAME, AND THE SCRIPTURE CANNOT BE BROKEN; SAY YE OF HIM, WHOM THE FATHER HATH SANCTIFIED, AND SENT INTO THE WORLD, THOU BLASPHEMEST; BECAUSE I SAID, I AM THE SON OF GOD?*" Phil. 2:5-8, "*LET THIS MIND BE IN YOU, WHICH WAS ALSO IN CHRIST JESUS: WHO, BEING IN THE FORM OF GOD, THOUGHT IT NOT ROBBERY TO BE EQUAL WITH GOD: BUT MADE HIMSELF OF NO REPUTATION, AND TOOK UPON HIM THE FORM OF A SERVANT, AND WAS MADE IN THE LIKENESS OF MEN:*

AND BEING FOUND IN FASHION AS A MAN, HE HUMBLED HIMSELF, AND BECAME OBEDIENT UNTO DEATH, EVEN THE DEATH OF THE CROSS." Please see how being extremely minute gifted Gods, who *can* forcefully create our ethereal quantum perceptions made of unknown particles and wave properties, yet cannot control the larger known particles that form our bodies or reality at large, and being the one vast FATHER above, are extremely different things on every level. Our souls are gifted on a principle of drawing our own *desired* and perceived reality, out of GOD'S provided aether; which permeates all quantum souls. Now true knowledge is only revealed by GOD to a soul in its natural order and moment of proper understanding. But mankind in rebellion, denies GOD, and runs far from the truthful order of reality. By going far afield, chasing their illusory imperfect desires, they see and hold to false suppositions and belief in physical miracles. Now they have come full circle through technology and rebellious curiosity, to peer into the quantum realm through both their soul's wonder, and in very small and slow ways, through their physical eyes. They refuse to see the breadth of their limitations, or that looking into the quantum realm for them, is looking through a dark distorted glass. And there is a vague image just on the other side of that dark glass they can't quite see. Mankind strains their imagination, trying to see the other side. Can the rebel mind of mankind overcome that dark glass, to see their soul looking back through that quantum mirror. 1st Cor. 13:12, "*FOR NOW WE SEE THROUGH A GLASS, DARKLY; BUT THEN FACE TO FACE: NOW I KNOW IN PART; BUT THEN SHALL I KNOW EVEN AS ALSO I AM KNOWN.*" In understanding that GOD knows us all as a soul, there is only one responsible truth that puts all things at peace. Now we have all seen projected holograms, created using scientific holographic principles. So, please see how we consciously generate representative three-dimensional patterns of creative force *quanta* within our soul, to project into the aether to build, or collapse if you prefer, vast numbers of lesser particles within a larger wave aspect we also create, that becomes our perceptions. Now in this next step of GOD'S scientific understanding, we expand the scientific Observer's Frame of Reference, and Theories of Relativity, to allow for all conscious size entities. To visualize this, we apply the forceful conscious creations of Dualism to the Universe. The entire Universe is a consciously created particle-wave based perception of far greater size and complex firmness, created by FATHER'S vast conscious soul. Now *deep within* FATHER'S vast wave-particle-based perception we call the universe, he eternally partitioned out from his vast greater soul, tiny distinct 3D spheres or spaces of forceful desirous controlling awareness. This is who we all consciously are as a soul. So, we add to the three physical dimensions, the three dimensions of each soul. One dimension is a soul's size and force; dimension two, applies to all minute gifted conscious entities alone, being the conscious effect of momentum in our traveling through GOD'S Universe; and three, is not time, but the moments or *size* and *rate* of each soul's created streaming perceptions. These three dimensions of each soul are relative to the three physical dimensions of each body's sensing ability. So, based upon the size, speed, *and* forceful reach, of each soul, there arises a spectrum of comprehension, as to the smallest to largest, slowest to fastest, particle-wave-based creations, that can possibly be perceived to exist by each size entity. Each lesser soul, is bound by size, momentum, and the laws of its relative perceived reality. Each differing size soul is grounded to, and looking out from one common center of the entire defined spherical spectrum of reality. Making every soul's perceived reality, a relative yet subjective physical creation and conscious experience. Beyond a soul's smallest and largest perceptive boundaries, the laws of perception are distorted. All the smallest, fastest, largest, slowest, ethereal and firm realities, cannot exist in any true sensed relative way to all lesser sized entities. All realities beyond the upper and lower sensing boundaries of the physical bodies we indwell, become a non-perceivable existence. To demonstrate this, greater size and speed, to a far lesser size and speed relativity. See how the soul and bodily personage of FATHER, is at minimum, seven or more times greater in points and force, than his entire created particle-wave-based perception of the universe. This makes FATHER'S conscious size and speed perspective, vastly different from mankind's. From being so *minute*, looking *up* and *out*, mankind senses only the slowest of all light particles. And by mankind's small slow perspective of reality, this allows a belief that *all* light requires billions of years to cross the universe. Yet FATHER, being vastly larger and faster, uses a far, far faster luminary particle to see the *entire* universe, in every moment from a top-down perspective of reality. FATHER could, using this far faster light, perceive moving his hand slowly across the entire universe in a tiny moment; making mankind's perception, perspective, and comprehension of the universe's age, paradoxical. By their great limitations, to mankind it can appear *as if* the universe has existed for far longer than FATHER'S perceived comprehension of the same. 2nd Peter 3:8, "*BUT, BELOVED,*

BE NOT IGNORANT OF THIS ONE THING, THAT ONE DAY IS WITH THE LORD AS A THOUSAND YEARS, AND A THOUSAND YEARS AS ONE DAY." Isaiah 48:13, *"MINE HAND ALSO HATH LAID THE FOUNDATION OF THE EARTH, AND MY RIGHT HAND HATH SPANNED THE HEAVENS: WHEN I CALL UNTO THEM, THEY STAND UP TOGETHER*". Isaiah 45:12, "*I HAVE MADE THE EARTH, AND CREATED MAN UPON IT: I, EVEN MY HANDS, HAVE STRETCHED OUT THE HEAVENS, AND ALL THEIR HOST HAVE I COMMANDED.*" This is FATHER'S objective observation; if it is not, what unintelligible things these are for FATHER to say in his ancient inspired Scriptures. This Conscious Entity, to Physical Relativity Principle, shows us *mankind's* claims as to what are the observed laws, times, and measured constants, governing the universe. Are only relative and subjective, yet *appearing* objective to mankind's small, slow, inside looking out perspective of space-time. The laws, constants, and theories, material science perceives and proffer, in no way, govern reality. These only relate to mankind's perceptions of reality. Further, their laws do not even govern mankind's perspective of reality; their laws are formed and governed by their slow small perspective. In this, we see why mankind's observance, cannot kill Schrodinger's cat, opening a physical box external to us. Which is part of FATHER'S created perception. Yet in our own *internal* created particle-wave perception of opening a box and finding a dead or live cat; we do instantly create a dead or living cat within our created perceptions of that occurrence. So, that paradox is eliminated. Accepting a 4D view of reality, in place of the true 6D, downgrades one's perspective from GOD'S top down, outside in, larger faster eternal perspective, to mankind's bottom-up, inside out, slow, small, temporary timely perspective. Our soul's awareness is our existence; so, as we create and perceive something, we consciously exist. If we perceive nothing, we temporarily cease to consciously exist. Our soul's generated and projected forceful desire, is the only conceivable patterned force that can explain our created perceptions. All physical complex creations are a biproduct of reality's two basic conscious and physical qualities, coming together to create a perceived complex firmness and relevance to all things then known. Do any deny that we *first* forcefully create our perception, and only then do we experience our real brightness, sounds, and feelings, that come from inhabiting and knowing our own perceived creation. How could any of us accomplish this without a form of projected conscious patterned force over substance? We could not. Circular reasoning asks, do our logic and desired forceful purpose and knowledge, form our every created perception; or do our perceptions self-create their own particles waves, logic, and awareness from no prior knowing desire or purpose? And if so, how and why? In pure Dualism, for any knowledge, understanding, or purpose, to exist *now* in certainty; it must be fully and completely known, understood, and desired in this moment by at least one soul. This truth is now a certainty, as it has surpassed every *probability* that it does exist. So, do not say the sciences have a built-up knowledge as firm undeniable proof, for all souls hold only their desired subjective perceived interpretation of knowledge. Hence all the changing secular theories and endless debating by scientist. If there were any undeniable slow small physical understanding of reality, we would see pure undeniable consonance and continuity. Yet science is riddled with constant re-evaluations and changes of direction. It is filled with paradoxes, greatly missed modeled predictions, and internal rising forms of disunity. We find that beyond spelling and simple math, mankind's understanding of reality grows only more confused in it coming from so many insufficient perspectives. All godless souls are in constant flux of forming their own understanding, and not one of their understandings is beyond being disproven. Accept and rest secure in GOD'S belief and understanding as it is given to us. Most souls are no more than a current created belief, devolving from their past created beliefs; it being the sum of who they spiritually are in a moment. A lost soul's awareness, exists completely within this process. If this alarms you, we are sorry; yet we welcome you to the pure truth. Now in that FATHER'S supreme conscious force has formed the universe; this universe is then consciously deterministic, compassionate, and a lively thing in that we find Conscious Determinism can be reasoned with. By GOD'S loving desire, it has his elements of mercy and forgiveness. Isaiah 1:18-19, "*COME NOW, AND LET US REASON TOGETHER, SAITH THE LORD: THOUGH YOUR SINS BE AS SCARLET, THEY SHALL BE AS WHITE AS SNOW; THOUGH THEY BE RED LIKE CRIMSON, THEY SHALL BE AS WOOL. IF YE BE WILLING AND OBEDIENT, YE SHALL EAT THE GOOD OF THE LAND:*". By GOD existing, the nature of the conscious term "*the complete truth*" means a full firmly formed conscious conception of perfect existential logic and understanding, formed in knowing *all* realities. *And all* being expressed only through their original desired and created intent. We see such a conscious concept, can be nothing born of the physical. Dualism is the only path to fully understanding a physical existence that incorporates our awareness of it. In pure Dualism, both components must interact to sense, build, and perceive the physical complexities and dimension that enable self-awareness. We all experience our nightly non-existence, called dreamless sleep. By sleep, we now also comprehend our longer state of non-existence, called physical death. Physical death lingers until our soul is resurrected to a new body, to again allow our soul's required physical sensed experience in the surrounding reality FATHER perceives and provides next. By our soul, still creating perceptions in sleep, or death, this reveals there are other-worldly, quantum, less physical, connections to be sensed by our soul. In sleep dreams, we most often see weak patterned information bringing intermittent muddled perceptions, that often make little sense. All external information, sensed from this world is sent to our soul through the body interface, being the central and vast extended nervous system. Our quantum realm souls, are loosely adjoined to the body, from within the quantum realm, via a cross-space, air-gap, quantum entangled connection. The electro-magnetic spectrum, that describes this universe to us, was also forcefully perceived into certainty by FATHER, utilizing the points that form his aether. Now the laws of conservation state the soul and body of FATHER, must be far greater in all ways and means, than the universe and its orders and mechanisms he has created. Again, we each create and indwell our particle-wave-based dreams, deep within the vast particle-wave-based dream or perception of FATHER. But the three physical dimensions are perceived in a way, relative to the size and forceful reach of each soul. These dimensions for mankind, consists of all the substantive things that can be perceived by mankind to have depth, width, and length. We see that if something is too small or large, too slow or fast, it cannot be perceived by mankind, and exists beyond mankind's sensed reality. The full three-dimensions FATHER perceives, are far beyond mankind's comprehension. So, as substance is broken down into its smallest aspects, or moves too fast or slow. At a point relative to the size of each soul and body interface, its physicality begins to blur before them. It becomes what can only be defined as waves or fields of a fluid ethereal thought-like intermittent phase. And in the next phase beyond that, for mankind, it will disappear altogether. Of course, it did not disappear altogether, it simply broke down into pieces too small or slow, or fast, for mankind to comprehend. Perhaps all the way down to the eternal single points of pure physical singularities that form FATHER'S body and external aether. The created three physical dimensions, cannot exist for any entity beyond their sensed comprehension of them. Now in this moment, project your desires to create within your mind, a bright vision of two trains going in opposite directions in space. Let's call these, the trains of chosen belief. See in your created vision, that one train is on a single track, going in the upward direction of wisdom, truth, and love of eternal life. See each soul, who by their deep desired acceptance of GOD'S truth, are wantonly and knowingly attached, and represented as one of that train's cars. And out in front is the most powerful forceful conscious source of perception, being the conscious locomotive that teaches, loves, creates, and pulls all cars along towards the oneness of an eternal peaceful loving existence. Now look in the other direction at rebellious individuals who have detached themselves from FATHER'S train. We see a track which constantly diverts and splits over and over into an ever-increasing number of darker colder descending tracks, with fewer and fewer cars on each track. This leads to the eventual eternal separation of descending single conscious cars on single utterly dark conscious tracks forever. This is the forever devolving path of individual souls, minus GOD'S omniscient guidance to oneness. Individualism leads all to ever-dividing confusion, and eventual loss of communication with all souls. Now, your soul's effortlessly *creating* those bright particle-wave-based ethereal images of two trains was miraculous, effortless, and even God-like, wasn't it? But your soul's aware understanding of those perceptions, was by many magnitudes, a far more amazing Godlike reality. How is GOD aware that they exist? It is by their soul. How are you and I aware of our existence? Is it in an entirely different way? Is there an entirely different way to experience existence? Is our soul, something truly different than GOD'S soul? No. They are of the same essence, ours is just a gifted minute measure. When we see a beautiful painting hanging on a wall, we become aware of its beauty. But again, we never see the actual painting, we are only aware of our created perception of it. The body only senses patterns of information coded in the waves and photons that pass between us and a picture. Our soul uses those sensed patterns to create our representative image, being a particle-wave-based perception. Only then do we become aware through our second-hand ghostly creation. But the secular sciences want us to believe this is all one single physical occurrence or event. They do not want us to see how we are aware of our perception, only after we create them. They want you to believe you see the painting on the wall, and not your created perception of it. They want us to believe the created perception *is* also our awareness. But *we* sense in every way, that we become secondarily aware of the image, that is our created perception. We witness our perceptions move and enable our streaming awareness. We can in no way feel or see this as a singular event. We know there is a true separation of qualities and even quantities in this process. Why do secular scientists, pressure us to accept their belief of faith? Why don't they want us to know how we forcefully create particle-wave-based perceptions? Why do they blend and blur these aspects into one pure physical event? They know, if they allow our soul's forceful creation of substantive perceptions, and our soul's awareness, into our beliefs, it causes us to see the true necessity of GOD, as the one who must have perceived the universe. Our path leads only to a confirming belief in GOD; that they deny exists. Ask yourself where is the possibility or logic, that a created image, and our awareness of that image, are one occurrence. I see particle entanglement throughout all physical realities; but I see no sane purely physical way to explain how this could exist. By simple desire and emotional denial, materialist believe that all perception and awareness are one event, and refuse to consider it further; as they don't like or want the alternative. Insanity means to lose touch with objective reality, and in seeing *all* individuals form their own subjective reality, they are each thereby detached from the one true reality of GOD. So, they are all by definition and degrees, insane. The politics of secular science has been about building emotional barriers and false understandings, designed solely to block out facts, and to allow souls more comfort in their ongoing rebellion against GOD. The one pure truth, is found only through conscious force determinism. Minus our forceful conscious perceptions of it, there is no knowledge or truth of anything. For any reality to exist, based upon and containing one single knowledge and truth, its completeness *must* be perceived and known by at least *one* soul. Our perceptions are completely known to us; just as FATHER'S perception we call the universe, is completely known to him. For *eternal* love and peace among many, GOD must separate individual subjective imperfect beliefs, thus eliminating speculation, supposition, and conjecture. Then souls see only GOD'S perfect truthful understanding of reality. To return to GOD, there must come within each fallen soul, a spiritual metamorphosis. Consider thoroughly that if you disagree with another soul, on most everything, conversing with them becomes impossible. As you are each speaking and hearing in a completely different belief of language. So, we see how belief becomes the greater underlying part of all language and communication. And realizing this, we must see that any reciprocal peaceful loving communication between us and GOD above, must require we share in his understanding and belief. Please see how all unintended and intended destructive actions of mankind, are born of imperfect understanding, and by accepting their self as being a temporary aspect. Once we know we are an eternal soul, we can see our Godly eternal responsibility between all souls. If souls see their selves as temporary individuals, life loses so many degrees of meaning among such souls. All misunderstandings are raised in the confusion of pride, envy, and fearful selfish preservation of one's individual imperfect beliefs. And in a complete opposite fashion to a flawed weak fragmented mankind, we see GOD'S omniscient supreme conscious force, and great eternal purpose and desire, having no fear or weakness, and no external physical needs to exist beyond their self. So, we see GOD in loving omniscience, are incapable of fearful self-protecting prideful envious physical desires arising within their souls. There existing no imperfect place, weakness, or error, or fear, or need within them, from which those might ever even arise. In this, we can see and understand the eternal truth and love of GOD'S omniscience. Individualism is observed to qualifiedly and quantifiably build differences and separations, and tear apart and destroy shared peace, life, and love. A strong spiritual culture, survives only until it becomes divided. Now, according to thermodynamics and empirical evidence. Once conscious disorder is introduced into perfect conscious stability, that disorder must continually increase. So, which existed first, perfect order, or complete disorder. Imagine pure dynamic complex *guided* motion and activity, minus all disorder; and add complex perfect dynamic orders of life. What does one tiny aspect of disorder do, when cast into that perfection. Now disorder has only one thing in common with light, in that it is seen in two ways; so, it must be explained in two ways. At its existence, disorder must be seen as both a cause *and* an effect. Disorder, or entropy, is simply the divisive destructive division of all dynamic and stable complexity. All the immense complexity we see, need, and love, can exist in a pure perfect order forever. But nothing complex can exist forever having one tiny single internal disorder. In our every understanding, we must rely on the fact that our own inner desires, are each divisive or cohesive. I hear many ask what defines the declared eternal perfection of GOD; and in essence, this is the same question as what is eternity. Eternal existence, is a concept impossible to define by any seen example within this current timely temporary dissolving existential theatre. Eternal is a term that can only pertain and apply to perfect things that are then without end;

of which, we see none in this deteriorating classical physical existence. In many ways, eternity means there is no past and future, there is only and always the ever-present expanding moment. Time can only be scientifically defined as the observed disillusion of all things between their beginning and their end. And eternity is defined as the absence of all times affects. I have heard many try to explain, define, or deny, the perfection of GOD. But always in their individual theological and moral subjective understanding. But GOD'S pure total perfection goes beyond all conscious aspects; yet no one seems to see the objective perfect physical side of GOD. We see this vast created universe is being stretched and consumed with confusion, entropy, and fission. These evil forces pushing the physical creation towards its final dissolution. So, we see another *Principle*, being that minus these timely evil forces, there *can* be a dynamic stable eternal shared physical reality for all who live within GOD'S omniscient ways. By GOD'S omniscient perfect truth, both consciously and physically, GOD is eternally bound immovable and immutable to that which defines them. GOD is then physically perfect, meaning without weakness or limitation over anything created below them. GOD being consciously and physically perfect, could never create a physical reality or any sum of realities that could ever physically overcome them. This is an objective fact that declares GOD'S perfection above us. All that ever exists for *us,* within both concepts of eternity or time, exists only in our current moment of perceived physical reality. So, in *not* ever experiencing or expecting physical or conscious sickness or confusion, or aging, hurt, loss, death, or doubt as to GOD'S truth; and in having eternal love for GOD and all perfect souls within GOD'S creation, we then hold the *objective* perception of what eternity and being eternal feels like. The feelings of both eternity or time, exist *before* us in between that which is created now, and that which will be created next. Eternity is comforting and lacks every threat of future loss of anyone or anything good. Most souls would embrace eternity, except for one thing, being that eternity is only found to exist in living within and among that eternal perfection of GOD. But most want to redefine perfection, and be a god over their own life. But individuals are not supreme, so they can never redefine FATHER'S perfection. The Materialist knows so little, they must forever wonder in their confusion about the physical structure and purpose and substance of creation. Romans 1:19-22, "*BECAUSE THAT WHICH MAY BE KNOWN OF **GOD** IS MANIFEST IN THEM; FOR **GOD** HATH SHEWED IT UNTO THEM. FOR THE <u>INVISIBLE THINGS OF HIM</u>*

FROM THE CREATION OF THE WORLD ARE CLEARLY SEEN, BEING UNDERSTOOD BY THE THINGS THAT ARE MADE, EVEN HIS ETERNAL POWER AND ***GODHEAD****; SO THAT THEY ARE WITHOUT EXCUSE: BECAUSE THAT, WHEN THEY KNEW* ***GOD****, THEY GLORIFIED HIM NOT AS* ***GOD****, NEITHER WERE THEY THANKFUL; BUT BECAME VAIN IN THEIR IMAGINATIONS, AND THEIR FOOLISH HEARTS WERE DARKENED. PROFESSING THEMSELVES TO BE WISE, THEY BECAME FOOLS.*" The experimental data that science uncovers, is never to be discounted outright, but it needs to be interpreted by HOLYSPIRIT. 1st John 1:20, "*BUT YE HAVE AN UNCTION FROM THE* ***HOLY ONE****, AND YE KNOW ALL THINGS.*" FATHER'S pure forceful conscious control, once rested upon all perfect motions and consistencies of creation. And upon the process of maintaining physical life, being what science calls the phenomenal forces. Scriptures describe how that first rebellion of imperfection entered the creation, and how FATHER'S hands which hold the universe, recoiled and pulled back. This stretched his creation out like we would pull on a curtain. So, it was no longer in its perfect stable design, as *one* imperfect conscious choice, introduced the growing element of physical disorder. Isaiah 42:5, "*THUS SAITH GOD THE LORD, HE THAT*

CREATED THE HEAVENS, AND STRETCHED THEM OUT;" Isaiah 48:13, "*MINE HAND ALSO HATH LAID THE FOUNDATION OF THE EARTH, AND MY RIGHT HAND HATH SPANNED THE HEAVENS: WHEN I CALL UNTO THEM, THEY STAND UP TOGETHER.*" Mankind has en-masse, consciously rejected and disrupted the perfection of GOD'S original design. But they cannot alter the personage of GOD, other than to disappoint, hurt, and destroy the love between themselves and GOD. *By three things*: one, GOD has twice, stretched and stopped, then stretched and continues to stretch the universe a third time; two, at least three vastly different speeds of light particles were created in GOD'S design; and lastly, mankind's limited slow small observational frame of reference. By these three, we see the age of the universe cannot be known by its current rate of expansion, measured using the slowest particle of light. This universe is at most, around twelve thousand years old. Thess. 2:10-12, says, most choose to believe a lie, so FATHER hides himself. And by mankind's subjective beliefs, those who don't desire GOD, form strong delusions of science, religion, morality, and all aspects of conscious belief. It can be said without offending GOD, or science, that all we see was first perceived in layers, during the first six literal rotations of the earth. But its original design is lost to those holding a small-slow-temporary belief and perspectives of reality. All who desire this strong delusion, run from GOD, chasing false hope. FATHER, in once being all alone in great loneliness, did perceive for himself an outward appearance, designed to share love. FATHER then cried out loudly, in all his conscious forceful desire of wanting the love of a SON. FATHER'S first WORD, carried a portion of FATHER'S own soul and body. And that eternal release of a portion of FATHER'S forceful desire to exist and his body, came forth to became self-aware, being the second of GOD. JESUS said in John 16:28, "*I CAME FORTH FROM THE FATHER, AND AM COME INTO THE WORLD: AGAIN, I LEAVE THE WORLD, AND GO TO THE FATHER*." JESUS said in Prov. 8:22-25, "*THE LORD POSSESSED ME IN THE BEGINNING OF HIS WAY, BEFORE HIS WORKS OF OLD. I WAS SET UP FROM EVERLASTING, FROM THE BEGINNING, OR EVER THE EARTH WAS. WHEN THERE WERE NO DEPTHS, I WAS BROUGHT FORTH; WHEN THERE WERE NO FOUNTAINS ABOUNDING WITH WATER. BEFORE THE MOUNTAINS WERE SETTLED, BEFORE THE HILLS WAS I BROUGHT FORTH:*" FATHER'S very first WORD spoken *forth* outwardly, became the *firstborn* of every creature. Col. 1:14-15, "*IN WHOM WE*

HAVE REDEMPTION THROUGH HIS BLOOD, EVEN THE FORGIVENESS OF SINS: WHO IS THE IMAGE OF THE INVISIBLE GOD, THE FIRSTBORN OF EVERY CREATURE:" John 1:1-4, "*IN THE BEGINNING WAS THE WORD, AND THE WORD WAS WITH GOD, AND THE WORD WAS GOD. THE SAME WAS IN THE BEGINNING WITH GOD. ALL THINGS WERE MADE BY HIM; AND WITHOUT HIM WAS NOT ANY THING MADE THAT WAS MADE. IN HIM WAS LIFE; AND THE LIFE WAS THE LIGHT OF MEN.*" John 1:14, "*AND THE WORD WAS MADE FLESH, AND DWELT AMONG US, AND WE BEHELD HIS GLORY, THE GLORY AS OF THE ONLY BEGOTTEN OF THE FATHER, FULL OF GRACE AND TRUTH.*" Before this first WORD, there was only FATHER and his *internal* knowing, desiring, and dreaming; with all perceived existence remaining within him all alone. There was only a growing love, belief, and understanding, between FATHER and SON as FATHER spoke in an ongoing sharing of his soul in love and truthful teaching between them. This back and forth forceful and substantive flowing form of truth, was a teaching and learning communion between FATHER and SON. This flow that existed in between them, grew in form and existence to become the moving HOLYSPIRIT, being the third soul of GOD. HOLYSPIRIT became aware, and has conveyed the love, belief, and understanding of GOD to all who would receive it. So, we see GOD become three souls, yet one in pure perfect loving belief and agreement. The Trinity then, were in agreement as FATHER spoke to perceive all layered external aspects of the universe. FATHER also whispered to gift all lesser creative souls, having their free-will to hear GOD and create their beliefs according to their desires. The Scriptures say that by GOD, do all things *consist,* meaning to hold together. We know all forces that hold all physical things together, and all forces of physical life, are actually FATHER'S willful projected desire. Now if we study the physical information that stimulates the body's neural system, we see correlating areas of increased neural chemical electromagnetic activity. And if we apply the very same form of study to a computer, we see the same types of correlations between all input information and all external conscious sources and physical devices; we see a computer's graphics card sends its information to a monitor, to create visual representations, and a computer's audio cards information is sent to speakers, to create audio representations. But images and sounds have no self-awareness, until they touch our souls. A neural system, has and exhibits the same informational correlations as a computer, as it interfaces between the world and our soul, raising our created awareness. The entire neural system is just a complex interface. Our self-aware sense of existence does not come by, or originate from those neural aspects alone. Our awareness is an intrinsic attribute of our soul's eternal desirous forceful conscious estate. Our soul's awareness, rests in all our emotion and desires. Our awareness of visualization, hearing, feeling, and emotion, are all unique enigmatic aspects of our soul alone. The body-brain sensing correlations, being composed of elemental matter and energy, are *never* aware of, and never feel or know anything. If visual awareness came by the eye's interaction with the patterns of information in physical light, how do we brightly dream in the darkness of night. How can we see, hear, and feel another reality filled with brightness, sound, touch, and our every emotion. Our created dreams range from obscure, to bright and vivid. In a most lucid vivid dream during sleep, I could smell the air. I felt my presence in standing and feeling the sunlight from one side, while a breeze came from behind, moving my hair. I could see in better brightness and detail than when I am awake. I could see more distant things in brighter details, yet I knew I was dreaming. It was so bright, so firm and real in every correct way. I could not find one weakness in my senses or emotions, compared to my wakeful reality. This realism in a dream creation is a rare occurrence in most people. Imagine if I did not wake up, and my dream continued in that realism in that place, I had never been before. Even sleeping and waking and running into the same people I had never previously met, as this continued for weeks and months. I would have to accept that my soul had ceased to exist in one created physical place, and I now exist in another. This is what it will be like to go into the long sleep of death, and awake to our afterlife. Except we hope to see those we previously knew and loved. If GOD changes our body and location, it is just like changing a TV station on a screen for us; we consciously existed on one channel, and then another. Now there are only two channels beyond this world. Don't leave this up to chance, because chance eventually loses every time. This is a free-will choice that must be made out of love, not chance. Lust is born of chance, but love is born out of free-will choice. Reality and our existence have *always* been seen, heard, felt, and desired within our soul; and all within the quantum realm. The one difference between wakeful and dream awareness, other than the strength and clarity of our awareness, is our belief. In being awake, most believe they know where they consciously are, when they don't. And in dreams, most believe they don't know where they are, when they should. Whether asleep or awake, we consciously exist only within our bright perceived created reality within our soul. This truth completely explains the depth and form of our full body and environmental experience, with its expanding and shrinking presence of qualia we feel. Visualize how through vastly differing sizes, speeds, densities, and consistencies, that within one aether, two created great dimensional realms of existence, cohabitate and overlay in the same space. At the fall of the physical realm of life, a gulf formed between all lesser conscious to physical perspectives, that subjectively govern their relative perceptions. We can never change our platform, and cross between these realms, unless GOD desires and causes it. We are greatly enlightened by this divide of perceptive ability between the eternal ethereal quantum realm, and the now temporary firmer classical realm. This, by design, is a necessity, this divide between the ghostly realm, and the much firmer physical realm, is a point and place where the conscious force of FATHER works within us, where he and his physical creations interact. The most basic qualities are forceful desire, and all pure physical points. Being minus the other, neither can firmly and soundly perceive, or be firmly or soundly perceived to exist in any awareness of complexity. Our awareness is enabled, only after the created dimensions of our internal perceptions. We exist as a bicameral phenomenon, choosing our conscious path, by two distinct parts, being the physical existence of anything complex and designed, and our consciously perceived knowledge and belief of its existence, in sequential order. Consider deeply, how secularism and humanism say there is no single perfect belief or way of life, and that there are only individual preferences, and an ever-changing controlling consensus. I wholly disagree. We watch the secular sciences measure and ponder the remains of FATHER'S original desired creation of the universe and all physical life. But they hold no desire to see these remnants as originating by FATHER'S perfect forceful creative desires. The marriage bond that exists between our every projected conscious forceful desire, and its patterned operation that creates our physical perceptions, are perfectly symmetric at their origin. So, imagine if you were the only one to exist, and you create your internal particle-wave-based dreams, just like GOD. But to not be alone, you desire, speak forth, and create an intricate external perfect physical spherical dreamscape of your desires, using your own physical properties. Then deep within your external creation, you partition out from your soul, minute conscious spheres, or souls of desire, to dream create their own perceptions; all having free-willed awareness, deep within your created dreamscape. But you must also provide them with an aether or medium, and a body to sense all moving relatable physical information. All this, in your hope they will share love and life with you in your perfect eternal ways. They could perceive and share in your creation, which is perfectly aligned to exist according to your desires forever. But what if they reject your perfect conscious and physical created ways, and choose to interject only multitudes of contrary patterns of desire. Your desire to find eternal love and perfect fulfillment might be destroyed. But wait, you could proceed using your designed propagation of physical life, gifting more souls, and find the fewer along the way who come to understand and share your desires. Then with these who came to love and know you, you could perceive a new loving eternal outcome with them. By this example, we see that though every individual's internal belief is perfectly aligned between their soul and their created perceptions. There is still the one greater required *external* perception that provides physical life, which we must share in the way *it* was perceived. FATHER'S desires for eternal love, as he gifts and creates life, are real; and are now mocked, and his physical creations ruined; yet FATHER finds *his* children along the way. We perceive and exist in our minute weakness, need, and accountability, to the Creator of this vast and great external perception of dynamic radiating flowing information that surrounds, enables, and sustains, our awareness. Around 2000 years ago, mankind saw just a small glimpse of the original conscious forceful desires in all highest perfection, who came again to walk the earth. We see the mind of GOD, who put off his eternal body to walk in the flesh of mankind. And by his conscious projections of force, many diseased, and even the dead were physically perceived anew. Broken bodies were seen to realign to the perfect desires of FATHER through CHRIST and HOLYSPIRIT. Our souls are quantum projecting forceful creators of particle-wave-based perceptions, but these required bodies are of the firm physical world. The Angel's differ, in that their souls, and ethereal bodies, reside in the quantum realm. FATHER can perceive any soul's current body to be translated between consistencies, realms, to any location, and in any size. Heb. 11:5-6, "*BY FAITH ENOCH WAS TRANSLATED THAT HE SHOULD NOT SEE DEATH; AND WAS NOT FOUND, BECAUSE GOD HAD TRANSLATED HIM: FOR BEFORE HIS TRANSLATION HE HAD THIS TESTIMONY, THAT HE PLEASED GOD. BUT WITHOUT FAITH IT IS IMPOSSIBLE TO PLEASE HIM: FOR HE THAT COMETH TO GOD MUST BELIEVE THAT HE IS, AND THAT HE IS A REWARDER OF*

THEM THAT DILIGENTLY SEEK HIM." Col 1:12-17, "*GIVING THANKS UNTO THE FATHER, WHICH HATH MADE US MEET TO BE PARTAKERS OF THE INHERITANCE OF THE SAINTS IN LIGHT: WHO HATH DELIVERED US FROM THE POWER OF DARKNESS, AND HATH TRANSLATED US INTO THE KINGDOM OF HIS DEAR SON: IN WHOM WE HAVE REDEMPTION THROUGH HIS BLOOD, EVEN THE FORGIVENESS OF SINS: WHO IS THE IMAGE OF THE INVISIBLE GOD, THE FIRSTBORN OF EVERY CREATURE: FOR BY HIM WERE ALL THINGS CREATED, THAT ARE IN HEAVEN, AND THAT ARE IN EARTH, VISIBLE AND INVISIBLE, WHETHER THEY BE THRONES, OR DOMINIONS, OR PRINCIPALITIES, OR POWERS: ALL THINGS WERE CREATED BY HIM, AND FOR HIM: AND HE IS BEFORE ALL THINGS, AND BY HIM ALL THINGS CONSIST.*" Acts 8:39, "*AND WHEN THEY WERE COME UP OUT OF THE WATER, THE SPIRIT OF THE LORD CAUGHT AWAY PHILIP, THAT THE EUNUCH SAW HIM NO MORE: AND HE WENT ON HIS WAY REJOICING. BUT PHILIP WAS FOUND AT AZOTUS: AND PASSING THROUGH HE PREACHED IN ALL THE CITIES, TILL HE CAME TO CAESAREA.*" John 6:20-21, "*BUT HE SAITH UNTO THEM,*

IT IS I; BE NOT AFRAID. THEN THEY WILLINGLY RECEIVED HIM INTO THE SHIP: AND IMMEDIATELY THE SHIP WAS AT THE LAND WITHER THEY WENT." To translate mass, means to change its form and place. There are many Scriptures in which GOD *translates* angel's ethereal body's to be small and firm on the earth, and back again. He translates Prophets bodies from the physical, to be ethereal in the realm of Heaven, and sends them back to a new physical location before the temporary aspects of time could move. In body translation, only a body changes and moves; the soul stays in the quantum realm, and is not translated in any way. A soul can experience a shock of understanding and belief, but no change to our soul's essence or quality in such a move. Now all things perfect and *sealed* are eternal, as they have no weakness that can allow disorder to enter. GOD'S original creation was perfect, but in placing untested unlearned free-willed souls, and one forbidden choice within it, it was not sealed. Seeing all eternal things must be perfect, GOD cannot bring eternal love and peace to any who refuse his created ways and choose their own imperfect destructive ways. We see the patterned forceful sustaining work of FATHER has never decreased or increased, but the ever-increasing physical confusions and destructions of mankind, renders that level of care more and more insufficient. FATHER will not provide additional care to all his creations. FATHER'S will, is moving daily away from mercy, and towards wrath for those dishonoring the Sacrifice of his SON. If your son died to help others, and they harshly and hatefully rejected and refused to consider *your* son's sacrifice, how would you feel? What if they even mocked and abused your son's name? FATHER is furiously heart-broken, seeing all of this, *screaming* my Holy SON suffered unimaginably and died to help you, and it means nothing to you! Having free-will, calls out and requires us each to know that love is the ruling force of desire over all reality. This places in us a call to obey all the loving words of GOD. *True* religion, is a term that applies only to the one pure belief and nature of GOD that they want us to share; and not what any rebellious soul wants GOD'S belief and nature to be. We find by the pollution of confusion and evil desires, that there is no end to imagined false individual beliefs. GOD'S one true belief allows no compromising of contrary hopes, desires, or traditions, for this world and the flesh. This truth declares there are endless individual religions, based on partial-truths, mixed with the desires and traditions of the flesh. Pure true religion, is to accept and work in ourselves, that which will bring us in our fullness, to be one in the family and belief of GOD. If any soul's religion, cannot reach this goal *while in this world*, that soul should in all fear, give great consideration to their fatal religious flaws. Now omniscience (omni-science) means all science, and only FATHER is omniscient, meaning knowing the true and complete science of GOD. So, omniscience and eternal love are of GOD and his Children alone. Do all souls love and value those who love them, above those who do not. Can FATHER work eternally to sustain the joy of those who do not love him, who forever offend and reject him. Think of a murderous molesting drug crazed addict, being an offensive form of life. Should you be forced to love and sustain that person eternally in your home, in their choices. If you say this is not the same, because you are not anything like that person. Then you must be *assuming* GOD is like you. Consider that in GOD'S higher beliefs, the *difference* between what you see as your goodness, and GOD'S true goodness, may be far greater than the difference you see between you and that murderous, molesting addict. We don't get to define what is good or evil; not even a little. We must learn and accept the real differences between GOD'S perfection and eternity, and fallen mankind's imperfect destructive desires and elements of time. GOD'S realities are incompatible in every way, with all imperfections. Any allowed imperfection, destroys pure eternal perfection.

There are endless degrees of imperfection in disobedience to GOD'S omniscience; but only one pure degree of perfection. Mankind stands at the door of truth and knowledge, and yet through their rebellious reasoning, they try to see many certain aspects of GOD'S work. But as Scripture says, they strongly desire to misinterpret what they vaguely see, to deny GOD. Romans 1:20-21, "*FOR THE INVISIBLE THINGS OF HIM FROM THE CREATION OF THE WORLD ARE CLEARLY SEEN, BEING UNDERSTOOD BY THE THINGS THAT ARE MADE, EVEN HIS ETERNAL POWER AND GODHEAD; SO THAT THEY ARE WITHOUT EXCUSE: BECAUSE THAT, WHEN THEY KNEW GOD, THEY GLORIFIED HIM NOT AS GOD, NEITHER WERE THANKFUL; BUT BECAME VAIN IN THEIR IMAGINATIONS, AND THEIR FOOLISH HEART WAS DARKENED.*" Scripture never once says GOD willed things into existence from nothing. From the moment their perfection was lost, mankind's individual evil desires have removed GOD from their sight. Evil souls can*'t* see why GOD allows so much evil to occur, raising their doubt about GOD'S existence. Free-will is absolute, it is the greatest requirement and necessity for love to exist. GOD never over-rules a soul's free-will choice. Every destruction in this fallen creation has occurred by free-will souls disobeying GOD'S created ways.

Many souls believe that physical chance controls *all* things, allowing them to see no need for GOD to exist; that being impossible. Please try to understand that if CHRIST ever tells you that you have been found wanting, and are sealed as lost forever. That this is an absolute that can never-ever change. Before GOD ever speaks, in their omniscience, they consider every possibility and probability of thought, desire, and action throughout all endless directions and eons. Remember, we never see or speak of a finite linear progression of reality; but an eternally fulfilling spherical reality. Point this world's most powerful telescope in in every direction, and you will see space and reality are there in a vast equal distance. This says the universe is not visually flat or linear, but visually spherical. And we do and must also accept the facts that this spherical physical reality is entirely aether based; and only secondly, is a certain part of it designed under GOD'S quantum laws, forces, reasons, and purpose. And lastly, the least part of this entire aether-based reality is more firmly designed under all god's laws, forces, reasons, and purpose. Yet for all of this, GOD'S entire spherical substantive reality, in all those same levels, ways, and forces, remains consciously deterministic. So, in the current picture, once beyond the aether, it exists in uncertainty, and currently in a temporary state. And in my mindful understanding of GOD, I cannot grasp how such a thing as woven space-time can exist. How can time and space be woven together, if one is spherical and uncertain, and one is said to be linear and certain. Also, if they are woven as one, how can we move up and down, and back and forth through space, but not time? Why does only momentum distort the equations of time? All concepts of time are described as flat and linear, yet the aether, space, and all the lesser firmer created parts of reality are all spherical. Now we know their linear descriptions and equations of time and reality are for instigating mankind's wrongful conceptual understanding. But why not use an expanding spherical visual for our ever-changing moments of conscious experience, to align with the spherical physical side of reality? Why choose a flat linear visual? Well, a personal experiential spherical visual of time and life, leans to, and opens doors, avenues, relationships, and questions of belief in the soul, which affect physical theories in deterministic ways that materialist don't like. GOD does not look backwards at a linear time-based past, or forward to a linear time-based future; but in every possible inner and outer spherical direction, dimension, and consideration, before ever making a final judgment. *And* if CHRIST ever says, welcome thou good and faithful servant; this too will be eternal, and can never change. We have touched upon substance, dimensions, energies, and forces. But what is *the* most powerful ruling known or unknown force? What is that greatest force of all, being the primary force that judges all reality? The force that desired and created all things, and that will stop all evil. It is the force of Eternal Holy Love. The only force known, that has the power to stay or slow the hand and wrath of GOD. This love is the highest governing force that rules over all creation by a singular pure eternal truth of purpose; ruling over all physical, and conscious, and all unknown realms. It has zero meaning, that FATHER and we are so vastly different in size. Love is love, being a desire that has no size; it only exists in degrees of strength and purity. Can one value the love of a bigger, over a smaller person. Can a bigger person hold more love. GOD can truly and eternally love us, and we can truly and eternally love them. And now the boundaries of the love and mercies of GOD, are beyond all scientific inquiry, as they exist in the immutable inalterable heart of GOD. Amen

THE LAST DAYS OF GREAT DECEPTION

We hope no one sees it as strange, that being deceived and choosing anything evil, and lacking Godly understanding or desire, go hand in hand. So, having the full true understanding desire of GOD and all reality, is our only defense against every deception. Why is so much false understanding of belief being propagated? And which deceptions do we need protected from the most. There are worldly deceptions all around us. Scams by evil souls out to use others for personal gain or pleasures. There are devious lies and sales pitches at every turn. But the most dangerous lies and miss direction, are set in all religious deceptions, born of individual Scriptural interpretations. We must be profoundly aware of Satan's work in religious belief, seeing false teachers and preachers offer delusional doctrines. This ubiquitous occurrence, makes the pure truth and wisdom of GOD, the most important knowledge we can acquire. How is it that so many souls, quickly and easily become deceived? What root causes can explain how so many are easily fooled? We see many causes for this, such as pride and lust. But one evil cause, as strong as pride and lust, never gets its share of blame; being curiosity. Remember how in hearing Satan's lie, that Eve and then Adam *wondered* about what he had said, and then *doubted* GOD. A curiosity was in their wondering what would really happen, if they stepped out of GOD'S defined ways. Our curiosity is born at the edges of our hopes and fears. It rises out of our limited knowledge and impatience for what unknown occurrences come next. This conscious aspect called curiosity is in its ways, related to confusion, as both can tempt one to step beyond FATHER'S created ways. Curiosity and questioning, can creep into our struggle between our need for the spiritual, and desire for the physical. The senses of the body, can tempt and trigger a soul's questioning curiosity. A Soul must then choose how to respond to that temptation of the body. This is a spiritual against physical battle, over which side controls our actions. Does our soul's love for GOD rule, or our desire for the sensations of this world? Curiosity always plays its role in souls getting lost; but Godly responsible faith keeps us safe. The term spirituality, refers to our soul's beliefs and expressed emotional attributes and desires. By all *these*, we see GOD sees our soul as being an evil spirit, or a Godly spirit. Devils and demons, are terms describing souls who have evil attributes and desires, contrary to GOD'S. Yet, so many souls believe they can *live* in sin, and still be in GOD'S grace. Why would GOD love and call us his children, if we emulate his enemies. All who act contrary to FATHER'S created ways, are devils. Only in our relentless, responsible, fervent, faithful, following in the ways of GOD, are we described as being of GOD. A curiosity towards sin, fill every vacuum left in a soul losing their desire for GOD'S ways. Because that is all that is left in their reality to want. Since mankind's fall from grace, their curiosity is for the flesh; it is to seek, feel, and experience, endless physical thrills and pleasures. A reprobate seeks only things of physical pleasure, security, comfort, and continuity. And just as all evil fantasy and curious desires are born of the flesh, all eternal questions and desires for eternal peace, joy, and love, are born only in souls searching only for GOD. The WORD says seek and you shall find; but what do we each truly seek and find? We seek whatever we desire, and we also seek it in the strength we desire it. We can desire to search for all of GOD, or just a certain desired amount; and the latter GOD rejects. By free-will, all who hold to a weak false religious doctrine, whether they know, or admit it, they were desiring and looking for and wanting that same weakness in GOD. Evil desires and curiosity are on our one side pulling; this is called temptation. If we want something evil, and see that evil thing is available, we experience strong temptation. If we remove all availability, and work to remove the desire, we have a chance. Our *eternal responsible* desire, is our Godly strength. And all curious desires for pleasure, come from weakness. Can't we all see this occurring before our eyes? Are most blinded to it, being lost in their own desired destruction? I sit across from, or walk past people, and I see deception and curiosity eat away their ability to ever know, believe, and live the created ways of GOD. We each must *fight* hard, knowing only GOD'S purposeful ways bring eternal peaceful understanding. Many want some of their ways, and some of GOD'S ways, just not all of them. Most want GOD in a lesser measure that allows for some worldly desires; and these try to have both worlds. Scripture says such a life, being lukewarm, brings the worst judgments. There is deception hiding around every turn in this world. But the sneakiest, most dangerous, least acknowledged deception of all, is *self*-deception. *Self*-deception is only and always born out of our desires for worldly things. It is at *enmity* (war) with GOD. The flesh, the devil, pride, and the lust of the eye, are found in wanting the creation, and having no greater desire for its Creator. We now see so many religious pagan beliefs, like at CHRIST'S first coming. False doctrine is prevalent among believers who are more focused on family, careers, fun, safety, and possessions, to the extent they haven't the time or desire to search for GOD'S greater truth. Most souls can believe GOD does not want or need much in return from them. These souls act like our love for GOD, is meant to be different from our love for family, friends, and our physical lives, and more like a reverence for GOD'S supreme abilities, and being thankful for all physical provisions. Quite like the entitled sense liberals have towards government provision. To clarify, whenever we say "*professed* Christian", we refer to weak confused uneducated believers in GOD, still caught in confusion, deception, and idolatry. On balance these souls live more, to much more for the flesh, and *on balance,* have a weak ungodly Spirit. Though they love GOD in a lesser way, they have a vain profession of grace, being minus HOLYSPIRIT and the eternal perspective. Matt 15:7-9, "*YE HYPOCRITES, WELL DID ESAIAS PROPHECY OF YOU, SAYING, THIS PEOPLE DRAWETH NIGH UNTO ME WITH THEIR MOUTH, AND HONOURETH ME WITH THEIR LIPS; BUT THEIR HEART IS FAR FROM ME. BUT IN VAIN THEY DO WORSHIP ME, TEACHING FOR DOCTRINES THE COMMANDMENTS OF MEN*." We must touch here on a few false doctrines. But we do not at all say that one's current weak investment in false doctrines, precludes one from *turning from* all such doctrines, and attaining the salvation of the pure truth. I also do not present myself, as invulnerable to this world; as we are all tried by the fires of this world. We all must strive daily in reaching for the heights of GOD; we must reject, and climb out from the depths of this fallen horrid world of hell and death. Please do not think I speak or write from a place safe from attack, or of spiritual comfort; the attacks against me are strong and relentless. We speak only the truth, and we call out and preach as much to our self, as to others in all things.

Scriptural truth is from no man, including me; it comes from GOD alone. HOLYSPIRIT'S words, are always for both the hearer and the orator. And we pray for Scriptural interpretation from GOD alone, Amen. The most prevalent *false* teaching today, is the doctrine of a pre-tribulation rapture, which has been around for about 150 years. It was ridiculed in nearly all churches, up until the 1970's. This false doctrine has grown fast along-side the *false* prosperity doctrine. These and all *false* doctrines feed off each other and have become sister doctrines that come from different directions, attacking the same weakness. Do any find it odd, that all new doctrines allow more and more physical possessions and more freedom from GOD'S commands; and they all ask less love from mankind for GOD in return. Scripture denies a pre, or mid-tribulation rapture. Matt 24:29-31, "*IMMEDIATELY AFTER THE TRIBULATION OF THOSE DAYS SHALL THE SUN BE DARKENED, AND THE MOON SHALL NOT GIVE HER LIGHT, AND THE STARS SHALL FALL FROM HEAVEN, AND THE POWERS OF THE HEAVENS SHALL BE SHAKEN: AND THEN SHALL APPEAR THE SIGN OF THE SON OF MAN IN HEAVEN: AND THEN SHALL ALL THE TRIBES OF THE EARTH MOURN, AND THEY SHALL SEE THE SON OF MAN COMING IN THE CLOUDS OF*

HEAVEN WITH POWER AND GREAT GLORY. AND HE SHALL SEND HIS ANGELS WITH A GREAT SOUND OF A TRUMPET, AND THEY SHALL GATHER TOGETHER HIS ELECT FROM THE FOUR WINDS, FROM ONE END OF HEAVEN TO THE OTHER." Rev. 6:9-11, "*AND WHEN HE HAD OPENED THE FIFTH SEAL, I SAW UNDER THE ALTER THE SOULS OF THEM THAT WERE SLAIN FOR THE* ***WORD*** *OF* ***GOD****, AND FOR THE TESTIMONY WHICH THEY HELD: AND THEY CRIED WITH A LOUD VOICE, SAYING, HOW LONG, O* ***LORD*** *HOLY AND TRUE, DOST THOU NOT JUDGE AND AVENGE OUR BLOOD ON THEM THAT DWELL ON THE EARTH? AND WHITE ROBES WERE GIVEN UNTO EVERYONE OF THEM; AND IT WAS SAID UNTO THEM, THAT THEY SHOULD REST YET FOR A LITTLE SEASON, UNTIL THEIR FELLOW SERVANTS ALSO AND THEIR BRETHREN, THAT SHOULD BE KILLED AS THEY WERE, SHOULD BE FULFILLED.*" Scripture says the Church is still here, enduring the tribulation, with more martyred brethren to come. Revelation 7:2-4, "*AND I SAW ANOTHER ANGEL ASENDING FROM THE EAST, HAVING THE SEAL OF THE LIVING* ***GOD****: AND HE CRIED WITH A LOUD VOICE TO THE FOUR ANGELS, TO WHOM IT WAS GIVEN TO HURT THE EARTH AND THE*

SEA, SAYING, HURT NOT THE EARTH, NEITHER THE SEA, NOR THE TREES, TILL WE HAVE SEALED THE SERVANTS OF OUR ***GOD*** *IN THEIR FOREHEADS. AND I HEARD THE NUMBER OF THEM WHICH WERE SEALED: AND THERE WERE SEALED AN HUNDRED AND FORTY AND FOUR THOUSAND OF ALL THE TRIBES OF ISRAEL.*" Here we see FATHER pour out his SPIRIT, and seal 144,000 prophets among the twelve nations. Again, servants, are the true Church. Revelation 7:13-17, "*AND ONE OF THE ELDERS ANSWERED, SAYING UNTO ME, WHAT ARE THESE WHICH ARE ARAYED IN WHITE ROBES? AND WHENCE CAME THEY? AND I SAID UNTO HIM, SIR, THOU KNOWEST. AND HE SAID UNTO ME, THESE ARE THEY WHICH CAME OUT OF GREAT TRIBULATION, AND HAVE WASHED THEIR ROBES, AND MADE THEM WHITE IN THE* ***BLOOD*** *OF THE* ***LAMB****. THEREFORE ARE THEY BEFORE THE THRONE OF* ***GOD****, AND SERVE HIM DAY AND NIGHT IN HIS TEMPLE: AND HE THAT SITTETH ON THE THRONE SHALL DWELL AMONG THEM, THEY SHALL HUNGER NO MORE, NEITHER THIRST ANYMORE; NEITHER SHALL THE SUN LIGHT ON THEM, NOR ANY HEAT. FOR THE* ***LAMB*** *WHICH IS IN THE MIDST OF THE THRONE SHALL FEED THEM, AND*

SHALL LEAD THEM UNTO LIVING FOUNTAINS OF WATERS: AND ***GOD*** *SHALL WIPE AWAY ALL TEARS FROM THEIR EYES.*" Throughout Scripture, all Prophets, Apostles, Martyrs, and our beloved LORD. All suffered great cruelty in this world. We see CHRIST'S Church, through the ages is brutalized, crucified, and killed in horrible ways by this evil world. But the current weak generation wants to hear *only* sweet comforts. These seek false preachers telling them how sweet and easy things will be for them. The pre, and mid-trib. also believe that after their false rapture, other souls will find salvation, but they are persecuted and killed for *their* faith throughout the tribulation up until the last trumpet. These pre, or mid-trib. rapture doctrines say, that *only* those alive now, who profess the name of CHRIST, in all of history, are chosen by GOD to not suffer any trying of their faith. Why would these very weak and rebellious souls, get unique treatment, even above CHRIST. What will a soul do if they believe they will be raptured out before any coming wrath and trying, and finds it not to be so? What devastation will come upon their faith, it being weak and worldly already. Every Holy Word of Scripture denies a pre, or mid-tribulation rapture. When asked by the Apostles, the LORD explained it in short fashion. Mark 13:4-27, "*TELL US, WHEN SHALL THESE THINGS BE?*

AND WHAT SHALL BE THE SIGN WHEN ALL THESE THINGS SHALL BE FULFILLED? AND ***JESUS*** *ANSWERING THEM BEGAN TO SAY, TAKE HEED LEST ANY MAN DECEIVE YOU: FOR MANY SHALL COME IN MY NAME, SAYING, I AM* ***CHRIST****; AND SHALL DECEIVE MANY. AND WHEN YE SHALL HEAR OF WARS AND RUMORS OF WARS, BE YE NOT TROUBLED: FOR SUCH THINGS MUST NEEDS BE; BUT THE END SHALL NOT BE YET. FOR NATION SHALL RISE AGAINST NATION, AND KINGDOM AGAINST KINGDOM: AND THERE SHALL BE EARTHQUAKES IN DIVERS PLACES, AND THERE SHALL BE FAMINES AND TROUBLES: THESE ARE THE BEGINNINGS OF SORROWS. BUT TAKE HEED TO YOURSELVES: FOR THEY SHALL DELIVER YOU UP TO COUNCILS; AND IN THE SYNAGOGUES YE SHALL BE BEATEN: AND YE SHALL BE BROUGHT BEFORE KINGS AND RULERS FOR MY SAKE, FOR A TESTIMONY AGAINST THEM. AND THE GOSPEL MUST FIRST BE PUBLISHED AMONG ALL NATIONS. BUT WHEN THEY SHALL LEAD YOU, AND DELIVER YOU UP, TAKE NO THOUGHT BEFOREHAND WHAT YE SHALL SPEAK, NEITHER DO YE PREMEDITATE: BUT WHATSOEVER WILL BE GIVEN YOU IN THAT HOUR, THAT SPEAK YE:*

*FOR IT IS NOT YE THAT SPEAK, BUT THE **HO'LY GHOST**. NOW THE BROTHER SHALL BETRAY THE BROTHER TO DEATH, AND THE FATHER THE SON; AND CHILDREN SHALL RISE UP AGAINST THEIR PARENTS, AND SHALL <u>CAUSE THEM TO BE PUT TO DEATH</u>. AND <u>YE SHALL BE HATED</u> OF ALL MEN FOR MY NAMES SAKE: BUT HE THAT SHALL ENDURE UNTO THE END, THE SAME SHALL BE SAVED. BUT <u>WHEN YE SHALL SEE THE ABOMINATION OF DESOLATION SPOKEN OF BY DANIEL THE PROPHET STANDING WHERE IT OUGHT NOT</u>. (LET HIM THAT READETH UNDERSTAND,) THEN LET THEM THAT BE IN JU'DAE'A FLEE TO THE MOUNTAINS: AND LET HIM THAT IS ON THE HOUSETOP NOT GO DOWN INTO THE HOUSE, NEITHER ENTER THEREIN, TO TAKE ANYTHING OUT OF THE HOUSE: AND LET HIM THAT IS IN THE FIELD NOT TURN BACK AGAIN FOR TO TAKE UP HIS GARMENT. BUT WOE TO THEM THAT ARE WITH CHILD, AND TO THOSE THAT GIVE SUCK IN THOSE DAYS! AND PRAY YE THAT YOUR FLIGHT BE NOT IN THE WINTER. FOR IN THOSE DAYS SHALL BE AFFLICTION, SUCH AS WAS NOT FROM THE BEGINNING OF THE CREATION WHICH **GOD** CREATED UNTO THIS TIME, NEITHER SHALL BE. AND EXCEPT*

*THAT THE **LORD** HAD SHORTENED THOSE DAYS, NO FLESH SHOULD BE SAVED: BUT FOR THE ELECT'S SAKE, WHOM HE HATH CHOSEN, HE HATH SHORTENED THE DAYS. AND THEN IF ANY MAN SHALL SAY TO YOU, LO, HERE IS **CHRIST**; OR LO, HE IS THERE; BELIEVE HIM NOT: FOR FALSE CHRISTS AND FALSE PROPHETS SHALL RISE, AND SHALL SHEW SIGNS AND WONDERS, TO SEDUCE, IF IT WERE POSSIBLE, EVEN THE ELECT. BUT TAKE YE HEED: BEHOLD, I HAVE FORETOLD YOU ALL THINGS. BUT IN THOSE DAYS, AFTER THE TRIBULATION, THE SUN SHALL BE DARKENED, AND THE MOON SHALL NOT GIVE HER LIGHT, AND THE STARS OF HEAVEN SHALL FALL, AND THE POWERS THAT ARE IN HEAVEN SHALL BE SHAKEN. AND THEN SHALL THEY SEE THE **SON** OF MAN COMING IN THE CLOUDS WITH GREAT POWER AND GLORY. AND THEN SHALL HE SEND HIS ANGELS, AND SHALL GATHER TOGETHER HIS ELECT FROM THE FOUR WINDS, FROM THE UTTERMOST PART OF THE EARTH TO THE UTTERMOST PART OF HEAVEN*." I read not one word about anyone being taken out before the tribulation. Rev. 19:19-21:1, "*AND I SAW THE BEAST, AND THE KINGS OF THE EARTH, AND THEIR ARMIES, GATHERED*

TOGETHER TO MAKE WAR AGAINST HIM THAT SAT ON THE HORSE, AND AGAINST HIS ARMY. AND THE BEAST WAS TAKEN, AND WITH HIM THE FALSE PROPHET THAT WROUGHT MIRICLES BEFORE HIM, WITH WHICH HE DECEIVED THEM THAT WORSHIPPED HIS IMAGE. THESE BOTH WERE CAST ALIVE INTO A LAKE OF FIRE BURNING WITH BRIMSTONE. AND THE REMNANT WERE SLAIN WITH THE SWORD OF HIM THAT SAT UPON THE HORSE, WHICH SWORD PROCEEDETH OUT OF HIS MOUTH: AND ALL THE FOULS OF THE AIR WERE FILLED WITH THEIR FLESH. AND I SAW AN ANGEL COME DOWN FROM HEAVEN, HAVING THE KEY OF THE BOTTOMLESS PIT AND A GREAT CHAIN IN HIS HAND. AND HE LAID HOLD ON THE DRAGON, THAT OLD SERPENT, WHICH IS THE DEVIL, AND SATAN, AND BOUND HIM A THOUSAND YEARS, AND CAST HIM INTO THE BOTTOMLESS PIT, AND SHUT HIM UP, AND SET A SEAL UPON HIM, THAT HE SHOULD DECEIVE THE NATIONS NO MORE, TILL THE THOUSAND YEARS SHOULD BE FULFILLED: AND AFTER THAT HE MUST BE LOOSED A LITTLE SEASON. AND I SAW THRONES, AND THEY THAT SAT UPON THEM, AND JUDGMENT WAS GIVENM UNTO THEM: AND I SAW THE SOULS OF

THEM THAT WERE BEHEADED FOR THE WITNESS OF ***JESUS****, AND THE* ***WORD*** *OF* ***GOD****, AND WHICH HAD NOT WORSHIPPED THE BEAST, NEITHER HIS IMAGE, NEITHER HAD RECEIVED HIS MARK UPON THEIR FOREHEADS, OR IN THEIR HANDS; AND THEY LIVED AND REIGNED WITH* ***CHRIST*** *A THOUSAND YEARS. BUT THE REST OF THE DEAD LIVED NOT AGAIN UNTIL THE THOUSAND YEARS WERE FINISHED. THIS IS THE FIRST RESURRECTION. BLESSED AND HOLY IS HE THAT HATH PART IN THE FIRST RESURRECTION: ON SUCH THE SECOND DEATH HATH NO POWER, BUT THEY SHALL BE PRIEST OF* ***GOD*** *AND OF HIS* ***CHRIST****, AND SHALL REIGN WITH HIM A THOUSAND YEARS. AND WHEN THE THOUSAND YEARS ARE EXPIRED, SATAN SHALL BE LOOSED OUT OF PRISON, AND SHALL GO OUT TO DECEIVE THE NATIONS WHICH ARE IN THE FOUR CORNERS OF THE EARTH, GOG AND MA'GOG, TO GATHER THEM TOGETHER TO BATTLE: THE NUMBER OF WHOM IS AS THE SAND OF THE SEA. AND THEY WENT UP ON THE BREADTH OF THE EARTH, AND COMPASSED THE CAMP OF THE SAINTS ABOUT, AND THE BELOVED CITY: AND FIRE CAME DOWN FROM* ***GOD*** *OUT OF HEAVEN AND DEVOURED THEM. AND*

THE DEVIL THAT DECEIVED THEM WAS CAST INTO THE LAKE OF FIRE AND BRIMSTONE, WHERE THE BEAST AND THE FALSE PROPHET ARE, AND SHALL BE TORMENTED DAY AND NIGHT FOR EVER AND EVER. AND I SAW A GREAT WHITE THRONE, AND HIM THAT SAT ON IT, FROM WHOSE FACE THE EARTH AND THE HEAVENS FLED AWAY; AND THERE WAS FOUND NO PLACE FOR THEM. AND I SAW THE DEAD, SMALL AND GREAT, STAND BEFORE ***GOD****; AND THE BOOKS WERE OPENED: AND ANOTHER BOOK WAS OPENED, WHICH IS THE BOOK OF LIFE: AND THE DEAD WERE JUDGED OUT OF THOSE THINGS WHICH WERE WRITTEN IN THE BOOKS, ACCORDING TO THEIR WORKS. AND THE SEA GAVE UP THE DEAD WHICH WERE IN IT; AND DEATH AND HELL DELIVERED UP THE DEAD WHICH WERE IN THEM: AND THEY WERE JUDGED EVERY MAN ACCORDING TO THEIR WORKS. AND DEATH AND HELL WERE CAST INTO THE LAKE OF FIRE. THIS IS THE SECOND DEATH. AND WHOSOEVER WAS NOT FOUND WRITTEN IN THE BOOK OF LIFE WAS CAST INTO THE LAKE OF FIRE. AND I SAW A NEW HEAVEN AND A NEW EARTH: FOR THE FIRST HEAVEN AND THE FIRST EARTH WERE PASSED AWAY; AND THERE WAS NO*

MORE SEA." In reading these Scriptures, and the entire Bible over and over and over many times, front to back. I have found not one word, taken in proper context, that says GOD'S people are taken out until after the tribulation. To the contrary, every word I have ever read in the full and complete text, says the elect are here and persecuted until the end. One last example, Thessalonians 4:16-17, "*FOR THE **LORD** HIMSELF SHALL DESCEND FROM HEAVEN WITH A SHOUT, WITH THE VOICE OF THE ARCHANGEL, AND WITH THE TRUMP OF **GOD**: AND THE DEAD IN **CHRIST** SHALL RISE FIRST: THEN WE WHICH ARE ALIVE AND REMAIN SHALL BE CAUGHT UP TOGETHER WITH THEM IN THE CLOUDS, TO MEET THE **LORD** IN THE AIR: AND SO SHALL WE EVER BE WITH THE **LORD***." I hear many preach a part of these verses, but most leave out that part where this happens only at the voice of the Archangel *and* the Trump of GOD at the end of the tribulation. Instead, they read from, "the dead in CHRIST shall rise first:" It is very *deceptive* to leave out the context in the first half of a sentence. So many preach partial verses, being Spiritual *deception*. Another new doctrine growing fast, is the prosperity doctrine. I will not list the endless Scripture, that deny such beliefs; but instead say if you have a red-letter Bible, read through the four Gospels. Reading only the words of JESUS and see the truth for yourself. Where time after time, JESUS says to take a vow of poverty, and not love or seek the possessions of this world. JESUS taught us to seek only the riches of GOD, being love and life eternal. The lively example of the early Church, and the acts of the Apostles, are of a good spiritual communal nature. We see they sold and share all things, being of one spiritual purpose and belief. This must only be a personal and spiritual act, it in no way comes by a government formed under fallen mankind, such as communism or socialism, that always restrict and deny religious liberty. The New Testament of CHRIST, says the true ways of GOD are not of this world. So, they are not of any monetary or physical nature. When we see the word "blessing" in the Bible, we should know all blessings from GOD are not for personal abundance, but for spiritual strength; being gifts of love, life, wisdom, patience, and lastly stewardship. All of GOD'S blessings promised in Scripture, after the suffering passion of our LORD, are surely of the Kingdom to come; having no part or share with this fallen world. The doctrine of personal prosperity is pure evil, as it leads one to focus upon what physical things they hope GOD will give them. Those who teach such things, turn the grace and love of GOD into lasciviousness. Another new false and dangerous doctrine is in the abandonment of the law, and our perfection coming through CHRIST alone. The end results of this doctrine are apostasy and hypocrisy flowing like rivers through the hearts of those who desire it. JESUS said "*BE YE PERFECT AS YOUR FATHER IN HEAVEN IS PERFECT*". Walk in GOD'S laws of love and life, and obey all the Commandments. I hear preachers saying we cannot be perfect or righteous, this is pure evil misdirection. We cannot be perfect and righteous in the same fullness as GOD, and we are not called to hold GOD'S highest perfection. But we are called to the flawless perfection of mankind in submitting to, and emulating the perfection of GOD. GOD asked Satan if he had seen Job, his righteous and *perfect* servant. There are many Bible references to faithful servants walking perfect before GOD. To say we cannot do the same is a false reason and license to sin, where none exists. Have you not read how many times FATHER said he was sick of repetitious burnt offering and blood sacrifices. Every time GOD offers a way back, most souls abuse it till it loses all true meaning. If we should rarely sin, then each time a price was paid by Messiah, for each new forgiveness. We need a way back to perfection and righteousness with GOD, but why does there need to be such a price? Because all sin is the destruction of love, life, and faith. Here and now, in our time of choosing, without great cost, there is no great value in anything. Without the possibility of loss, there is no treasured having or holding. Without the lost, who are the saved? And why are the few saved, and the many not saved? There is GOD'S definition of sin, being all things divisive and destructive; and sin cannot be allowed. This is why the cost of GOD'S eternal Kingdom must be high. GOD gave all souls the ability to see the destruction and nature of imperfection, and they teach us *why* sin is evil. This to provide a way to redeem our hearts back to seeking a perfect love in true sincere repentance of all sin. The animal sacrifice was meant to be a severe heart felt thing, watching an animal life's force poured out, for the sole reason that they sinned. It was their fault an animal which lived, had to die without purpose. This was to make souls witness and know the destruction of sin, which is always death. And by this sacrifice of lesser life, hopefully bringing to that heart the strength to not sin again. But mankind grew numb, and turned it into the legalism of monetary pay to play. Reducing their penance, to no more than paying a fine as a societal function. It was not meant to be that if you had a vast heard of animals, this allowed you to sin more than others, or as often as you could afford to offer an animal's life. They turned it into a futile exercise of trying to purchase redemption. They lost all meaning, repentance, sorrow, pain, and reason; they turned it into a license to sin. This is the true definition of legalism. *Never* do this with the Sacrifice of our LORD. Legalism is not in following the laws of love; legalism is acting like a lawyer to create false physical loopholes in the law, and deny the spirit of the law. This all makes GOD'S heart break and grieve. For again FATHER tried to have a loving relationship with mankind; yet as in all times before, mankind replaced the true strength of the sacrifice, with their own weak desires. So, what did FATHER do when the animal sacrifice failed, did he just destroy all mankind as they richly deserved? No. GOD made the ultimate sacrifice. Do you have a desire to understand; can you know the deepest broadest highest eternal aspects of love, without the indwelling HOLYSPIRIT? Do you think you can comprehend life truly without GOD'S help? What in all of creation could be the ultimate sacrifice by GOD for love. For knowing this, we have been given a Holy Book of unspeakable value. For it holds the free true understandings of love, and conscious treasures that can never be purchased by us, their price being beyond all things. We receive HOLYSPIRIT, to discern all written things and more. In these Holy Gifts, we find only the most wonderful beautiful truths, which paint the picture of GOD'S ultimate sacrifice. A picture, so loving and beautiful, that most cannot begin to comprehend it. We see in this picture a loving FATHER, and a rebellious dying creation, refusing *his* truth, and choosing death. We see FATHER try to teach and show the gifted and created, having free-will, what eternal love and truth really are. Yet mankind most always finds this world and themselves to be the greater love in their heart. These choose their own ways, which lead to hell and death. So, after every other effort and means failed, FATHER knew an eternal example of loving sacrifice is the greatest picture of love to bring and give a greater strength to souls. FATHER in great pain, sent the Holy image and example of the far, far Greater, coming to lay down his life for those who are far, far less; followed by an offer of his indwelling spirit. For the center of this greatest picture, it could not be just anyone, not just a decent man, not a Prophet or an Angel, nor all of them together. FATHER sent his only begotten SON, who is the far greatest part of his heart. We need to all ask ourselves the most important question we could ever ask. Did FATHER do all of this, to give us an example of perfection to follow and emulate as he always has; hoping we find a loving strength to overcome all things of the flesh, as he tried to do many times before. Or, did FATHER give up on our Holiness, and give that most painful example, just to give us a license to sin against him as we please. I admonish all souls, in all the righteous love, fear, and anger I can feel, in knowing the eternal horrors that surely come to all holding such beliefs. Do not once again turn to legalism; *this* Sacrifice, is not to be unrighteously used or abused. Do not *trample* on the suffering Sacrifice of JESUS. If we ever call on the Blood of CHRIST, let our heart break in pieces, and turn to him in shame each time. We are claiming the Holiest Sacrificial Love and the Holiest of Covenants. So, it must be given reverence above all Creation. It is not to be called upon in any feeling of right or entitlement, but in great fear. I warn you for your soul, that GOD does not forgive *un*repentant ways. GOD, please show us mercy in our *shame*. Amen. This leads us to another horrific false dangerous doctrine that is spreading, but not as fast as the others. It is the belief, that all our past and future sins were forgiven in our first prayer to CHRIST. The once saved always saved doctrine is a doctrine of hell. Do any wonder, how and why these new doctrines start and take root, being contrary to Scripture. It seems the height of all arrogance, to say that for thousands of years, everyone from the Apostles forward got it wrong, until this last weak entitled generation, finally got it right. This is the greatest apostasy, idolatry, self-indulgence, and self-deception since Sodom and Gomorrah. This false doctrine says it now turns out that GOD doesn't hate sin and its destruction as much as once thought. GOD now just wants to get along with us all, and make us all rich and safe while mankind goes insane with sin and physical perversions. I am profoundly afraid to think such thoughts. The WORD, is in all ways a two-edged sword. Cutting both ways in speaking of the granted things from GOD, being mercy and forgiveness from above; and our responsible faithful mature stewardship and selfless protection of the Kingdom of GOD. Most pulpits have put down this two-edged Sword, and raised in their hand a new sword, bearing one edge. It is such great sadness to see souls adopt any such false religious beliefs; knowing these must someday see the truth that they were actually Satan's front guard, who helped weaken and lure lost searching souls into darkness. How in that great day, shall these hold themselves up, finally hearing GOD'S pure truth, and learning their countless un-forgiven sins are in league with Satan. Being found as a soldier among the murderers of GOD'S Children.

THE LAST DAYS OF GREAT LOVE AND FEAR

We have all heard the saying "love is what you make it". And if that were true for love, it would be true for fear as well. And so fallen souls, all have at least a slightly different description and definition of what these two emotions mean to them. Yet again, there is GOD'S one highest definition of love and fear. FATHER holds the highest purest eternal love, and he alone knows the form and level of love and fear we need to hold towards him to endure eternal, reducing all of mankind's fallen weak definitions of love and fear to lies. Scripture says GOD is love, which means beyond all question, every aspect of true eternal love is perfect and holy. And that if we do not have the indwelling HOLYSPIRIT, we can hold no measure, or know the eternal love of GOD. Eternal love is a desired selfless faithful responsibility. Eternal love is not in any way a temporal temporary thing. *Eternal* love has not one single shared aspect with physical desire, or physical pleasure. The eternal spiritual love of GOD is perfect in its ways. Worldly loves are tragic, as they all fade, die, and end in complete loss. All loves born of this world are imperfect, and all this world's strongest of loves, slowly fade and die. All the lost souls *of* this world, will lose all the love they have known; it will all come to naught in the end. So, we see love has its temporary and eternal perspectives. There is our eternal love for GOD first, that redefines our love for self and all lesser souls as commanded. And our eternal love for all lesser souls, must be seen and known through our eternal love for GOD. Can or do many souls see this order within themselves? I do not find this to be the case, as most souls don't know, and are unconcerned with eternal love. Yet these two created perspectives, divide the saved from the lost. Most believers think it's who believes, and who does not believe in GOD that draws this separation between the saved and the lost; that is not so. The fallen angels surely believe, and are lost in their sin. Now do you believe in love at first sight? We tell you in sincerity and all confidence, that instant worldly attraction has not one thing to do with the eternal love of GOD. Any sudden attraction among mankind, is physical lust and nothing more. We must resist this world's playing with the definition of GOD'S love. The difference between GOD'S love and worldly lust, is eternal love is born of GOD'S perfection, growing in our soul as a tiny seed, never bringing its fullness of joy at the first. A seed that grows into a selfless spiritual peace in serving GOD'S perfect Kingdom. It's a growing desire to bring increase to GOD'S Kingdom, and to remove every offence. At sensing these spiritual aspects within our soul, our eternal love has begun to grow. We must study to nurture this love and find the wisdom of GOD, and do all good things which allow love to grow eternally. After much learning and impassioned effort, our greater love bears even greater joy. But with physical lust, being worldly fleshly desire, we see the opposite order. Worldly lust brings its greatest joy at the first moment two souls eyes lock in desire, and experience their first great physical excitation. But from that first experience, their desire and excitation will slowly or quickly fade until you sense the thrill is gone. Examples are, you see a new car and you say you love and acquire this car. From that moment going forward, your joy and satisfaction will diminish until you find you want another car. When two people see each other physically, and there is instant attraction. How can that be or have any part with FATHER'S love, when his love is found only in spiritual ways. Love for GOD grows with knowing them, as it is an intimate knowing thing. In lust, at first sight they physically know each other, being only what they have seen, and not what their soul has known. A soul can call their desire for physical relationships with souls and the things of this world "love" but they are not of *GOD'S* love. Worldly souls for the most part, have no clue what GOD'S created eternal love is, so they cannot give or receive it. It is like all things good and spiritual from GOD that are replaced in this world and in the lost with a cheap physical imitation. Love is a force and a commandment of GOD; so, it is spiritual and abstract, yet very real. It is the creative sustaining force of all rising eternal perfection. True loves forceful brightness is sensed by GOD, flowing from and to its intended. Love is the opposite of lust, in that if we love, our life force flows from us to another, being sacrificial. Lust seeks, draws, and absorbs, the life forces of physical comfort and pleasure from others, to one's self. The love *of* GOD, in Corinthians, is defined by its pure spiritual aspect called charity; meaning sacrificial selfless giving. In Godly giving, there is hope, need, and a now low expectation of return. Love is an effectual Spiritual force of life's work, given from one to another for Spiritual wellbeing. Think of eternal love as the true force of eternal life itself, because it is. And do, or will you give away, that which keeps you alive to others if you love them? Yes. In loves pure form, it is free of thoughts or desire for one's self, and highly focused on the Spiritual needs over the fleshly needs of its intended. If you have this true real love in you, then you have GOD in you, and you are in GOD. Why do so many question whether GOD is loving? Because they don't recognize or understand GOD'S love, it not being like theirs. To give in pure spiritual intention, is a rare most valuable force. For a soul to know that GOD and their Children have this love for us, is a true pure strength, and a greatest of eternal spiritual satisfactions in all reality. Our eternal love exists in spiritual giving, riding upon our physical sacrifice. 1st Cor. 13:1-13, "*THOUGH I SPEAK WITH THE TONGUES OF MEN AND OF ANGELS, AND HAVE NOT*

CHARITY, I AM BECOME AS SOUNDING BRASS, OR A TINKLING CYMBAL. AND THOUGH I HAVE THE GIFT OF PROPHECY, AND UNDERSTAND ALL MYSTERIES, AND ALL KNOWLEDGE; AND THOUGH I HAVE ALL FAITH, SO THAT I COULD REMOVE MOUNTAINS, AND HAVE NOT CHARITY, I AM NOTHING. AND THOUGH I BESTOW ALL MY GOODS TO FEED THE POOR, AND THOUGH I GIVE MY BODY TO BE BURNED, AND HAVE NOT CHARITY, IT PROFITETH ME NOTHING. CHARITY SUFFERETH LONG, AND IS KIND; CHARITY ENVIETH NOT; IS NOT PUFFED UP, DOETH NOT BEHAVE ITSELF UNSEEMLY, SEEKETH NOT HER OWN, IS NOT EASILY PROVOKED, THINKETH NO EVIL; REJOICETH NOT IN INIQUITY, BUT REJOICETH IN THE TRUTH; BEARETH ALL THINGS, BELIEVETH ALL THINGS, HOPETH ALL THINGS, ENDURETH ALL THINGS, CHARITY NEVER FAILETH; BUT WHETHER THERE BE PROPHECIES, THEY SHALL FAIL; WHETHER THERE BE TONGUES, THEY SHALL CEASE; WHETHER THERE BE KNOWLEDGE, IT SHALL VANISH AWAY, FOR WE KNOW IN PART, AND WE PROPHECY IN PART, BUT WHEN THAT WHICH IS PERFECT IS COME, THEN THAT WHICH IS IN PART SHALL BE DONE AWAY. WHEN I WAS A

CHILD, I SPAKE AS A CHILD, I THOUGHT AS A CHILD, BUT WHEN I BECAME A MAN, I PUT AWAY CHILDISH THINGS. FOR NOW WE SEE THROUGH A GLASS, DARKLY; BUT THEN FACE TO FACE; NOW I KNOW IN PART; BUT THEN SHALL I KNOW EVEN AS I AM KNOWN. AND NOW ABIDETH FAITH, HOPE, AND CHARITY, THESE THREE; BUT THE GREATEST OF THESE IS CHARITY." There is no Prophet, Scholar, Linguist, or Poet in all reality, dead or alive, or yet to be born, who can say this chapter does not speak of the pure form of selfless sharing of our spiritual tangible life force that GOD calls love. Charity is not in what you physically give, it is in what, why, where, when, and how, you spiritually give all things. Giving all you physically have to feed murderous drug-addled reprobates, is not GOD'S Spirit of Charity. This *love* that GOD extends to all who return this same love, must flow, or die; it is like light, and all the great moving gifts of GOD. Love enlightens and perpetuates all Spiritual life as it moves. Eternal love cannot be received by or from this world, or held on to as lust can. Eternal love when flowing to us, must flow back immediately or it loses all its purpose and promise, and it is dead. Love must be fed, yet it comes and goes without wanting or asking; so, our life should be about giving and sensing eternal love.

We are all created to be eternal conduits of GOD'S healing love; nothing else compares to this calling. To eternally love, we live and give all of ourself *spiritually* in GOD'S name and purpose, but only for the reason we sense a Godly need. If a worldly murderer says he needs money to buy a knife to kill someone, do you charitably give what they physically need in that moment? Of course not. We can't physically give, to eternally ease any perceived need. Eternal love is our way of life, it is our first nature and who we are. In our projecting true love towards the lost, we hold no expectation of love in return. GOD'S love is not natural to fallen mankind, so their charity is physical, and falls short. Every moment we're not in GOD'S eternal circular flow of love is our wasted experience, gone forever, never to be counted for anything. Our every wasted moment, will be burned at our judgment as unrighteous work. We are to feel the love from GOD flowing into our soul, and so now ours; and commanded to give away that entire love, exactly as GOD does. What then of the hate GOD has for evil. Many Scriptures say, we are to always hate what GOD hates in all our righteous anger, with the pure perfect understanding he gives us about these evil divisive things. Revelation 2:2, "*I KNOW HOW THOU CANST NOT BEAR THEM WHICH ARE EVIL.*" Revelation 2:6, "*BUT THIS THOU HAST, THAT*

THOU HATEST THE DEEDS OF THE NICOLAITANES, WHICH I ALSO HATE." Our righteous anger sees every tiny sin's, tiny evil destruction, yet we resist physical force against it. Instead, we separate ourself and flee, lest our zeal consume us. Most don't see how it is far harder for mankind to hate all the things GOD hates, than to love all the things GOD loves. If each soul had a list from GOD of all the things they each do *not* hate, that GOD *does* hate, they would be shocked. The Church body has much sin and destructiveness to overcome. Please study the Revelation of CHRIST, chapters 2 and 3; which is JESUS telling the Churches what good he sees them doing; and more importantly, warning what good they are not doing. Hear his fearful warnings of the consequences to come upon those who don't correct themselves. A loving eternal GOD, created us to fulfill an eternal loving purpose; this is our calling. Do any think our calling is something less? Tell me why? What is a perfect love from mankind towards GOD? To love GOD with our entire mind, heart, and soul, means what? It means to seek GOD completely and intimately, to learn all we can about living in GOD'S ways, to pray and fellowship with GOD in a spiritual state of love, wonder, and fear. To believe the Scriptures that say our possessions *in* this world, turn us into slaves and possessions *of* this world. Once we truly know GOD is there, what could there be in this world to hold our attention above our Creator. Everything physical loses its luster in that comparison. When you see a great work of art, like a 500-year-old Painting. Do you, in seeing the beauty of a painting immediately go in your mind to the Painter's Spirit? Do you wonder what they thought and felt, and why did they paint it? What is its intended meaning and purpose? What was their motivation? Did the Artist agonize in sadness over it? Or was it in joy? Or do you just stay on the beauty of the Painting? When you see a beautiful ancient Cathedral, do you see the many decades of pain and suffering of conscious and physical anguish and death spent in its building? Do you see its spiritual purpose and passioned reason for its existence? Do you sense the many sacrifices of the desperately poor, going without to help fund its building? Do you feel and hear the cries of pain of the multitudes of peoples who have fallen right there to their tired scarred knees before GOD over hundreds of years, crying out in their great suffering losses in this world? And can you hear the echoes of their cries, and sense the pain of the many that GOD has heard and felt there? Or do you in your mind, just enjoy a very old beautiful physical building? If we don't know and care far more for the Creator and the creative intent, we can hold no true eternal understanding or appreciation of any creation. GOD says this physical creation is now a vapor, and warns all against living in a physical state of being blinded by the flesh and holding a different love and belief, than that of GOD. We must grow in FATHER'S ways and belief, as we know his sustaining force touches every aspect in his creation at all times; yet most souls never recognize his healing gift. They are blind by lust, envy, and pride of physical possessions. The true eternal love, defined by GOD, *verses* all desired, realized, or imagined, temporary worldly lust, are the two factors found in every decision made by everyone. Most souls think this true love from and to GOD, is for each individual to define for their self. By this, declaring they also define GOD for their selves. For such souls, GOD becomes more and more a reflection of who they want GOD to be, rather than who GOD actually are. If you think it's ok for individuals to define truth and love for themselves, you are saying we are all on our own, and we are not our brother's keeper; saying you figure out your own way to GOD, as I am busy finding my different way. GOD and their words will harshly judge this. We know there is the one way of truth, and the eternal love delivered by GOD. And this we seek as one spiritual body, not in individual worldly ways. While lust is individually born and defined, GOD'S love is a singular eternal choice, flowing through

GOD'S eternal perfect order, defined by them through us. If you love your home, if you love how something or someone makes you physically feel? These are lust and self-fulfilling. Lust is all about how it makes one feel in the flesh. Can we all be honest, and see that if you choose someone in a relationship for how they make you feel. Do they make you laugh? Do you enjoy and have worldly fun with them? Do you feel they are attractive, and a catch for you? Do they excite you physically? And do you feel physically comfortable, safe, and proud, with them at your side? These are feelings of the flesh, they are self-serving, just as with a new car, boat, or home. Now we can surely have *measures* of both love and lust for someone. But if you want to marry, you must seek someone of the *opposite* created sex for their Godliness, over the flesh. And daily, after giving your greater loving part to GOD, you remain spiritually and physically responsible and faithful to the one you are espoused in all your ways, acts, and thoughts, for your entire life on earth. This is why GOD ordained the Holy Institution of Marriage. GOD has ordained man and woman to become one-flesh in a physical union, which is to contain all physical closeness and propagation, but not for lustful recreation. Holy Marriage is different, in that two are made one flesh in GOD'S blessed permission. It is GOD making and seeing two as one.

If we are two halves of one flesh, and treat the physical subject of marital congress as propagation and our containment of temptation as according to Scripture, not giving it any undo control over ourselves. We are then not physically lusting after another, because we are one flesh. But going beyond Biblical boundaries of marital procreation and coming together to stay temptation, to enter into pure recreational physical congress, even in marriage, is lust. And it breeds the Spiritual division of having a mixture of lust and love in your marriage. When lust fills part of your heart, love can only fill the rest. Love and lust do not dwell peacefully, there is a constant conflict of each growing and waning in battle for supremacy in our fight to control the flesh. One will always be winning, and one losing. We must make sure our Spiritual understanding is always in strong control. And lovingly obey GOD, while holding no inappropriate desires for the world, or the things therein, is to be pure, strong, upright, and perfect in GOD'S eyes. Can we attain this state in this world? Yes. But only in our knowing passionate love for CHRIST, and receiving HOLYSPIRIT'S strength, wisdom, and all spiritual blessings from above. Sin is never the more powerful, unless we desire it and allow it to be in control. In the past strength of our flesh, it may at times of our life have been the more powerful. But that was in our spiritual weakness and rebellion. By a Holy Anointing comes wisdom and a force that strengthens our desire to put away all lust. It is truly possible to grow while in this world, and walk perfect before GOD. To say this is impossible for mankind, is to have a form of Godliness, yet deny the power of GOD and love. And to deny JESUS has overcome the world to show us the way. A soul *can* say it is impossible for them, because they don't love GOD enough. But that is all a soul can honestly say. A Minister who says the perfection of mankind is impossible, does so out of self-delusion, because they too, are trapped in sin and ruled by the flesh. How wonderful in Heaven CHRIST must be, and how he must reign in glory. JESUS is worthy for overcoming the flesh and showing and preparing our way to FATHER. Should we all desire to see CHRIST'S unsurpassed beauty, like a lifelong treasure hunt, adding knowledge with study, prayer, and fasting, getting closer and closer to our desire. Life's only real value is to truly and *intimately* know the one true GOD. Look at what you surround yourself with. Do you realize that when you desire physical beauty around you, it goes to self-worship. If an acquired thing is owned for any reason beyond a physical-requirements to exist. If you have chosen a spouse, largely or partly for physical appearance. If you have a beautiful home, nice cars, clothes, jewels, and whatever, for the sole reason we physically desire it all. What do you think this does to you, why do you desire these things? Why do such things bring *physical* pleasure and satisfaction? And if you don't have these things, yet you want them, you are found just as guilty. These things let the flesh feel elevated and safe above others who don't have them. This adornment and security, is called pride of self, and we must struggle against this. By attaining desired possessions, we can sense others admire, envy, and desire to be like or with us, and desire to have what we have. In other words, they physically elevate us in this physical world; and GOD says this is called worship. This is not living the truth of our purpose, but a lie of our purpose as we are not created to be worshipped. Within all worldly fame, rests the desire to be seen and known by others for what one looks like, what one can physically do, and the worldly possessions and access one has acquired. This world does not see the soul, it holds the celebrated flesh up, and idolizes them for beauty, possessions, accomplishments, and desires. The flesh thrives on attention; so, we see a self-destructive probability equation in the heart of all free-will souls. As when Satan said he will exalt his throne above the stars of GOD, he was saying, I want all to see, desire, and worship me, in my beauty, intelligence, and ability, just like

GOD. Satan, after acquiring dominion of the Earth from Adam through deception, asked JESUS to bow down and worship him to be given these things. And in self-loving delusion, Satan thought JESUS should admire him. If one gathers worldly treasures, one seeks to be worshipped by others to some degree. When JESUS said don't covet anything your neighbor has, he was not saying do not have any desire to steal what your neighbor has. CHRIST was saying don't want what you see a neighbor has at all. In the end, physical lust leads to un-fulfillment and eventual misery, at the realization it is all temporary. These can never satisfy or meet any eternal desire or need. The only guarantee you get, is that it will vex your spirit, and you will lose all you have. All physical beauty turns to a semblance ravaged by gravity and disorder. The joys of great possessions, turn to a burden of who will inherit your curse. Eternal happiness comes only by fulfilling our created spiritual purpose; so, we cannot define any blessing or cursing by increase or decrease in worldly possession, ever! If you want to know if you are truly blessed by GOD? Look and discern the Spiritual World around you. Look left and right, examine and consider those going about lacking the love and fear of GOD. Consider the whole world. Do you see deception and confusion everywhere? Do you sense a spiritual emptiness as the masses blindly march off the edge? Do you see horror, sorrow, hell, and death, overflowing the whole of the earth? Do you always see temptations and lies at every turn? Do you see the souls lost in the flesh, temporarily smiling, singing, dancing, and not seeing their eternal doom right before their eyes? If you see all these things in your great fear for GOD, then you see reality as GOD'S Holy Angels see it. And you are truly blessed in this, because you see the world as a Spirit of GOD, and no longer in the flesh. The flesh walks through the world seeing only where it goes, seeking out every available pleasure and comfort, lacking any true strong fear of GOD. FATHER knows that if we truly see the great GOD he is, that our great fear of him is natural. And he knows we need a *strong* natural fear for the love GOD will take away at the last from all sinners. We need both our love for GOD, and fear of its loss to keep ourselves always. Isaiah 66:22-24, "*FOR AS THE NEW HEAVENS AND THE NEW EARTH, WHICH I WILL MAKE, SHALL REMAIN BEFORE ME, SAITH THE LORD, SO SHALL YOUR SEED AND YOUR NAME REMAIN. AND IT SHALL COME TO PASS, THAT FROM ONE NEW MOON TO ANOTHER, AND FROM ONE SABBATH TO ANOTHER, SHALL ALL FLESH COME TO WORSHIP BEFORE ME, SAITH THE LORD. AND THEY*

SHALL GO FORTH, AND LOOK UPON THE CARCASES OF THE MEN THAT HAVE TRANSGRESSED AGAINST ME: FOR THEIR WORM SHALL NOT DIE, NEITHER SHALL THEIR FIRE BE QUENCHED; AND THEY SHALL BE AN ABHORRING UNTO ALL FLESH." Here we read that even in Heaven, we will go out the Gate of the New City Jerusalem, to see the lake of fire and those roiling in it. Why? Because we need to always be grateful for love, and fear losing it. Sadly, love alone is not strong enough in all lesser gifted creatures having eternal free-will, without the fearful consideration and knowledge of possibly losing it. This is why *worldly* godless pastors, falsely teach to not fear GOD, as they do the devil's work. We have touched a little on the eternal sacrifice and responsibility of love. But what is the ultimate Godly fear. Imagine it is our day of judgment, and even if you are GOD'S Child, I pray you know that when you come before the Throne of GOD, you will be filled as never before with a fear that steals your very breath away. Emotions of your great love will be overcome in a great fear and awe you never dreamed you could feel or know. I hope you also know you will not be standing before GOD, as none will have the strength to stand. On that day, as Scripture says, all will be on their knees. You will tremble and cry with eyes as wide as possible, emotion so strong it shakes the whole body. None will run to hug JESUS sitting above all on his throne, and surely none will boldly curse him as some say. What a horrid realization, to in this world think one's self saved by crying out with *only* one's mouth, and not with one's life to declare their love for GOD. And by their choices and actions in their heart and life, to be found untrue and denied. What kind or level of horror and fear is a soul capable of holding, feeling, and comprehending in a moment? Scripture says there will be wailing and gnashing of teeth, at such hard realizations. None in this world have ever felt or imagined such fear. There is a fear on a level that cannot be known by any other way, than being before GOD'S Throne, and be denied by GOD in one's greatest horror. It is wise, and we do well now and always, to hold as great a fear as possible for that coming day. And for all who crucify their flesh, having their commandments written on our hearts and minds by our faith in CHRIST. In coming to that day, we, *and* the lost, will feel the same absolute fear and humility. But *we*, the found, will have our saving eternal responsible faithful love. Many strange false beliefs and practices reside in those who profess CHRIST with their mouths, but not with their life. One strange belief is, they live like they have been given into their own hands, the power to be saved. Thinking salvation rests in their power, to in CHRIST'S name ask for forgiveness and mercy. Believing by this alone they are saved; thereby making it up to them alone. According to Scripture, this is not true. We all have the power to ask for forgiveness, and all have the power to believe they are forgiven. But only GOD grants forgiveness upon seeing true repentance, which means to turn from *all* our sins. Matt 7:21-27, "*NOT EVERY ONE THAT SAITH UNTO ME, LORD, LORD, SHALL ENTER INTO THE KINGDOM OF HEAVEN; BUT HE THAT DOETH THE WILL OF MY FATHER WHICH IS IN HEAVEN. MANY WILL SAY TO ME IN THAT DAY, LORD, LORD, HAVE WE NOT PROPHESIED IN THY NAME? AND IN THY NAME HAVE CAST OUT DEVILS? AND IN THY NAME DONE MANY WONDERFUL WORKS? AND THEN WILL I PROFESS UNTO THEM, I NEVER KNEW YOU: DEPART FROM ME, YE THAT WORK INIQUITY. THEREFORE WHOSOEVER HEARETH THESE SAYINGS OF MINE, AND DOETH THEM, I WILL LIKEN HIM UNTO A WISE MAN, WHICH BUILT HIS HOUSE UPON A ROCK: AND THE RAIN DESCENDED, AND THE FLOODS CAME, AND THE WINDS BLEW, AND BEAT UPON THAT HOUSE; AND IT FELL NOT: FOR IT WAS FOUNDED UPON A ROCK. AND EVERY ONE THAT HEARETH THESE SAYINGS OF MINE,*

AND <u>DOETH THEM NOT</u>, SHALL BE LIKENED UNTO A FOOLISH MAN, WHICH BUILT HIS HOUSE UPON THE SAND: AND THE RAIN DESCENDED, AND THE FLOODS CAME, AND THE WINDS BLEW, AND BEAT UPON THAT HOUSE; AND IT FELL: AND GREAT WAS THE FALL OF IT." If CHRIST says he *never* knew these souls, then they were *never* forgiven. They obviously asked for forgiveness and believed they had been, but never were. They believed in and coveted the physical provisions of GOD, but did not desire to live in GOD'S loving ways of eternal life. Until we hear JESUS say *enter* "thou good and faithful servant" we wait in our loving faith and hope of his promise, and our obedience to his ways of eternal life. Beyond our rendering to GOD our true love, faith, hope, *and* obedience to their every spoken word, forgiveness is out of our hands. Scripture repeats a multitude of times that we are called to study and show our selves approved. To crucify ourselves daily, and come to the fullness of love. Many Scriptures speak of CHRIST rejecting many who were called, and who did ask for forgiveness, but are not chosen because they asked without true repentance. It is solely within our power, to look at this world around us and say we see nothing eternal to be desired except GOD, his ways, and his Children. To see all that is here physically, is but a vapor that soon disappears before our eyes at FATHER'S desire. Allow *nothing* to be between you and GOD. This is the free-will power given to us, to love and walk with GOD in our perfection every day; which is our eternal and great love for GOD, and our eternal great fear of GOD. Luke 9:23-24, "*IF ANY MAN WILL COME AFTER ME, LET HIM DENY HIMSELF, AND TAKE UP HIS CROSS DAILY, AND FOLLOW ME. FOR WHOSOEVER WILL SAVE HIS LIFE SHALL LOSE IT: BUT WHOSOEVER WILL LOSE HIS LIFE FOR MY SAKE, THE SAME SHALL SAVE IT.*" Matthew 10:34-39, "*THINK NOT THAT **I** AM COME TO SEND PEACE ON THE EARTH: **I** CAME NOT TO SEND PEACE, BUT A SWORD. FOR **I** AM COME TO SET A MAN AT VARIANCE AGAINST HIS FATHER, AND THE DAUGHTER AGAINST HER MOTHER IN LAW. AND A MANS FOES SHALL BE THEY OF HIS OWN HOUSEHOLD. HE THAT LOVETH FATHER OR MOTHER MORE THAN ME IS NOT WORTHY OF ME: AND HE THAT LOVETH SON OR DAUGHTER MORE THAN ME IS NOT WORTHY OF ME. AND HE THAT TAKETH NOT HIS CROSS, AND FOLLOWETH AFTER ME, IS NOT WORTHY OF ME. HE THAT FINDETH HIS LIFE SHALL LOSE IT: AND HE THAT LOSETH HIS LIFE FOR MY SAKE SHALL FIND IT.*" Titus

1:16, "*THEY PROFESS THAT THEY KNOW* ***GOD****; BUT IN WORKS THEY DENY HIM, BEING ABOMINABLE, AND DISOBEDIENT, AND UNTO EVERY GOOD WORK REPROBATE.*" Titus 2:11-15, "*FOR THE GRACE OF* ***GOD*** *THAT BRINGETH SALVATION HAS APPEARED TO ALL MEN, TEACHING US THAT, DENYING UNGODLINESS AND WORLDLY LUST, WE SHOULD LIVE SOBERLY, RIGHTEOUSLY, AND GODLY, IN THIS PRESENT WORLD; LOOKING FOR THAT BLESSED HOPE, AND THE GLORIOUS APPEARING OF THE GREAT* ***GOD*** *AND OUR* ***SAVIOUR JESUS CHRIST****; WHO GAVE HIMSELF FOR US, THAT HE MIGHT REDEEM US FROM ALL INIQUITY, AND PURIFY UNTO HIMSELF A PECULIAR PEOPLE, ZEALOUS OF GOOD WORKS. THESE THINGS SPEAK, AND EXORT, AND REBUKE WITH ALL AUTHORITY. LET NO MAN DESPISE THEE.*" Stand strong in GOD'S Faithful Responsible Authority. Galatians 5:24, "*AND THEY THAT ARE* ***CHRIST'S*** *HAVE CRUCIFIED THE FLESH WITH THE AFFECTIONS AND LUSTS.*" Galatians 6:14, "*BUT* ***GOD*** *FORBID THAT I SHOULD GLORY, SAVE IN THE CROSS OF OUR* ***LORD JESUS CHRIST****, BY WHOM THE WORLD IS CRUCIFIED UNTO ME, AND I UNTO THE WORLD.*" Romans 6:6, "*KNOWING THIS, THAT OUR OLD*

MAN IS CRUCIFIED WITH HIM, THAT THE BODY OF SIN MIGHT BE DESTROYED, THAT HENCEFORTH WE SHOULD NOT SERVE SIN." The fullness of the true love for GOD and his ways, is such a miraculous thing. Scripture says continual growth, brings a complete change to all our desires, knowledge, and every thought of our heart. A new Spiritual creature is born and grown in deed and all actuality. Yet more and more Churches preach doctrines that allow the lust for this world and the flesh, and ask very little to nothing of souls for GOD above a verbal profession of love. The spiritual message today for lost souls, who before hearing it, believed they were a good person to start with in not killing, doing drugs, drinking to excess, smoking, cussing, stealing beyond insignificant things at work, and being physically faithful to their spouse. These hear in most sermons that they do not need to make any great changes. They hear they can do all they did before, only now they do it all in the name of the LORD. Just come to church, tithe, pray, be honest by your own definition, and give thanks for your physical abundance. These souls walk where they walked before, doing most things they did before. These continue in covetousness and love for this world and the things therein, just as before. But now they claim it is all a physical-blessings from GOD. 1st John, 2:14-17, "*I HAVE WRITTEN*

UNTO YOU, YOUNG MEN, BECAUSE YE ARE STRONG, AND THE WORD OF GOD ABIDETH IN YOU, AND YE HAVE OVERCOME THE WICKED ONE. LOVE NOT THE WORLD, NEITHER THE THINGS THAT ARE IN THE WORLD. IF ANY MAN LOVE THE WORLD, THE LOVE OF THE FATHER IS NOT IN HIM. FOR ALL THAT IS IN THE WORLD, THE LUST OF THE FLESH, AND THE LUST OF THE EYES, AND THE PRIDE OF LIFE, IS NOT OF THE FATHER, BUT IS OF THE WORLD. AND THE WORLD PASSETH AWAY, AND THE LUST THEREOF: BUT HE THAT DOETH THE WILL OF GOD ABIDETH FOR EVER." Isaiah 5:20-21, "*WOE UNTO THEM THAT CALL EVIL GOOD, AND GOOD EVIL; THAT PUT DARKNESS FOR LIGHT, AND LIGHT FOR DARKNESS; THAT PUT BITTER FOR SWEET, AND SWEET FOR BITTER! WOE UNTO THEM THAT ARE WISE IN THEIR OWN EYES, AND PRUDENT IN THEIR OWN SIGHT*!" All who profess false repentance, and hold to false belief, see the same darkness as before, but now call it light. They hear the same lies, but now call them GOD'S truth. They still walk in the flesh, but now call it the spirit. They still desire the things of this world, but now call them blessings from GOD. They hold to the bitter desires of their rebellious flesh, and call them the sweetness of

GOD'S love. HOLYSPIRIT is not in any part with this world. One cannot claim to have and drag HOLYSPIRIT through this sinful world. HOLYSPIRIT is not a stamp of approval, or something to lean on if one feels the need. HOLYSPIRIT is to be worshipped as our teacher and leader, not as our follower. GOD does promise to meet our basic simple physical needs in this world. Which means keeping us alive if he sees a Godly purpose in us. But the Church today, replaces the words want and desire, with need. Does a true servant of GOD say; I *need* a bigger house in a better neighborhood, or I *need* a new car, or I saw a new closet organizer and I *need* one of those. Or the thousands of things we see that can make our flesh more comfortable, and safe. That does not mean we *need* it; it means we covet it. What we *need* for the flesh in this world, are sturdy clothes to cover us from the weather and hide our nakedness, we *need* air, water, and enough simple food to keep us alive, we *need* shelter from the harsher environment, and the healing that comes from GOD, and not just doctors and surgeons. FATHER will never let a *true* child of his die for the lack of these. 1st Tim. 6:8, "*AND HAVING FOOD AND RAIMENT LET US BE THEREWITH CONTENT.*" We need the eyes and ways of GOD upon us in truth and love. We need correction, guidance, rebuke, chastisement, and we need HOLYSPIRITS strength, patience, wisdom, and knowledge. Are you a slave to a mortgage and car payment, how about insurance and an employer? What about rebellious family members? We cannot be consumed in all Spiritual matters, if we are consumed in fleshly matters? How can one be consumed twice. We cannot stand strong and surrender at the same time. If we are busy enhancing and maintaining our physical life in this world, how can we be eternally learning, changing, and preparing our heart for the next. If we are busy hearing the things of this world, how much are we missing about the wonder of the next. Have no doubt FATHER'S commands are not just read or heard, his commands must be loved, discerned, and sought out in ever-more abundance and lived and experienced. Never think you cannot hear Scripture in the flesh, and receive a false worldly interpretation. When GOD speaks of false doctrines, they speak of all worldly fleshly interpretations. To hear the one truth from GOD, requires a humble effort, born of a pure desire of our heart to hear and live in the ways of GOD. After JESUS spoke a parable, being a moral allegory that requires an effort to interpret or misinterpret. The Scribes and Pharisees asked JESUS to speak plainly, and not in parables, as they had no desire to try to discern what he was saying. CHRIST knew *their* desires, *their* truths, lusts, and beliefs, were set already.

Their life was full, and there was no desire or room for change. They had been wrong about GOD all their life, so, they heard what JESUS said, but had no desire for what he meant. Like liberal politicians, they spun his words to mean what suited their own purpose. We see how a simple parable will separate many, raising in them an excuse and a desire to mock. Such reprobates hear foolishness in their conceited hearts. Have you ever known someone, living in an abusive relationship? And it is not always physical or verbal abuse. Sometimes there is the great abuse of neglect in not providing the things of love needed by the other. If two come together for only what they get from the other, there will always be abuse from both sides. Neither will see their own abuse towards the other, but each will feel their needs are not a priority of the other. If one person sees and tries to meet some needs of the other, but gets less in return, we see greater abuse from one side. If we come to FATHER, hearing what he promises his Children, and not hearing what he needs and asks from us, we abuse his love. FATHER created us with needs as children, and he comes wanting to meet our needs as a loving FATHER. But he has his loving needs also. A good FATHER will receive great love, honor, *and* obedience from a loving child, but not from a rebellious child. Until we love GOD in wanting to live their ways, and know and meet their needs; wanting no more, than to see FATHER'S smile of love and approval in return. Then we come not in love, but lust; not in light, but in darkness; not in sweetness but in bitterness. GOD cannot love forever, a soul trying to come to him in any false way. It is such an easy thing to love JESUS if we truly know him. And it is impossible to truly love JESUS if we know him in our own false way. If a soul loves GOD in their own false belief, they love not the actual true GOD, but a figment of their conscious creation. They love a fantasy god, that suites their desires and reflects their imagination. The first commandment says we are to love GOD more than all reality. But if we don't learn all the truth about FATHER, how can we love him above everything? It is impossible to love with all our heart mind and soul, someone we don't know. And we cannot just say or think we are Godly. GOD'S ways, are in all ways contrary to the ways and desires of the flesh. When you have a guest, do you desire to make them feel at home? Do you offer them things that please you? Would you offer these to JESUS, if he came to you in physical embodiment? Would JESUS standing in front of you, desire such worldly things? Do you say you would not offer these to JESUS, because you know these are desires of the flesh, and GOD holds contrary desires? But, if JESUS is not *physically* standing in our presence, does HOLYSPIRIT go with us, out of sight and mind, while we enjoy these same things? Or, does HOLYSPIRIT wait at home, or outside a restaurant or theatre? Does HOLYSPIRIT close his eyes and stop his ears? Do you think that while fulfilling the covetous desires of your flesh, HOLYSPIRIT is not offended? How great eternal love must be for GOD to endure his painful search.

THE LAST DAYS OF GREAT SELF INDULGENCE

These last days could also be called the days of self-rule, instead of self-indulgence, as people assume all freedom from GOD in their use of the body and the environment GOD has provided. Most souls now make their own truth, definitions, rules, and love for life as they go along. Their ways are seen in how they understand, use, and abuse themselves, others, life, love, and FATHER'S vast created reality. Most everyone now draws their own line as to what is good and evil in their own eyes. Most souls today are too absorbed in their physical lives to even hear what Scripture says about how we are designed and meant to love, learn, and live eternal. These play with fire, not seeing the horror they bring on themselves and others. These think their desired ways are benign and of no consequence to GOD, others, or creation. Now at *what* farthest point, can a soul still turn back to GOD. Turning can only occur at a point where a soul can possibly hear and understand why they must obey every command to live in GOD'S eternal righteous ways. But in that most souls will never do so, we see those who can't, say no one can. These choose, take, and play a role not theirs. Free-will means two things: one is, we are capable of convincing ourselves and believing whatever we strongly desire; and two is in so many ways, we are then slave to our own strongest created chosen desired beliefs. Souls can easily obey every word of GOD, if they lovingly, deeply, and strongly, desire to do so. And no soul can obey every word of GOD, if they have no desire to do so. This is why materialist want to believe their desires are biological and beyond their control; denying free-will exists. Rebellion is when any soul assumes control of FATHER'S created ways of life. Being the body, the extended creation, and his created perfect way that adjoins us all in eternal love. Rebellion is such a fearful foolish thing, being the non-acceptance and non-conformity towards GOD lively creation. Can the rebellious know in any moment the evil they do, seeing they hold no desire to know this? No. Only those who truly love GOD, know, care, or consider, what they do for or against GOD can know such things. Evil souls hijack FATHER'S creation. Picture a movie scene as someone jumps in and steals someone else's car and floors it. They look in the rear-view mirror to see if they have gotten away with it. They sense a stolen privilege, possession, and power of feeling freedom from their reality. They feel a control and exhilaration that the flesh loves. In this world no one will call the police for anyone denying the one true Creators rights, or taking control of *his* rightful property. For the greatest crimes in the universe, these thieves never hear an accusation while in this Kingdom of Satan. Yet this crime is real, and I too have been guilty of this crime. But I was caught and tried in my heart, feeling the horror of my convictions, thrown face down in the dirt of my own life and emptied out. I saw I had no eternal perspective or purpose, and was void of all that is good. I was filled with the despair of a captured criminal, knowing I deserved a horrible eternal punishment; yet in my barren hollowed out emptiness, a messenger came speaking unique words that held a renewing force. His words burned deep into my heart, giving birth to my renewed enduring faith. I felt GOD'S great love, and I saw a beautiful created way out, I thought was gone for me. Not a worldly physical way, which deceives the eye and betrays and fades, this was a pure and bright spiritual beauty. This world's physical pleasures, can satisfy a soul for a short moment; but they are all weaken us in the long term. Temporary relief from having no eternal purpose, creates a physical dependence, or addiction. Now please, do not think we can only be addicted to things like sex, drugs, alcohol, or food; we become addicted to all the things that fill our life. Am I spiritually addicted to GOD in that I need them more each day, and more than anything else? Yes. But is an eternal spiritual dependency upon GOD good? Our every form of desire, comes with our dependent need for that desire to be met; this is the nature of all desire. We can desire *only* to receive what FATHER gives; or we can desire to receive from, and give back to FATHER in all his loving created ways of life. We can also, be void of GOD, and turn in desire to take and give in our own temporary ways of life that are not real. We are eternally dependent upon FATHER; this is the truth of reality. CHRIST our King, paid the price for every returning child, in suffering their physical debt. He was savagely beaten, cut, and torn for the sins of those who truly repent and return to GOD in fullness and all loving perfection. And CHRIST'S loving scarred out stretched arms, say come live with GOD and the Holy Angels in our ways of eternal love. So, I never want to self-indulgently steal control of this physical life again. It's not mine, and can never satisfy my eternal purpose or desires. This is such a needful lesson that only the fewer learn, a lesson few realize they need to learn, and one most do not want to learn. So, how can one live filled with HOLYSPIRIT, and deny the flesh, if they can't properly see, know, and separate the soul and flesh? How can HOLYSPIRIT be one with us, if we live after the flesh? Of course, this can never be. The ways of the flesh, and the eternal life found with GOD, are opposed in every way; but they are not at all difficult to spot or understand. Difficulty only arises in those *not wanting* to see or accept GOD'S full truth. Souls can easily refuse to see the difference between the fallen flesh, and a Godly Spirit, if they individually interpret Scripture to blur and blend the Spiritual and fleshly attributes together. By this, they form self-indulgent false beliefs, and a personalized god for their life. These confused souls can't see that the term Pagan, means to hold an individual religion that exists strictly between one's self and a god of one's own interpretation. If we wait and allow HOLYSPIRIT to interpret and reveal all things, we *can* receive what is true. But we must then live accordingly. The difficulty each will encounter in doing this, is determined by how deep and strong, are their desires for this world and the flesh; and how much their soul recognizes the deceitful allures of physical comfort, beauty, and the ways of this fallen world. And if they can see it all, in contrast to the fullness of GOD'S ways and promise. But the greatest factor is our own strength and purity of love for GOD. And again, we're not talking about thankfulness for what FATHER has done and promised to do for his children, we are talking about our love *for* GOD in all they are and do, and all they say. This means we love and emulate CHRIST in who he truly is, and the way he exists; and not in a way we wish CHRIST to be. We desire to be and give whatever GOD needs from us above all things. In our fervent desire we turn away from *this* world to find JESUS. Let's examine these fleshly attributes we war against, being the flesh holds an evil covetous for worldly beauty, comfort, prosperity, security, all physically exciting things of this world, and the power of being an individual. The flesh thinks life is their own; so, it desires vengeance, wanting retribution if it feels wronged, and it finds satisfaction in the same. These are a few of the easy to see attributes of the flesh. But *we* must reject all such temporary things and feelings, found through the temporary ways of this world. The Spiritual attributes are a strong need to know everything possible about GOD and their forceful creative ways on an intimate level. We spiritually see that any desire to stay in this world, for any reason beyond GOD'S work, arises from within the flesh. We spiritually see we must love GOD back, for who GOD is, and not what GOD gives. We never see ourself, outside of GOD, and focus on the Gospel and our work to help increase and sustain the Kingdom of GOD. We see the flesh and this world as the least important elements, and we treasure the souls of GOD'S Children, as and above our own. We have a profound discernment and recognition as to the subtleties of evil in this world, and an aversion to its over-all evil state. We see the true magnitude of sin that permeates all creation. We have a fearful understanding and hate for the temptations and dangers of this world that work to divide us. We have all heartfelt knowledge that we are Eternal Spiritual Entities, and that we abide in a physical world in which we do not belong. We feel a strong desire to be one together with GOD and all GOD'S Children, having a prayerful longing for the coming Kingdom. We know only true spirituality brings our home and family. We see and pray for the ways and means to come out of, and gain complete victory over this world. These are some of the attributes of our Spirit of Godly love. If we are ruled by the Spirit of GOD'S love, how can unrighteousness be desired, allowed, or ever accepted. Can a child of GOD say in their heart, it does not matter if *we* are not one and together in the ways of CHRIST. Can we say it should not or does not matter to us what the members of Churches do when they go home. Thinking that is between them and GOD, having nothing to do with us. Can we say GOD does not require Holy-unity in one Spirit. If these are things you can accept in your heart, it is then hypocritical to admonish a preacher who teaches and practices all deviant forms of perversion, and immorality. Can we see any righteous judgment in accepting denominations that allow different levels of rebellion, and ask different levels of submission to GOD'S every command. Every sin is sin, and every rebellion is rebellion. The only question that can then be asked, is does GOD allow any willful measure of these? Or has FATHER commanded, and given us the power to walk perfect before him, if we have the same true love to do so. And if any believe FATHER *does* accept a certain level of sin, who judges that acceptable level? If any believe this, I am sure their drawn line, has much to do with allowing all the sins they desire most. Self-preservation makes it hard for any to see themselves as rejected by GOD. The mostly *apostate* Church of these last days, says and believes that the only righteousness we can attain before GOD, is in asking for forgiveness, and not in our loving obedience to every word of GOD under grace. Such souls, turn righteousness inside out, and on its head, and it is a great offense to the truth of GOD. In all such compromised ways, we see a subtle antichrist religion. HOLYSPIRIT shows us why we cannot desire worldly things we don't require in order to live, without losing the spiritual strength we require to live eternal. We cannot fill our soul twice with both evil and good. There is a true three-dimensional conscious space and size to our spherical soul as we exist in GOD'S reality, and evil seeks to fill it up. These last days of self-indulgence now reach deep into the Church. A most fruitful approach by Satan, is attacking mankind at their greatest weakness, being the lust of the eye and the law; just as Satan did with Eve and then Adam. Such a simple subtle lie that mankind falls for over and over without end. This world brings moral abandonment, as many temptations are being laid out before all in this world. In times such as this, those who truly believe in GOD must ask themselves. Do I truly seek only Spiritual things? Or the things of this world? Or worst and most dangerous of all, do I self-indulgently seek to have both? If we seek the truth and power of GOD *only*, we must accept that Scripture is revealed over the course of our life at times and in measures by GOD. And it is not, and cannot be given in the beginning in fullness. Each soul first requires a new shared foundational understanding to support this new and higher understanding. Then we have our ever-expanding moment to come to hear and know our ever-building true understanding. Each step in its order, must be properly known before one can properly learn and know the next step. Learning GOD'S truth, like learning anything, is a step-by-step process. We must build from the beginning, or we have no true foundation. And this truth can only be found while walking in loving obedience. A soul having a weak impure love for GOD, cannot wait years for this instruction and correction; so, they tire and fall away. Or in thinking themselves special, they believe in their hearts that GOD gives them all things in their fullness in the time of their asking, and then in error, believe their own desirous thoughts are sent from GOD. Such impatient weak souls claim something to be true, that is not, and easily accept false doctrine. There is a time for all things in the will of GOD. If we are out of GOD'S time and place, we are then out of his will. And this is not just in our receiving, but in our giving also. To give in GOD'S love, will, time, and place, are our only true works of righteousness. Therefore, if we give, but are out of the love or will or time or place of GOD, we are not being led; and all such giving is an unrighteous work to be burned. Do you think the charitable giving of non-believers is a righteous work of GOD? Now our zeal for GOD is needful and good, but it can be a wild force in the hands of a babe in the LORD. So, be slow and sure as you move. In this time of great self-indulgence, we see most physical charitable works done in GOD'S name, are not done in his love, will, time, or purpose. A zealous mistake not seen by young weak believers, is made in not knowing righteous works are done for mostly to purely Spiritual, and not physical reasons. If we help to clothe, feed, or mend the flesh of others, it must not be meant for the benefit *of* the flesh. An elder who has lived a long-life emulating CHRIST, will most often offer a different help, than that of a new child in CHRIST. Scripture repeatedly and often instructs *all*, and surely the young in CHRIST to try every spirit; and to hear those who have found and walked the *narrow* path for a time sufficient to bear wisdom and fruit. Of course, such leaders are scarce, I find these to be most rare in my life. The further we go into these last days, and coming tribulation, the more we need to come out of this world. Temptations grow faster than the appetite of mankind can consume them. If a soul admits they need and seek worldly things, my only admonishment is to seek a new eternal love and direction. And any soul that seeks both the things of GOD, and the abundant things of this world, should tremble for the confusion within them. GOD guards their truth, and how impossible it is to find, if we don't seek it from its source, and in GOD'S way. The one true interpretation is not laid out for all to be picked up like stones strewn about. Scriptural interpretation is a Holy gift, decreed by FATHER, paid for by MESSIAH'S Blood, and delivered only through our Baptism of HOLYSPIRIT. When it comes to *truth*, we hear it said *in this world,* there are two sides to every truth. All such statements are born of the selfish worldly arrogance that fills the flesh; and in this world alone are they accepted. But if we hear the truth of GOD, wherever we see two sides of a worldly disagreement, dispute, or misunderstanding between two souls, we see both sides are always wrong. I ask you to see that in this world, there is almost always three sides to the truth. There is a false perceived truth of one soul, the false perceived truth of the second, and then there is the truth of GOD undefiled, being the author of all such things. This holds true throughout Heaven and Earth, from the tiny squabbles between little children, to the monumental irreconcilable differences between countries that bring only war and death. GOD'S Creation has no provision for individual truth. The truth is one, and GOD is its author. None can, or has any power or right to alter it. By all evidence, most all souls in this world have committed grievous, heinous, abominable acts against GOD'S truth. *And* we know there have also been those souls of men, women, *and* children who in loving GOD, have accomplished perfect works. FATHER admonishes mankind to walk perfect before him; so, this we can and must do. And understand that our required lesser perfection of mankind, is not equal to the full perfection of GOD. We are created for and called to a lesser purpose than GOD. And to fulfill our lesser purpose, is to be our lesser perfection before our LORD. The perfection of mankind we are called to by GOD, has no self-indulgence or self-rule. The perfection of mankind is to come in faithful love and live in all the created ways of GOD. We must answer the love of GOD by becoming one in CHRIST. But as the selfishness of this world grows stronger, and all true love for GOD grows colder, this brings only sadness. Most souls live in a temporary mode of reality. What does this world eternally search for? What will they eternally achieve? Where is their eternal gain? Theirs is an illusion of a darkened eye and a cold stilled spirit. Followers of false doctrines, look at their physical possessions and see a sign that GOD is pleased with them. And by that belief, they lose all they think they have, and all they might have ever truly had. This prosperity doctrine is responsible for building most mega Churches; and in these, one hears messages designed to please those who covet the same. JESUS told the man who would be perfect, to sell all he had, give it away, follow him, and he would have treasures in Heaven. On earth, JESUS was a man of sorrows, acquainted with grief, owning nothing of this world, having nowhere to lay his head. This world is a boiling cauldron of death, and the endless sinful desires for it, permeates the flesh that simmer therein. The greatest act of self-indulgence in the heart of mankind, is to say there is no GOD. In this one simple statement, lies the heights and fullness of self-indulgence. By this statement, one claims absolute rule over one's own beliefs and actions; it is an attempted coup, and denies the evident existence of the Spiritual world. Another deception we see, is the feeling of salvation by grand association. It is basically a type of fanaticism or fan-based action. Think of a sports team's fans, which at some level deep inside, feel they are part of the team. How else could they share joy with the players in victory, and their pain of defeat. A team cannot exist without fans. I have heard from many, how they derive their stronger feelings of connection to GOD when they attend church, and feel weaker the rest of the week. This feeling is found in a person not living by the WORD. Some may feel they are on the right path by joining a group that to them, appears to be on a good path. And the bigger the group and grander the worship structure, the more assured some may feel. There is an element in their search, born of the old herd mentality adage that there is safety in numbers. When a church reaches a certain size, and can put their services on television, such churches in a short moment can explode in membership. People will drive passed many small neighborhood churches to get to a mega church; yet most can't see the vain cause for this phenomenon. Has GOD ever said, to each his own work, if they say it is done in GOD'S name? No. Has GOD said, to each his own ways and beliefs, if they say they are done and professed in GOD'S name? No. GOD has truly said by the fullness of Scripture, that he created man, and then woman, after he created the true strait Spiritual ways, works, and belief in which we should live eternal. But mankind, being lost in their great self-indulgent pleasures of this world, now believe Heaven must be a buffet or democracy. When discussing the acts of self-indulgence, we shine more light on those professing false Judeo-Christian doctrine, and less on those who profess no belief. Having a form of religion or godliness without GOD'S power; from such we are told to turn away. These souls, by every word of GOD are self-indulgent in their religious practice and doctrine of belief. 1st Tim. 6:5, *PERVERSE DISPUTINGS OF MEN OF CORRUPT MINDS, AND DESTITUTE OF THE TRUTH, SUPPOSING THAT GAIN IS GODLINESS: FROM SUCH WITHDRAW.*" 2nd Tim. 4:3-4, "*AFTER THEIR OWN LUSTS SHALL THEY HEAP TO THEMSELVES TEACHERS, HAVING ITCHING EARS; AND THEY SHALL TURN AWAY*

THEIR EARS FROM THE TRUTH, AND SHALL BE TURNED UNTO FABLES." Oh GOD, grant us mercy, Amen.

THE LAST DAYS OF GREAT IDOLATRY

There are the two great facets to what Scripture defines as idolatry: one is the reason of desire within one's heart for why one idolizes; and two would be the object or personage that is idolized. In the past, before people worshipped the scientific technical and medical manipulations of the physical reality we now see. Most souls believed in GOD, or a god. And when souls did not want to believe in, or obey the one true GOD. As now, they could and did in self-deceit, turn to a personally created GOD. And back then they could project their god's onto things carved of wood, or a molten image. They molded personal house gods, that reflected their individual beliefs, in not wanting to be molded into GOD'S image. They attributed deity to physical aspects of reality, which allowed them to project their own desires, as commands coming from their false god. This practice in effect, makes an idolizer, the creator and false god over their life. The goal of idolatry, is always to replace GOD and their ways with one's self and one's own ways. In these last days, just as long ago, idolatry steadily increases; and it will, up until our LORD brings his Throne to earth. What do we each accept as the true and full description of idolatry? And does it matter as an aspect of our salvation? Many definitions of idolatry are floating around, and as with all things, only GOD'S definition is true and right; so, it is vital. Now most current individuals see idolatry as a small simple thing, thinking it means the act of molding or carving an image, and looking upon it in ancient superstition, believing it brings a good life. But this is a vague surface definition. To idolize, is a confused evil act in one's soul that goes very deep. Idolatry is what is going on inside one's heart, where one's greatest trust and desire dwell. It is any partial or complete replacing of GOD over our love, truth, and life. Idolatry is a need to live more after objects and people, and surrendering to a great fear of losing the same. We must understand idolatry in its fullness to be able to avoid it, and not offend our love for GOD. Idolatry occurs in our soul's highest place, designed to house our highest love and worship, if we put anything or anyone there in that place meant for GOD. GOD must always be first and foremost in our heart, or we will begin to destroy ourselves and others. This is part of the design of all creation. There is one created purpose leading to eternal life, and each and every *contrary* purpose leads to death, meaning separation from GOD. GOD did not gift souls and create reality on a whim, having no Holy Purpose or Righteous Knowing of what is good and evil. And then as an arbitrary afterthought, declare certain things are now good or bad. All things were created to function eternally with all forethought of purpose according to the eternal good desires of FATHER, to fulfill his Righteous need for an eternal loving family. Praise GOD for their love and mercy by seeking to fulfill their purposed desires; for which we are made and given the great gift of life. If *this,* we say we cannot do? Then why? It is only our soul that can desire to replace the supremacy of GOD with people and objects, and only for one's own temporary physical sins and pleasure. The fallen desires of the flesh are in us all, in our being born in sin, not knowing what is good, and being separated from GOD. Now a soul can come to love and trust GOD, but the corrupted flesh of mankind can never trust in GOD, because it has no salvation. We must keep the flesh in our strongest subjection. Ezekiel 20:10-16, "*WHERFORE I CAUSED THEM TO GO FORTH OUT OF THE LAND OF E'GYPT, AND BROUGHT THEM INTO THE WILDERNESS. AND I GAVE THEM MY STATUTES, AND SHEWED THEM MY JUDGMENTS, WHICH IF A MAN DO, HE SHALL EVEN LIVE IN THEM. MOREOVER ALSO I GAVE THEM MY SABBATHS, TO BE A SIGN BETWEEN ME AND THEM, THAT THEY MIGHT KNOW THAT I AM THE* ***LORD*** *THAT SANCTIFY THEM.*

BUT THE HOUSE OF IS'RA-EL REBELLED AGAINST ME IN THE WILDERNESS: THEY WALKED NOT IN MY STATUTES, AND THEY DISPISED MY JUDGMENTS, WHICH IF A MAN DO, HE SHALL EVEN LIVE IN THEM, AND MY SABBATHS THEY GREATLY POLLUTED: THEN I SAID, I WILL POUR OUT MY FURY UPON THEM IN THE WILDERNESS, TO CONSUME THEM. BUT I WROUGHT FOR MY NAMES SAKE, THAT IT SHOULD NOT BE POLLUTED BEFORE THE HEATHEN, IN WHOSE SIGHT I BROUGHT THEM OUT. YET ALSO I LIFTED UP MY HAND UNTO THEM IN THE WILDERNESS, THAT I WOULD NOT BRING THEM INTO THE LAND WHICH I HAD GIVEN THEM, FLOWING WITH MILK AND HONEY, WHICH IS THE GLORY OF ALL LANDS; BECAUSE THEY DESPISED MY JUDGMENTS, AND WALKED NOT IN MY STATUTES, BUT POLLUTED MY SABBATHS: FOR THEIR HEARTS WENT AFTER THEIR IDOLS." We must know that eternal love is only from and for GOD. It is in no way, a part of this fallen realm. And in this lost world, eternal love can only grow inside the hearts of those who desire all of GOD'S ways. By our every action, word, and thought, we answer GOD'S call of who or what do we idolize. Idolize meaning to supremely worship, love, and live for. Do we worship physical beauty, comfort, and security, in seeking beautiful homes, vehicles, expensive array, and other physical pleasures of this world. Scripture says any desire for the things in this world, above our true needs to live, are lustful covetous. To most, this sounds far too harsh, but this is a mild admonishment compared to all other forms of idolatry. We now must confront the toughest idolatry questions; being do we idolize people we love. This is always a harder question, because love is part of worship; and we are commanded to love our brothers and sisters in CHRIST. We are to hold a general hopeful love for salvageable enemies. We are to give our love from GOD to everyone who can receive it. But our love for people opens the door to a most dangerous form of idolatry; being a form we are all susceptible to. We are not, and cannot be sorry if this offends anyone, in it being the truth. But most professed Christians love their family not only more, but far more than they love GOD. Even as they deny this to themselves, and declare to all that they don't. If we ever place family, even our newborn child above GOD'S love and ways in our heart, whether we know or admit we do, we offend GOD by idolatry. Matthew 10:37-39, "*HE THAT LOVETH FATHER OR MOTHER MORE THAN ME IS NOT WORTYHY OF ME: AND HE THAT LOVETH SON OR DAUGHTER MORE*

THAN ME IS NOT WORTHY OF ME. AND HE THAT TAKETH NOT HIS CROSS, AND FOLLOWETH AFTER ME, IS NOT WORTHY OF ME. HE THAT FINDETH HIS LIFE SHALL LOSE IT: AND HE THAT LOSETH HIS LIFE FOR MY SAKE SHALL FIND IT." The act of elevating anyone, or *their* needs or desires above those of GOD in our heart, is idolatry. In idolizing another, one can still have and feel a strong but *lesser* measure of desire for GOD in their heart. Yet GOD is displaced of his *rightful highest* place, and moved down. This is idolatry. No Gifted Conscious Entity has an unlimited capacity to love as our forceful soul is finite. And if we live and love more for this world, our soul then has less love and life for GOD. By all the laws of reality, our soul must always be filled to its capacity with all our loves and hates. And if we fill our hearts with the desires for this world, our family, and self, we see we will not have the place or room to give GOD the love they deserve. Each day and each minute, give to GOD first, that greater part which is theirs, and the remaining to those around you, and you will never err. The two easiest to see signs of idolatry are: do we live more for GOD, or for certain people and possessions. And the second sign rests in the possessiveness of our thoughts for family. If we are not careful, we can believe family members are *ours*, and not GOD'S. One can believe too strongly that a soul is truly *their* spouse, and that these are truly *their* children, siblings, or parents. Psalms 24:1, "*THE EARTH IS THE **LORD'S**, AND THE FULNESS THEREOF; THE WORLD, AND THEY THAT DWELL THEREIN.*" These are not literally ours, neither were they ever ours. It is a near impossible thing for *professed* believers being still in the *flesh*, to think of their Godly spouse and children as being their closest brothers and sisters in the family of GOD. Most see these as above that standard, and as far more. Yes, as to this world, we have the most love and life invested in these, yet they are not now, nor ever will be ours; they are GOD'S. Our love for them can seem so strong, natural, and good, but keep this love from becoming idolatry. The word *natural* has become a great false bridge to accept worldly things by the Church. Remember this world's use of the word natural, means of the physical world, and not the eternal soul. Scripture says there are *only* two eternal families, containing all lesser forms of families. There is the eternal family of GOD that includes all his Children; the other is Satan and his children. All souls are originally FATHER'S gift; and we are called to love those near and do all in our power to make sure they come to and live in FATHER'S ways for their eternal provision and life. We must love and treat others the same way we treat ourselves before GOD, in giving and calling for the same perfection and obedience we demand of ourselves. This is the love we are commanded to have for GOD'S people, and it should always be so. If a temptation and opportunity to sin is there before you; yet with all our hearts-desire you do not want to sin against GOD. Can the sin nature of the flesh take control over your soul and force you to sin? No. Only when you have a worldly desire to do so, can you be unfaithful and irresponsible, and sin and hurt GOD. Many, not understanding idolatry, try to get family members and friends to love them more and more. Yet in this way, they compete against GOD for a soul's greatest love. If we unintentionally or intentionally, get any soul in our life to love and need us more than GOD, we have become their idol, and broken their spirit. JESUS, when told by bystanders in Mark 3:32, that his mother and brethren where outside wanting to see him. JESUS said in 3:33, "*WHO IS MY MOTHER, OR MY BRETHEREN*" Looking around the room at those who sat listening, he said, in 3:34-35, "*BEHOLD MY MOTHER AND MY BRETHEREN! FOR WHOSOEVER SHALL DO THE WILL OF* ***GOD****, THE SAME IS MY BROTHER, AND SISTER, AND MOTHER.*" And JESUS did not say *FATHER*, for we have one FATHER in Heaven. We see a huge difference between people who have added an amount of GOD into their life in various ways and measures, and those for whom FATHER and his ways have become their life. Where does our love reside, and what and who do we idolize? Do we idolize things and flesh that are a vapor and soon gone? Or do we idolize the great eternal loving GOD of love, truth, and life? We have discussed idolizing worldly things, and family and friends, and even evil itself. Is that all, or is there another form of idolatry. We now come to the very center and vane of our existence, to again, look into a mirror. Do we see our Idol in the mirror, or a true servant of GOD in Heaven? Do you know if you place yourself above GOD? Does your belief tell you GOD will always love you, no matter what you do? That GOD will never leave you, no matter what you do? That GOD wants to give you all the desires of your heart? That GOD follows and goes with you, no matter where you go? This sounds like a soul who has lost all humility, and put themselves in that highest place in their heart, who now sees GOD as a servant. This is the same self-idolization born in Satan. Remember, GOD loves those who humble themselves, and obey GOD. GOD says to turn from sin, or they will turn from us. GOD says they want to give us only the good things of their Kingdom. FATHER says for our good, we need to blindly trust his omniscient love. How does idolatry, being contrary to GOD'S truth get started and grow in churches? Idolatry *in* the professed Christian Community, in many ways and measures is accepted, taught, and even *enforced.* People are taught by preachers to pray for things GOD will never give them. Even if they ask for something good and eternal for GOD'S Kingdom, most can never wait for the right time to receive it in GODS will. What happens most often in their impatience, is they turn to the flesh for help, sensing GOD is not responsive, or there for them. Many place those more responsive, higher in their heart for faster comfort and assurance, like doctors, family, and friends, that they sense are quickly and visibly responsive to their prayers for physical need and desire. Mankind has always struggled with knowing the line of separation that exists between physical need and want. Most think that if *something* will temporarily ease their physical burden, or increase physical comfort, pleasure, or security, they need that something. Is this how GOD defines our physical needs? When I read about the lives lived by GOD'S beloved Prophets, who trusted in GOD. We find two groups: one is a group of stewards, entrusted with worldly sustenance and authority to lead, provide for, and bond many people under leadership of that Godly soul; such as Righteous Kings and Patriarchs. Two, are those sent as Prophetic bearers of coming Judgment such as Elijah. Many see these Kings and Patriarchs, thinking GOD has given them *personal* wealth because they are righteous; this is not true. What has been bestowed is stewardship and responsibility over many. In the second group of Prophets, we see servants, sparsely provided for physically. Have I yielded to GOD, or do I gather much for myself? Using these wisdoms, I ask *myself* has GOD made me a responsible steward over many souls, or a lowly servant relying on GOD for simple need and provision. James 5:10, *"TAKE, MY BRETHEREN, THE PROPHETS, WHO HAVE SPOKEN IN THE NAME OF THE* ***LORD****, FOR AN EXAMPLE OF SUFFERING AFFLICTION, AND OF PATIENCE.*" These servants are the highly loved and favored of GOD. Our LORD himself was physically frail, lowly, and had nowhere to lay his head. When we look at the poorest countries, do we say GOD has not met their wants and needs, so GOD must love us more? Or do we come to the truth, that our true needs consist of HOLYSPIRIT, bringing us the one true interpretation of love and strength in Scripture. I would choose to be a poor HOLYSPIRIT filled soul, eating simple clean foods, others discard, over being the richest lost soul on earth. It is reflective suffering and chastisement that teach us wisdom of GOD, needing only enough simple food water and covering to keep us alive to do their will. Mankind struggles to know the line of separation that exists between spiritual wealth and physical poverty. The more the flesh has, the more it needs. And this is a great problem in that most think they're not rich until they have a mansion, millions of dollars, luxury cars, yachts, and a jet. We can never compare ourselves to those who have more; we are called to compare ourselves to those who have nothing. The average person in the U.S., that falls under this country's definition of poverty, is far closer to being rich than poor in the eyes of over half the people in the world. What do we possess, and how do we sustain this life in the U.S.? Through borrowing tens of trillions of dollars, we can never pay back; that is how. How many outfits of clothing, and how many rooms are in our home? Do we have a car, range, washer, dryer, television, and refrigerator? How many things do we own that we could literally live without (meaning not die)? We do not get to judge if we are guilty of storing up riches. GOD will judge these things. Idols are anything anyone loves more than GOD. So, most souls are filled to their brim with idols. In the very last days, Scripture says the world at large will replace GOD altogether with a coming idol of all idols, and the greatest of all counterfeits. There are so many examples of idolatry in the things we see and might physically desire around us each day. But the world will soon see a new thing, on an immense order of magnitude not ever seen before. In great mass the whole of the world, including most professed believers will fall away from the true GOD, and turn to the worship a false spirit, son, and false father. This being the greatest mass commission of idolatry ever committed, coupled at the same time, to the greatest form and the most heinous and hurtful act of idolatry against GOD of all time. As the greatest Idol of all will be placed in the Jerusalem Temple, and then indwelled by the soul of Satan to speak as GOD, to which the world will bow in their greatest act of idolization. Such a painful thought that we must witness this world's mass rejection of the greatest love, truth, and mercy of GOD. It often makes me ill to be here, having this knowledge, and witnessing how this world always vehemently turns on GOD. They dishonor and reject GOD'S great work, sacrifice, and loving offer of truth, hope, and life. I cannot understand a heart that rejects such righteous beauty and purity which out-shines the whole of creation. This world rejoices in turning away without shame, claiming in religious pride, their love and desire for their idols. I don't often quote outside Scripture, but it has been said no better than by James Lowell, "The idol is the measure of the worshipper". This is the most profound quote on the subject I have ever heard outside of Scripture. A rebuke to all who by their small measure, choose insignificant illusory things to worship. And when idols are held up, against the great light and measure of wisdom, love, and truth, of all who truly worship GOD, we see the true measure between light and darkness. How filled are we with idols of family, friends, self, houses, vehicles, clothes, food, pleasure, comfort, doctors, musicians, sports teams, politicians. There are endless idols one can fill GOD'S place with. Col. 3:5-6, "*MORTIFY THEREFORE YOUR MEMBERS WHICH ARE UPON THE EARTH; FORNICATION, UNCLEANNESS, INORDINATE AFFECTION, EVIL CONCUPISCENCE, AND COVETOUSNESS, WHICH IS IDOLATRY: FOR WHICH THINGS' SAKE THE WRATH OF* ***GOD*** *COMETH ON THE CHILDREN OF DISOBEDIENCE:*" We have said that the Fourth Commandment regarding the Holy Sabbath is all but forgotten and it is not alone; the Tenth Commandment also is forgotten. Each commandment has been weakened and forgotten by the church in differing measures. In asking others to define covetous, most say it is a thought leading to an act of theft or adultery; but these are extreme examples. To covet is to want anything you don't need or have. If you want anything beyond the required needs to serve GOD in this world, you covet that thing. Eve and then Adam coveted a piece of fruit, thought to enlighten; why would a soul trade their perfect eternal life and dominion over the earth, for a lie about a piece of fruit. Because they coveted something they were told they could not have. 1st Timothy 6:5-9, "*PERVERSE DISPUTINGS OF MEN OF CORRUPT MINDS, AND DESTITUTE OF THE TRUTH, <u>SUPPOSING THAT GAIN IS GODLINESS</u>: <u>FROM SUCH WITHDRAW THYSELF</u>. BUT <u>GODLINESS WITH CONTENTMENT</u> IS <u>GREAT GAIN</u>. FOR WE BROUGHT NOTHING INTO THE THIS WORLD, AND IT IS CERTAIN WE CAN CARRY NOTHING OUT. AND HAVING FOOD AND RAIMENT LET US BE THEREWITH CONTENT. BUT THEY THAT WILL BE RICH FALL INTO TEMPTATION AND A SNARE, AND INTO MANY FOOLISH AND HURTFUL LUSTS, WHICH DROWN MEN IN DESTRUCTION AND PERDITION.*" Before the end comes, I wish someone could, for me and the world at large, put into words a full good explanation for why many choose pleasure and comfort from the things of this world, and close their heart to pure eternal life found in GOD'S Spiritual ways. Why do they take this Holiness offered by GOD, being a priceless, pure rarity unparalleled, and under self-inflicted delusion, toss it aside. And beyond all eternal love, open their soul and take of the lust of the eye and the lust of their heart to fill their life with the foul vapor of things having no eternal value. Things which soon vanish as if they never were. Our tears should fall at seeing this occurring everywhere without rhyme or reason. We are so ashamed before the GOD of mercy.

THE LAST DAYS OF GREAT CONFUSION

Let's start with the question, what exactly is confusion? Simply put, for all intents and purposes, each experiential article of confusion that occurs within each and every soul, is defined as one not grasping to differing degrees, the true and full understanding of any fact, choice, or point of reason, set before them in a moment they need to hold and exhibit such understanding. Confusion occurs in its ongoing ways in every lost or incomplete soul, not seeing GOD'S full truth of reality. For individuals, reality is falsely seen through their individual preferences, bias, and false beliefs. Using this true and full definition, we see most souls are even confused as to what are love, joy, pleasure, and even life. We have heard Preachers say, we *can* live in the joys and pleasures of sin, but unless we come to GOD, we will lose these in the end. This teaching itself is confused and superficial at best. What is *true* joy, and what is true pleasure. The true state of anything, rests only in its eternal definition, and never a temporary definition. GOD says all true love, joy, and peace, are eternal. In FATHER'S wisdom, he sees a thing is not good, unless it endures faithful, true, and eternal. But we see all so-called pleasures of this world are brief and vanish. Souls spend most of their existence searching and waiting in between their moments of temporary gratification. The longer anything endures in purity, and we are talking millennia as a starting point, the less of a lie it is. What lost souls and flesh feel, is not *true* joy, it's a temporary illusion of joy. This world offers illusory gratification, in place of joy. The love and joy of eternal life, make all temporary imitations look like lesser suffering. Most seek and experience such temporary delusions of false hope in their ways of life, and feed off the souls around them as they deny they are lost and diminishing. There is a complete distinction between this world's moments of forgotten or reduced distress and disorder that delude, and must be replaced constantly, and true eternal joy. This world's desired temporary elevated sensations that accompany sin, just fill the moments in between the endless intermittent dissatisfaction and despair that overflow the lost. In a life of sin, bitterness and loss do not come at the end only, but all throughout. Souls train their selves to not see this, while seeking those momentary delusions. They don't see, it is all a temporary lie as they chase one temporary gratification after another, and never find true lasting joy. The lost, savor false physical sensations of prideful achievement, held in acquiring things lifted up by their fallen peers in a godless immoral culture. They live and run with a Pack who are all dead and in hell. And Preachers who say there is joy in sin, unknowingly spread confusion. In 1974, I wrote a paper for the Pentecostal Church I attended titled "Confusion, a non-Christian disease". I laid out how diseases are contrary influences finding their way into the processes of our lives where they do not belong. These bring disruption, destruction, and eventually an end to physical life; and they are passed along to others. So, what is an affective description of confusion? It is a conscious disease, infecting and affecting one's perspective. But why is it a non-Christian disease? The sense and occurrence we call being confused is a symptom, that unless caused by cerebral damage, presents as a soul's inability and ignorance to properly deal with the choices in one's situation. So, being burdened by the choices they are continually confronted with, fallen souls get caught in a feeling of losing control of all that is occurring. These confused feelings range from barely noticed and short-lived over things of little importance or consequence. But according to the level of *importance and consequences* of not knowing something, these symptoms increase up to a state of great fear and panic. As a consequence, daily confusion drives souls to seek solace in a consensus of group belief. But herd mentality leads to beliefs in vague theories and assumptions that ease the broader shared conscious symptoms; and by this they can never fully align. So religious confusion in professed believers in CHRIST, points to their rejecting the full truth of GOD, raising their Spiritual confusion. To accept a false belief about GOD'S reality, may temporarily calm, but it leaves one in the same questioning confused state, as the unbeliever. Titus 1:16, "*THEY PROFESS* <u>*THAT THEY KNOW*</u> *GOD; BUT IN WORKS THEY DENY HIM, BEING ABOMINABLE, AND DISOBEDIENT, AND UNTO EVERY GOOD WORK REPROBATE.*" 1st Tim 4:1-2, "*THESE HAVE BEEN SPUN FROM ALL TRUTH AND ARE GIVEN OVER TO A REPROBATE MIND. NOW THE SPIRIT SPEAKETH EXPRESSLY, THAT IN THE LATTER TIMES SOME SHALL DEPART FROM THE FAITH, GIVING HEED TO SEDUCING SPIRITS, AND DOCTRINES OF DEVILS; SPEAKING LIES IN HYPOCRISY; HAVING THEIR CONSCIENCE SEARED WITH A HOT IRON;*" 1st Tim 4:6, "*IF THOU PUT THE BRETHREN IN REMEMBRANCE OF THESE THINGS, THOU SHALT BE A GOOD MINISTER OF JESUS CHRIST, NOURISHED UP IN THE WORDS OF*

FAITH AND OF GOOD DOCTRINE, WHEREUNTO THOU HAST ATTAINED." How do souls arrive at various states of confusion; and again, we speak of Spiritual confusion, as to the truth of GOD. If we allow and accept into our hearts and minds, any teachings of the flesh, it will manifest as confusion when it becomes relevant to any line of spiritual thought or conversation. So, one then says, I don't know, or I'm not sure. But how and why can one not know and be sure? Only by confusion. Scripture calls us to seek with all our strength, and accept the one full pure truth of GOD, and to *walk in it* without fail. Having HOLYSPIRIT, endows us with the full truth of GOD'S love, purpose, and knowledge. GOD'S truth and love *always* provide us the right answer and course of action. This world has become so confused as to the truth of reality; and by evil design it grows more confusing to more souls each day. People everywhere are confused about what they should do concerning their daily needs and desires. They make decisions every day in their lack of surety, wondering if they made the right choice. I hope you clearly see, all of mankind's ways are fluid and born out of endless multiples, variables, and constant change. While GOD'S ways are born of unchanging eternal truths. The ways of GOD, and mankind, share nothing in common at any point. Mankind's ways should be discarded as the foolishness they are; their ways move in the winds of an ever-changing godless consensus. GOD'S ways are built into the fabric that governs reality. If you trust in mankind's ways, you will sense here and there, things that temporarily satisfy your fleshly desires. But in GOD'S ways, we know we are not of this world, yet we must for a moment endure, being out of our place. Through guidance from all higher sources, we know without confusion, that what comes after this world, is highly preferred. I hope you can agree, that by continually walking and living in GOD'S truth and love, you dispel all confusion of knowing and choice, making it a non-Christian disease. How much confusion do we see in this world? How much confusion will come before the end? Every new choice that arises for mankind, increases the world's confusion. Have you ever heard a person, or even yourself say in a store's cereal aisle, that all these choices are confusing. Cereal is such a small matter, but as an example it is true. The choice of the registered Christian denominations has grown to over 250 as of this writing. Are there 250 different confused Christian truths to choose from? If GOD'S one true belief is to be our belief, how does this occur? Does GOD have 250+ different beliefs? No. These are 250 groups of confused souls. Think of the number of choices available to each soul every day, and multiply that by the number of cultures, age groups, personality types, and beliefs. Multiply that by the number of persons in the world. Then multiply that by the number of interactions between all souls in a day. This is no true formula to calculate the level of world-wide confusion caused by all the wrong choices of belief, made daily. But a way of showing that ever-increasing individuality and ever-changing beliefs, create more confusion. Mankind was not designed or created for this process. Look at the rising levels of anxiety and stress, causing the masses to breakdown spiritually and mentally. We wake up making decisions, we fall asleep making decisions, we make decisions even in our dreams. Confusion has so many causes and side effects, being spiritually and physically destructive. One affect, is people escape into a shell of self, in being exhausted from all choices. There is often no strength left to hold onto each other. People sink under their confused decisions, while their love for GOD is drowned out. People more and more are confused about right wrong, good evil, love hate. They grow spiritually confused about politics, religion, country, sexuality, wealth, families, their rights and reasons for life, and their selves. People without the one pure truth, wander through life, never being sure of anything. They think the truth of reality is an ineffable thing; yet it is knowable. To hold onto a false semblance of sanity, souls can convince themselves they are sure of certain things, which helps them to cope with life. But then they are disproven, so again, they must find and convince themselves of new false certainties to cope. This is worse than a dog chasing its tail, it is like a dog chasing its tail, that has no tail. We hear in this world the sounds of quiet and loud desperation and self-deception. We hear the sounds of not wanting to know GOD, in a fear of not wanting what they find. So, we see mankind rejecting GOD'S truth, living by their desires towards the narrow physical evidence before their eyes, they each filter and interpret through their bias. These search for something to console themselves, while smiling at each other, assuring each other there is no need to worry. These, by a temporary perspective, define their own reality as it comes each day. That is, up until that day they find out they are wrong. It is a great burden of sadness to see such confusion afflict so many. Those far away and close by, on the right and left. Will any desire to escape it? We escape this world by GOD and our pure and fervent desire to do so. We see true light and witness GOD'S great pure truth. But most will not see it, they stare past it *and* us, far off as though deaf and blind to it all. If they refuse to hear GOD, why would they hear us. Our second greatest purpose after seeking, knowing, and loving GOD, is to be a conduit for that love and truth of GOD. To be passed from GOD through us to others in words and emotions of spiritual love and truth. And yes, we do fulfill all physical charitable needs that ease true suffering, for this softens the hearts of the confused. It is not only our obedience to every word of GOD, but even more *our* great desire to do so, which is the true and exact measure of our love for GOD. And so, the volume and sincerity in our sharing GOD'S pure truth with others is the exact measure of our love for them. We surely do not discount the temporary work to give food, shelter, and ease pain and suffering. But the *pure* truth, shared on GOD'S behalf, is the eternal need to be met. To bandage physical wounds, and feed the weak flesh, is momentary and necessary, bringing rest from the distractions of fallen life, so one might hear GOD'S truth. But to feed and mend only the deteriorating body which disappears forever, and not feed and mend the soul more so, lacks eternal purpose. Great confusion is born out of believing a multitude of lies. How can we lead one out of deep confused bewilderment. This depends on our mutual love for GOD, and the depths of confusion into which they have fallen, and how much do they want out. What words can they hear? What words can we say? Only the pure truth and eternal love of GOD are our guide, and these must become theirs as well. There is no other path to find and follow, there is no individual way for each soul; there is only one way for all. And yes, the Church rejected my paper. But there can be a perfect moment for each soul to hear. A few will hear the truth, but most refuse it; as this is the essence of fallen free-will rebellion. Can we ask and expect GOD to break through to our loved ones with his truth, to remove their free-will blindness, deafness, and defiance? But how can GOD do that? Can or will GOD alter their desires by force? No. Will GOD change them against their will? No. All HOLYSPIRIT can do for all free-will souls, is look for the right moments of need and hope, and call out in love. Can GOD forcefully affect a change in a backslider, a non-believer, or a person angry with GOD? How far can GOD go? What can GOD do without over-ruling our free-will? What can GOD do without subverting the Holy integrity of his loving purpose for all creation? How can GOD over-ride any soul's free-will, required for their love to exist? GOD will not easily return to call someone, after finding no desire in that person, having tried many times already. Many ask GOD to please give a lost loved soul, an undeniable sign; yet it is hard for many to accept in our heart that when someone has resisted all prior attempts by GOD, a sign rarely helps. If the one we pray for, still believes in JESUS, and has *some* true love and fear for GOD; yet is desperately uncertain of GODS love towards them? GOD may give *them* a sign. But for one so far away, whose heart is hardened and scarred over, by rejecting GOD'S love many times; if this is your desperate prayer, you ask for a very hard and most uncertain thing. I do not say don't go and ask, but go in nothing less than purity of heart, humility, great fear, with fasting, self-denial, and crying for GOD'S mercy and compassion. Then maybe GOD *will* call on them again. I would pray; GOD, please let every soul of truth, near to our loved one, be there with the words of your perfect strength at every perfect moment. And at every time of their every need and searching for someone or something to hold onto, let your perfect light of love be there whenever they might look up and still see. Please LORD if it be your will. Amen. If all else fails, and GOD sees your loving obedience and consuming love for his ways; and if this is your lost child or spouse. You may hope in GOD'S promise of grace by extension of your house in the Millennial Reign. It is sad, but we can't remove the false belief in anyone, only they can. A person will cling to whatever they have, holding it dear, being all they have for their hope of understanding. A soul fears giving up their former reality of faith, before receiving the new. How do the right words of love entering through the ears, soften the heart and open the eyes to see the new and the good. This confusion of worldly choices and free-will, being the ability to choose our belief for our-selves is powerful. And most souls seem prone to choose self-destruction over living according to GOD'S ways. Most of mankind, made in GOD'S image, and having free-will, have chosen self-destruction; as did even one third of the Angels. In prior writings, we touched on the subject of fallen Angels. We see many souls, having our gifted Godlike attributes, who just cannot bow to any eternal superiority. So, it is an amazing point that deserves to often be pondered. By what desire, purpose, and goal, and even logic, can souls who reside in Heaven, beholding GOD, hearing his voice in all his Glory. How do they let themselves be tempted and confused about the truth of reality; being that they were gifted and created by GOD for his familial desires born out of love. Why would a free-willed creature so close to the truth, choose to believe and live a lie? Why choose destruction over eternal love and life? This *warns* me, and I hope it warns *you* of the great powers of self-deception and confusion that comes with and against our free-will desires. This is why I cling tight to my fear of losing GOD. I adore and treasure both my love *and* fear. I walk in my love *and* my fear in all joy. I know love is the good and far greater power; but we see a probability equation of weakness exists in eternally securing the love within all lesser souls below FATHER. This according to their forceful polarity of free-will desires. I do believe FATHER with all his heart, wishes it were not necessary for us to fear him. I believe GOD would so prefer that love alone were all that was needed. But among most lesser dependent creatures GOD provides for, and has gifted with the free-will power of choice, there has arisen rebellion. And rebellion has shown itself to be a contagion. If allowed its way, it destroys all GOD gifts and creates; so, it must soon be stopped and separated. And by this soon appointed separation of the rebellious, and loss of GOD, the fear of GOD born. This natural fear gives strength and wisdom for the gifted to endure, who truly love GOD. FATHER in great sadness accepts our fear, seeing it needful, good, and useful to save and hold our souls to life. We are commanded to keep our fear with our love close for our well-being. The fear of losing any love, is a great true part of every love. I say this with all love and desire to see you live eternal in joy. As GOD has said, be very afraid for your soul. We know every force of sin adds destruction to GODS creation; always has and always will. GOD will not allow sin into his coming New Kingdom. So, the only way reprobate souls can ever be saved, is for FATHER to create for each one a world of their own, and make them the GOD of it. Which would be against all righteousness and Holiness. We should see this power of free-will gifted us by GOD, as the power to carelessly eternally destroy ourselves. GOD gave us choice, to see what we would desire to do with it. This power was never meant for us to do with as we want forever. By all truth, we are meant to choose to give that control back to GOD, who knows our eternal needs. We are *in*sufficient, and the truth of life will always be FATHER'S, by his great forceful wisdom. Remember that Biblically and Spiritually, we can go only one of two ways: either we surrender ourself to be interpreted by GOD'S WORD, or we will interpret GOD'S WORD to suite ourself. We can choose self by intention or uncaring or false knowledge or ignoring knowledge or apostasy or idolatry or confusion or deception or lustful self-indulgence. But not by FATHER'S true eternal love. Romans 1:24-25, "*WHERFORE* ***GOD*** *ALSO GAVE THEM UP TO UNCLEANNESS THROUGH THE LUSTS OF THEIR OWN HEARTS, TO DISHONOUR THEIR OWN BODIES BETWEEN THEMSELVES; WHO CHANGED THE TRUTH OF* ***GOD*** *INTO A LIE, AND WORSHIPPED AND SERVED THE CREATURE MORE THAN THE CREATOR, WHO IS BLESSED FOR EVER. AMEN.*" Do not be confused, CHRIST has not suffered in vain, and mankind has fallen from grace. Our LORD came to redeem and restore those who love GOD, back on high, from which we all fell. GOD has *not* suffered, so we can remain in the state into which we fell by the desires of the flesh. The fallen state of being an individual, is in hells confusion, separated from GOD'S truth, love, and ways of life. Individual's, exist unto them self in a belief of life, separate and apart from GOD. CHRIST has born such great suffering and paid such a great price to free us from that bondage and separation. We are called to become *one* again in FATHER, CHRIST, and HOLYSPIRIT as a child. There is confusion in the Church, as it bows to worldly pressures. Consider the Holy Ancient Religious Institution of Marriage, ordained by GOD, thousands of years ago. The souls of this world have usurped GOD'S rightful rule, and the Church allows it. A Minister who is licensed by a recognized religious institution, is allowed by law to marry those who have a government license to do so. Ministers refuse to marry without a license from their State, as it carries a fine, and is a misdemeanor in some states. And by that same unrightful Government control over marriage, we suffer rampant no-fault divorce, and abominable forms and defilements of marriage. I stand before all souls, and declare only a man and a woman as GOD has made them, can in a ceremony among brethren, declare their solemn vows to GOD to honor one another. And by that, and only by that, will GOD make them one flesh. If you are not married in the eyes of GOD, you are not in a truly sanctified marriage. Marrying without a license is not recognized by Government; but in bowing to government, it is recognized by GOD. Does GOD recognize unholy forms of marriage? No. and so, neither do I. Does GOD honor the words, by the power vested in me by the state of such-n-such, I declare you man and wife? No. And neither do I. And the list just gets more perverse, and vile from here. So, I stand with GOD, in not endorsing such; I also have zero concern or desire to debate the matter with anyone who disagrees with GOD. Amen.

THE LAST DAYS OF GREAT APOSTASY

The term "*apostate*" was coined back around the fourteenth century. At first, it was a strict narrow term, referencing a soul who'd abandoned all actionable faith in GOD. That definition was later expanded in the fifth book of the Decretals of Pope Gregory the IX, to include the broader term "*apostasy*". The broader definition of apostasy, then referred to each act of disobedience to every lawful command of GOD. Each willful disobedience to GOD is then a *measure* of one's apostacy. And anyone unable to overcome a certain sin, is a servant to that sin; living in a certain apostasy. In Scripture, we read how in GOD'S Children, being the church body, apostasy crept back in after every correction brought upon them by GOD. It seems to be a cycle of GOD punishing the rebellious till they turn back to him, and then they quickly lose their zeal for obedience and faithfulness, and inch back into their desires for the flesh and the world. Can mankind, through any *individual* desired beliefs, hold any true substantive faith in the one true GOD? No. True faith is born in, by, of, and towards GOD'S truth, belief, and eternal purpose. So, no soul can hold true faith in the true GOD, born of their own imagined belief and purpose. We must all strive to live in GOD'S belief and ways, and never try to weaken or redefine GOD'S ways. We must share in GOD'S belief and purpose to have a true viable faith in them. Mankind easily finds faith in false beliefs they have ascribed to GOD. Any soul who does not know and agree with GOD'S one true belief and purpose, yet has faith in a self-defined god; their faith rests in a god created by their desired imaginations. Apostasy meaning sin, brings division among believers in the forms of religious sects, splinter doctrines, and denominations. These can only appear by a lesser imperfect understanding of GOD'S commands, statutes, and guidance. In all these divisions of belief, there is a growing spectrum of accepted levels of apostacy. Do not take offence at my words unless you can show me why FATHER, having one absolute truth, would want his children divided into many different sub-truths. 2nd Timothy 3:16, through 4:4, "*ALL SCRIPTURE IS GIVEN BY THE INSPIRATION OF* ***GOD****, AND IS PROFITABLE FOR DOCTRINE, FOR REPROOF, FOR CORRECTION, FOR INSTRUCTION IN RIGHTEOUSNESS: THAT THE MAN OF* ***GOD*** *MAY BE PERFECT, THROUGHLY FURNISHED UNTO ALL GOOD WORKS. I CHARGE THEE THEREFOR BEFORE* ***GOD****, AND THE* ***LORD JE'SUS CHRIST****, WHO SHALL JUDGE THE QUICK AND THE DEAD AT HIS APPEARING AND HIS KINGDOM; FOR THE TIME WILL COME WHEN THEY WILL NOT ENDURE SOUND DOCTRINE; BUT AFTER THEIR OWN LUST SHALL THEY HEAP TO THEMSELVES TEACHERS, HAVING ITCHING EARS; AND THEY SHALL TURN AWAY THEIR EARS FROM THE TRUTH, AND SHALL BE TURNED TO FABLES*." There is one GOD, holding his one doctrine of desire and mercy that saves, and there are many weak believers desiring to alter his doctrine. The defining bright line between false doctrines, born of apostasy, and the doctrine of FATHER is this; it is crossing the line where one no longer believes that mankind is called and able to walk perfect in all his commands. But instead, believe that not only can we not do this, but FATHER accepts that the saved cannot do this. With all GOD'S spiritual love, sacrifice, gifts, and strength, offered to us; it is abominable to falsely stand before GOD as if clean, if we are not filled with their power over sin, born of eternal love. And any further descriptions from this point on, can only describe the severity or level of one's apostasy. There is no end to the weakening changes to the commands of GOD, one can apply and accept within their heart and mind, once they accept an impure weak foundation. Satan told Eve and Adam something less, and that they did not have to align with GOD in belief and truth; yet they would not surely die. Accepting *any* sinful belief in one's self, forces a soul to accept all other souls who have their own individual sinful beliefs. Apostates must then accept that *all* false doctrines are under GOD'S grace, or be hypocritical. Who accepts this? And if you accept this, how extreme or minimal can your acceptance of apostasy be according to your belief? And why do you think it will be accepted by GOD? Let's visualize the far end of apostasy. Most older professed Christians have heard a profession of faith in JESUS from the mouth of a man or woman, who say by their professed belief alone they are going to Heaven. But beyond hearing a verbal profession, they saw a person who hurts, takes, threatens, and abuses. They saw one unfaithful to all that soul said they love, being a foul-mouthed drunkard and drug abuser, lacking any compassion, a soul feared and avoided by their own family. Here is a professed believer, claiming salvation. Is this level of apostasy accepted by FATHER in your belief. 2nd Timothy 3:1-5, "*THIS KNOW ALSO, THAT IN THE LAST DAYS PERILOUS TIMES SHALL COME. FOR MEN SHALL BE LOVERS OF THERE OWN SELVES, COVETOUS, BOASTERS, PROUD, BLASHEMERS, DISOBEDIENT TO PARENTS, UNTHANKFUL, UNHOLY, WITHOUT NATURAL AFFECTION, TRUCEBREAKERS, FALSE ACCUSERS, INCONTINENT, FIERCE, DESPISERS OF THOSE THAT ARE GOOD, TRAITERS, HEADY, HIGHMINDED, LOVERS OF PLEASURES MORE THAN LOVERS OF* ***GOD****; HAVING A FORM OF GODLINESS, BUT DENYING THE POWER THEREOF: FROM SUCH TURN AWAY*." CHRIST'S love, sacrifice, reason, purpose, and commands, call us to overcome all sin; and we are given the power thereof, they deny. To an apostate, their profession alone is their only form of godliness. What about a professed believer who is quiet and polite who would not hurt anyone physically, and is in church every week; but has a problem with alcohol, drugs, lying, and cheating souls and government out of money to survive? Is that level of apostasy acceptable to FATHER? What if a professed believer has only one fault they refuse to confront and overcome in their love for GOD; is that level of apostasy acceptable to GOD? If you say yes, does it depend on the fault? Surely if they can walk perfect in all the commands except one, being thou shalt not kill; do we expect GOD to welcome all otherwise good serial killers? Do not play games with FATHER'S word, as all sin is destructive. We must learn and understand why and what sin destroys, and put away all offense to walk perfect according to our love for GOD. What causes all faults of weakness and apostasy? It is always one's desires for the flesh that manifest in pride, envy, and unrighteous anger, and a spiritual weakness of love for GOD'S ways. These words apostasy and apostate, do not appear in Scripture; but their meaning does. Their meaning arises, just outside the called, chosen, and faithful children of GOD; and define every sin and sinful believing and non-believing soul that exists. Every accepted and believed subversion to the full truth and love of GOD in the hearts of individuals, creates, defines, and is, a false pagan belief. We see a spectrum that runs from false professed Christians, to the most ardent atheist. A soul could extensively study the Bible in the flesh, and have in their mind a firmly held understanding born according to their individual interpretations and desires. They could live out that belief to the letter, every minute of their life; yet gain nothing. Their belief was formed in playing a god, it was *they* who defined the life entrusted to them. Most souls desire and trust the world more than FATHER for their needs; so, they seek preachers of like mind. These claim grace in physical impurity and a false love for GOD. Isaiah 30:1-3, "*WOE TO THE REBELLIOUS CHILDREN, SAITH THE* ***LORD****, THAT TAKE COUNCEL, BUT NOT OF ME; AND THAT COVER WITH A COVERING, BUT NOT OF MY* ***SPIRIT****, THAT THEY MAY ADD SIN TO SIN: THAT WALK TO GO DOWN INTO EGYPT, AND HAVE NOT ASKED AT MY MOUTH; TO STRENGTHEN THEMSELVES IN THE STRENGTH OF PHA'RAOH, AND TO TRUST IN THE SHADOW OF EGYPT! THEREFORE SHALL THE STRENGTH OF PHA'RAOH BE YOUR SHAME, AND THE TRUST IN THE SHADOW OF EGYPT YOUR CONFUSION.*" In Scripture, Egypt and Pharoah are analogous with worldly rule and provision; and it is as true today as then. If any seek, trust, and hope, in worldly church leaders and Government for their truth and provision; they are confused and hopefully can still feel shame. If you cover yourselves with false unsound doctrine, you are covered with false hope. If any seeks the light from mankind, and not HOLYSPIRIT, they will remain in darkness. Do you believe we all can disagree about many things of faith, and all go to Heaven? One aspect of GOD'S punishment for apostasy, seems a bit unusual, in that when GOD foretells of punishing apostasy, he reserves his harsher disdain for hypocrites. If one totally denies every word of GOD and curses and calls those who believe in GOD fools, all the days of their life. GOD says the eternal anguish of such who honestly and openly hate him, will be less than those who try to get only as close to GOD as is physically comfortable for them; these act as if GOD is vain and a fool. If we are commanded to go a certain way and a certain distance to find eternal life. We cannot go another way or less distance. This is an added insult to CHRIST'S mercy and Crown. Outright unmitigated rejection of GOD is less offensive, being of one mind. Where dwelling in forms of religious apostasy, reek of insane contradictions and double standards. GOD sees each foolish attempt in these souls to deceive themselves, GOD, and others. Scripture says, be ye hot, or be ye cold, (good or evil), but do not be lukewarm (part good, part evil). Do we see levels of apostasy in work-mates, friends, family? How about Christian friends in your Church, or your Pastor? How about in the mirror? If we call this being judgmental, refusing to see it; how can we ever overcome it? The good doctrine of judging, knowing, avoiding, and excluding sin in our lives is commanded in Scripture. But all final judgment (meaning sentencing for sin) was taken from mankind because of their own sin, and given to CHRIST, and reserved to GOD alone. Judge not, lest ye be judged. JESUS said, let he who is without sin cast the first stone. JESUS did not say, do not see or know or care that one has sinned against GOD. The apostate church has reversed the truth, teaching that you cannot judge whether something is a sin, or if someone is a sinner; this is insanity. Does GOD see acceptable levels of apostasy within our hearts. If so, what might be acceptable…10, 20, 30% apostasy? How about 49%? This is at least two percent more-good than bad. But again, this is the Luke-warmness we must not allow. We cannot dwell in, or accept any level of apostasy. Apostasy grows inside a soul if allowed. No one can deny that apostasy, or sin, is the effect of the lust for this world which wars against our Soul. Where can, or how does GOD'S true eternal love, accept the willful hurt to others? Which is what every tiny sin does. It hurts and destroys; and that is why GOD can never accept it. The difference between the one pure truth authored by GOD, and all weak false reinterpretations of his truth, is as great as the difference between a candle that has been blown out, and the sum of light filling the whole of creation. We see from the fall of Adam in Genesis, to the last trump sounding in Revelation, there are pure consistencies; being that GOD needs to love and to be loved equally, and tries to empower us to overcome all desires and temptations towards the flesh and the world. And we see that in doing this, we do not hurt GOD, each other, or his creation. We hear a consensus among the professed believers who preach a false divide between the Old and the New Testament. These declare there has been a change of direction, instruction, and intention, from GOD; meaning GOD has changed. I see no change in Scripture, or in GOD. I see only continuity in all things with a faithfulness to eternal love. GOD has continually given more and more, until they have offered and given all they are and have. We are to walk perfect according to GOD'S ways, by which, all gifts and creations were accomplished for GOD'S pleasure. The Revelation of JESUS, says in 21:5-8, "*AND HE THAT SAT UPON THE THRONE SAID, BEHOLD I MAKE ALL THINGS NEW. AND HE SAID UNTO ME WRITE: FOR THESE* ***WORDS*** *ARE* ***TRUE*** *AND* ***FAITHFUL****. AND HE SAID UNTO ME, IT IS DONE. I AM* ***AL'PHA*** *AND* ***O'ME-GA****, THE* ***BEGINNING*** *AND THE* ***END****. I WILL GIVE UNTO*

HIM THAT IS ATHIRST OF THE FOUNTAIN OF THE WATERS OF LIFE FREELY. HE THAT OVERCOMETH SHALL INHERIT ALL THINGS; AND I WILL BE HIS ***GOD****, AND HE SHALL BE MY SON. BUT THE FEARFUL, AND THE UNBELEIVING, AND THE ABOMINABLE, AND MURDERERS, AND WHOREMONGERS, AND SORCERERS, AND IDOLATERS, AND ALL LIERS, SHALL HAVE THEIR PART IN THE LAKE WHICH BURNETH WITH FIRE AND BRIMSTONE: WHICH IS THE SECOND DEATH.*" GOD accepts no sin. In the above, all things new, means the coming New Heaven and Earth, yet GOD does not change. Scripture says and teaches us over and over that GOD cannot change from their shared perfection. Yet many say GOD has changed towards evil, by accepting evil, and no longer expects us to be perfect before him. This is refuted over and over by the Apostles. Scripture shows that our brothers Mathew, Mark, Luke, John, Paul, Peter, James, Jude, and the rest, in so many ways profess the continuation and fulfillment of all prior Scripture by JESUS. From that first act of creation, FATHER taught his Children that our obedient faith towards his truth is our foundation of love for him. True love and faith are the foundation that perpetuates eternal life. The early Israelites, weakened in their love towards GOD; but tried to hang onto GOD'S promises, by relying on only one of three strong elements given them by GOD. They drifted into loving less, in a loose worldly faith; and relied on physical obedience to the animal sacrifice to receive GOD'S promise; but we read FATHER was sorely displeased by that. The LORD came to call all Israelites back to all three: faith, love, and obedience. And our Savior elevated, perfected, and fulfilled, the Blood Sacrifice to its new unknowable heights. This to give *us* a great strength and love, and to bring an end to the abusing the animal sacrifice. But the Children are now once again backslidden by their own lust. And lured astray into once again, making the same mistake of moving from a loving faith to live every word, and again hanging onto and abusing the Blood Sacrifice alone; and such an abominable abuse it is. When we see how it displeased FATHER, that mankind abused the animal sacrifice, to keep committing sin. Who can imagine the level of righteous anger FATHER will have for souls who abuse *his* SON'S Sacrifice to sin again. Scripture teaches us from early in the book of Genesis, through to the end of the Revelation, that true faith is foundational in pleasing GOD. And that our faith in GOD is only evidenced in our loving obedience. And then and only then is our faith and love furthered in our good works to increase GOD'S Kingdom. But every weak lacking faith, is evidenced by one's acceptance of apostacy. Scripture teaches us a work is only good if rooted in all three: faith, love, and obedience, to GOD. By this we see physical works alone, not born of loving obedience, are not good or accepted by GOD. Cane, in his sinful pride, gave his best physical gift, and GOD rejected it. Abel gave his best physical gift in his greatest love and humble obedience; and GOD received it. So, faith and love without works through our obedient love, are dead and untrue. But physical works, without loving truthful obedience, are also dead. JESUS showed us the wonders of Heaven in his obedient fulfillment of the Law and Prophecy. He did this to reverse the evil that had crept in by mankind's traditions and lustful worldly temporary perspective. CHRIST brought to the true Children of GOD, all the forgiving power, love, and the most perfect example that can be given. And he brought us his command to emulate his life, and overcome and endure until the end. JESUS fulfilled, corrected, and brought the perfect truth and power to all who can hear and desire it. The time and time again, of mankind's return to self and sin, we read of throughout Scripture, makes us wonder why GOD continues trying to save as many as possible from destroying themselves. All but eight souls were completely consumed by apostasy and vile lust, prior to the flood around four thousand years ago. But in GOD'S love, finding just eight righteous souls, he was moved to do miraculous things. FATHER repented gifting and creating mankind, yet did not give up on them. Scripture *is* the greatest love story of all time; and as in all love stories, it is filled with tragedy, betrayal, and eternal loss, for such is the greatest cost of eternal love. If reality were a Hollywood movie, it would surely have a happy ending, with everyone seeing their error, and turning from their evil ways; but this is not a movie. By the unimaginable sacrifice and unknowable heart break on the part of GOD, we each desire and choose to be a true Child of GOD, or a Child of Satan. But again, GOD accepts no mixture of the two. A deceitful heart will minimize and justify its own sin by all subtle means available; many weigh it against what pain will come upon them and others, if they do not lie. Some see it a necessity to lie to live and keep peace in this world. So, half-truth, or hiding unfaithful deceit, is seen as a good standard for now. We can somewhat agree with that premise; but only *as* it written. The important words in the first sentence are "*trying to live and keep peace in this world*." This is why we are told to come *out* of this world, and be a separate and peculiar people. So, we do not partake of her sinful ways. If you *very* rarely sin, and each time you do, it brings you great true long-lasting shame, and you strive too never sin again; forgiveness awaits. But that forgiveness does not come in making allowance for the lust and sin souls do not wish to turn from. Apostasy finds many doors into the soul: there are doors of refusing to surrender pride, refusing to surrender traditions, a door of refusing to surrender physical luxury and abundance. We see a big door of refusing to avoid people and things of evil influence, and a refusal to reject the fallen world at large. These all fall under the refusal to live within GOD'S eternal created ways. Many enter ongoing apostacy, because they do not yet know what they firmly believe. Do Apostates believe they are not under the law by love and faith? Do any believe they need not obey GOD? Do any not believe the laws are the commandments of GOD? Do any believe we must live by all 11 commandments or 10, or 9, or are most down to 8, 5, 3. Are most in their own *form* of Godliness? I ask all to consider, how could a perfect GOD, be fulfilled with children who do little more than believe he exists? Who give GOD a lesser place in their lives as they gather to sing and give thanks in pursuit of physical comfort, pleasure, and security, lacking all Godliness.

THE LAST DAYS OF THIS DISPENSATION

In true Judeo-Messianic-Christian parlance, we use the formal term of "dispensation" to set apart the 3 great eras seen as defining three great interventions of correction by GOD towards mankind. And GOD never once weakens or alters or compromises their belief and intention, throughout these three dispensations. With each of GOD'S subsequent interventions, we see great sacrifice and work on the part of GOD, trying to reclaim as many souls as possible. All according to the great sin by mankind, and their true repentance. Scripture defines three great eras of around two thousand years each, and one soon coming era of a thousand years; appearing in that order. Let's first define an important aspect of our third current era, known as the era of CHRIST'S Grace. All interpretations of this grace, are at the heart of all denominational differences. And with the term grace, we define another Christian parlance, being legalism. Grace *and* Legalism; what do these have in common in Scripture. Well both terms are most often greatly mis-used, mis-interpreted, and have become great weapons *against* uneducated minds, seeking GOD. If you believe in GOD; and yes, all fallen angels, now being eternally lost, do truly believe. So, if you believe in GOD, but love yourself, family, friends, or the world too much, having a lesser love for GOD; yet you obey every word of GOD without fail, believing that will get you into Heaven, that is legalism. But in fulfilling the physical aspects of the law alone, we are not saved. We were *born* in sin, and salvation is not obtained by works *alone*, it is truly a sacrificial gift from CHRIST. A gift given to those who return GOD'S love in their highest obedient measures. Grace is a gift CHRIST gives to those *lovingly* obey GOD'S laws and ways. It's not given for a work of following the law alone, or asking forgiveness alone, but for our loving desire to live for GOD. Do not be confused on the matter of eternally being under GOD'S laws of love. Let no one confuse you about the last days of this third dispensation, or your spiritual preparation *for* these days. Scripture warns and commands us to prepare for the worst things imaginable. Most professed Christians agree in words and thoughts that we are now in the last days, nearing the Tribulation; but these do not live like they believe. Many believe in error, they will be taken out of this world prior to the Tribulation. Scripture says, in the last days, nearly all souls existing on earth at that time, become lost in their endless iniquity, and earth will be a thoroughly polluted kingdom, set for the fires of destruction. But Scripture says GOD'S final reaping of earth, occurs just prior to the Battle of Armageddon at the end of the Tribulation. Scripture warns that throughout the last days, we will see false doctrines and their practitioners on an unimagined scale. What would best show GOD the great loving desires of your soul? Is it prayerful acceptance of all others? Or is it in the Godly ways in which you live? Can you say you believe in, and love MESSIAH, yet not obey him? If honest, you must see how this falls far short of any eternal loving belief, and it is a form of legalism. Can purposed thoughts and spoken words out of our mouth alone, declare our true love and belief? If you believe a child walking towards a cliff is about to fall to their death; is your belief evident by saying aloud that you believe it, or in your desperate act to stop it? Scripture teaches we must search for and live out the full truth of GOD. If any think GOD'S holy righteous ways of life are *not* a pre-requisite to the promise of eternal life, and believe that all that is required are prayerful words of thanks and belief, and a profession of CHRIST'S name over their ongoing endless sins, they are truly deceived. Eternal life is a Spiritual journey, and not a physical one, in which HOLYSPIRIT teaches and corrects our understanding of reality. We must circumcise our heart, mind, and body, in our Spiritual growth of building a Temple of HOLYSPIRIT. How is calling ourself a Christian, and attending Church, minus our great desire for personal holiness and righteousness enough? Is our salvation determined by what we say we believe; or in our living to emulate CHRIST? Does the power of salvation rest within us and our words and physical works?

Are we all saved by hearing and answering GOD in our own individual ways? Or does grace come at a true calling, a true hearing, and being truly chosen by CHRIST, who sees true repentance in our turning from and overcoming every sin? It is not in saying we are sorry we are constant sinners. Rev. 17:14, *"THESE SHALL MAKE WAR WITH THE* ***LAMB****, AND THE* ***LAMB*** *SHALL OVERCOME THEM: FOR HE IS* ***LORD*** *OF LORDS, AND* ***KING*** *OF KINGS: AND THEY THAT ARE WITH HIM ARE CALLED, AND CHOSEN, AND FAITHFUL*." Here we see those with the LORD, described by *three* terms. Being many are called, but fewer chosen, and fewer still remain faithful. Are all that are called coming with JESUS. Mathew 22:14, "*MANY ARE CALLED, BUT FEW ARE CHOSEN*". If we truly believe in GOD, and that JESUS has the loving right to save us. Then we hear the call to repent, and are called. But we see those coming with JESUS are not just called; *all* that are called, are *not* saved. Of those called, there are those who are chosen for GOD'S righteous holy reasons of eternal love. Then we see a third shared attribute, being they are faithful. CHRIST'S Twelve Apostles, were *chosen* from among all those *called* by GOD. And we know JESUS said, I have *chosen* you twelve, yet still one of you is an unfaithful devil, being not faithful to the end. We see one named Judas Iscariot, who was both *called* and *chosen*; yet he was not *faithful* to the end, and fell away. If you think this does not apply to us all, you are wrong. Only by a faithful obedience to GOD'S every word, can HOLYSPIRIT be effective through us in his work to sow and increase GOD'S Kingdom. By every false belief and disobedience, GOD'S Kingdom is hurt and diminished. Our every disobedience, means someone may not be saved. How can one live in ungodly ways, knowing they are seeds of destruction? I want to live in ways that perpetuate *and* increases GOD'S Kingdom. Please, please Children, live in GOD'S eternal ways, love GOD above all things so you will be chosen, and continue in loving obedience, being faithful. Do not live according to any ways of this world. Seek holiness and righteousness, and GOD will be pleased. Every tiny sin is in league with death and hell itself. If we know these truths, and know there is only one Holy GOD, we do well. Scripture says all indwelled by HOLYSPIRIT are CHRIST'S, and only they are heard by HOLYSPIRIT, FATHER and JESUS; and like-wise, GOD spiritually communes with them. FATHER reveals and teaches us so many things through CHRIST and HOLYSPIRIT and Scripture. JESUS said to Pilate in John 18:37-38, "*EVERY ONE THAT IS OF THE TRUTH HEARETH MY VOICE.*" Pilate answered, "what is truth".

Being the second greatest existential questions ever asked; number one being, is GOD loving? FATHER alone has authored the truth of everything, in that he defines his created purpose and reason for all reality. So, by HOLYSPIRIT'S indwelling, we earnestly see and hear and understand eternal life through our new eternal perspective. And we lovingly accept and live in the holy righteous ways in which FATHER has brought his creation and us in his divine purpose. We have all heard that GOD *can* do all things; but what does this mean exactly. Surely it does *not* mean that GOD *will* do all things. Scripture says GOD *cannot* sin, being contrary to who they are. GOD cannot cause all souls to stay in their good ways and desires, once he gifts us as creative free-will entities; unless he rescinds free-will. Free-will, means a complete and free choice of desire; not partial or sometimes, but always. Can FATHER say let there be up, or right, and yet no down or left; where then is any meaning or purpose. Each one is defined and exists, and has meaning, only by its opposing contrast to the other. Can GOD define a good and loving way for all free-will souls, and there not be free-will contrary evil ways, appear on their own? No. How can we be defined as eternally good, except by never having an evil desire towards GOD'S ways, beliefs, and highest form of love. We see among FATHER'S Creation, and gifting many free-will souls, that there can never be love, without hate also arising; or light without darkness; or truth without lies; and we see all these do exist by free-will contrary desires. Can FATHER gift souls to exist within his shared physical reality, and find something as great as love, without free-will? No. So what do we conclude? We see FATHER defines and knows all that is true and good, who in his essence, is everywhere, seeing his universe he made out of his spoken forceful substantive words. FATHER is unimaginably more forceful than all other souls and creations combined. FATHER is omniscient, omnipresent, and omnipotent in all good things and ways pertaining to us eternally. FATHER was before all things, so, it was he who gifted and created all conscious and physical things. And FATHER can end all physical things as he sees fit. Please see why FATHER'S conscious and physical laws of reality, must apply to him and his gifts and creations as well. When FATHER was one and alone, the laws of all reality were within him alone. But the conscious uncertainty laws that arise between many free-willed souls, now differ. Yet, FATHER'S immutable laws of love, truth, righteousness, and holiness, can never change. FATHER operates within all his created laws without fail. Can we know the full restrictions of free-willed reality that FATHER yields to, being the many-things he will never do, and so cannot do; seeing they are the boundary of all good realities and creations. We may best say GOD can never do certain things, because they are divisive, destructive, and not of the eternal loving desires and attributes of GOD. So, we all can learn and know and eternally rely on these set boundaries of GOD'S unchanging goodness. Have no doubt, FATHER was the source and giver of your soul, and he allows us our every free-will choice in this life. So, let's understand exactly what Scripture means, when it says GOD can do all things. Let us always learn, pray, respect, and expect the good things of GOD, but only in his Holy ways, times, and Righteous abilities. And only in their proper eternal fashions and contexts, according to GOD'S loving purpose and desires. Beyond this, all our desires and expectations arise out of our error and apostacy of pride and envy. We are thankful for FATHER'S gifts of love, truth, and life, above all things, in that he sacrificed to gift and create us out from within himself. Do you see how FATHER forcefully perceived his great firm works of creation *for us*? The bodies to which we are adjoined, are fashioned not from the substance of stars, as science says; these bodies and the stars also, are formed from the forceful spoken substance forming FATHER'S every word. We are his to love if we obey all his created laws of returning love; and his to not love and separate if we choose individual destructive ways. And his to despise most, if we try to be and have both. Now, is there something more beyond this creation from earth to heaven and to FATHER himself. We can only be sure that if there is, FATHER is there also. GOD you are our love, our truth, and our only pure light. You are our soul's great true desire and only good hope. Our found everlasting treasure, our true compass, and our greatest need. Our unlimited strength to stand and endure. You alone, are our life's very beginning and our only desired outcome. GOD you alone are our all, and our Holy perfection in your highest love. Have mercy LORD, show us loving kindness. Amen. There is an end coming to all we physically see and have known. And obviously, something new and wonderful for GOD'S Children will start. Revelation, 10:5-7, "*AND THE ANGEL WHICH I SAW STAND UPON THE SEA AND UPON THE EARTH LIFTED UP HIS HAND TO HEAVEN, AND SWARE BY HIM THAT LIVETH FOR EVER AND EVER, WHO CREATED HEAVEN, AND THE THINGS THAT THEREIN ARE, AND THE EARTH, AND THE THINGS THAT THERIN ARE, AND THE SEA, AND THE THINGS WHICH ARE THEREIN, THAT THERE SHOULD BE <u>TIME</u> NO LONGER. BUT IN THE DAYS OF THE VOICE OF THE <u>SEVENTH</u> ANGEL, WHEN HE SHALL*

BEGIN TO SOUND, THE MYSTERY OF ***GOD*** *SHOULD BE FINISHED, AS HE HATH DECLARED TO HIS SERVANTS THE PROPHETS.*" What mankind calls time, there and then loses all its former obscure meaning. And all whose names are not written in the LAMB'S Book of Life, are cast with their new eternal bodies into an eternal unfeeling prison for all imperfection, called the Lake of Fire. By receiving GOD'S fullness of grace in growth, we find our ongoing change from apostasy to righteousness as we learn and turn from all evil. And upon the completion of its purpose, all the ways, burdens, and cost, of this third dispensation will be set aside as fulfilled. All souls not firmly set in their choice, will be tried further in CHRIST'S Millennial reign. This will not include those who refused to decide, as by a desire to not choose, their decision against GOD was made. But the truly unknowing undecided will enter the final Millennial dispensation, to then and there, sense, consider, and choose between GOD or themselves and the world to be set and everlasting. No one in Heaven will miss the imperfect bondage called time. It has been a cruel master, a chain, a tempter, a cage, and a confused prison. The delusion called time has been a big part of the process of hell and death in the lost, a process from which we are freed by JESUS. HOLYSPIRIT teaches us the mind of CHRIST, and from our first search for truth, we should be filled with fear; not the fear of this world or of evil, but the fear of any weakness or lie within our self. And fear GOD'S final separation, reserved for those who fall into the temptations and ways born of all such weakness. We must fear losing HOLYSPIRIT; without whom, we have nothing. Outside of seeking for GOD'S truth, love, and mercy, we ask most for GOD'S help to not be deceived, and lead us to our strength and patience to wait for his wisdom. Many Scriptures say HOLYSPIRIT is given as our teacher of the mysteries of reality, and interpreter of GOD'S every word. Scripture warns of souls desiring to hear a path filled with compromise that allows their physical desires; these many variations in the church, are not of GOD. So, as commanded, we are to in fear, settle in our mind and heart very early that we will not be instructed in any final way, by any man, in waiting on GOD'S Truth. As a witness, it has been more than a few decades since I received my first revelation. And I do certify, that no deceitful words from any false science, pulpit, or government, has found entry into my understanding of, and from GOD. The world has not raised a single question or doubt in the perfect foundational truth that has been taught to me by HOLYSPIRIT. It has only been built upon and made stronger. It surrounds me as an impenetrable fortress. Prov. 2:1-9, "*MY*

SON, IF THOU WILT RECEIVE MY WORDS, AND HIDE MY COMMANDMENTS WITH THEE; SO THAT THOU INCLINE THINE EAR UNTO WISDOM, AND APPLY THINE HEART TO UNDERSTANDING; YEA, IF THOU CRIEST AFTER KNOWLEDGE, AND LIFTEST UP THY VOICE FOR UNDERSTANDING; IF THOU SEEKEST HER AS SILVER, AND SEARCHEST FOR HER AS TREASURES; THEN SHALT THOU UNDERSTAND THE FEAR OF THE ***LORD****, AND FIND THE KNOWLEDGE OF* ***GOD****. FOR THE* ***LORD*** *GIVETH WISDOM: OUT OF HIS MOUTH COMETH KNOWLEDGE AND UNDERSTANDING. HE LAYETH UP SOUND WISDOM FOR THE RIGHTEOUS: HE IS A BUCKLER TO THEM THAT WALK UPRIGHTLY. HE KEEPETH THE PATHS OF JUDGMENT, AND EQUITY; YEA, EVERY GOOD PATH.*" GOD speaks here of the spiritual equity of hope, not worldly possessions or outcomes. This is not an unusual scripture, it is like a multitude of others that say if we seek GOD'S truth with great excitement and desire in our Righteous-Holy-walk, and do not seek the treasures of this world, HOLYSPIRIT will guide us to all great mysteries and truths. Scripture says our true Baptism, Spiritual Resurrection, and indwelling of HOLYSPIRIT, are not gifts received upon our profession of faith or belief, but upon a higher standard of our loving faithful responsible obedience. Before receiving HOLYSPIRIT'S strong and sure indwelling. Early in our true repentant work, we sense HOLYSPIRIT'S distant calling and drawing. Our stronger recognition by GOD, comes only at our true repentant turning from our sin and accepting CHRIST as our Master. In this we are called, drawn, observed, and helped by HOLYSPIRIT, who comes closer and closer in his gradual growing trust in us. The indwelling HOLYSPIRIT, comes only at our strong point of knowing obedience to eternal love. Now we should see clearly, and learn quickly that a devils most fruitful place to do his work of deception, is not in pubs, taverns, or brothels, or dens of mass murdering terrorist. Satan's fervent efforts, are in misrepresenting the words of GOD in centers of worship in all corners of the earth, just as he did with Eve. And devils always find hearts willing to water down, soften, and corrupt GOD'S truth; who then share, or sell it to weak covetous buyers. HOLYSPIRIT does not find nearly so many to stand up, restore, and affirm the purity and strength of CHRIST. Professed Christians seek out doctrines of belief from unreliable sources, accepting great contradictions to Scripture. We are commanded to speak out when we see this, *and* must warn all according to Scripture. Ezekiel 3:17-21, "*SON OF MAN, I HAVE MADE THEE A WATCHMAN UNTO*

THE HOUSE OF ISRAEL: THEREFOR HEAR THE WORD AT MY MOUTH, AND GIVE THEM WARNING FROM ME. WHEN I SAY UNTO THE WICKED, THOU SHALT SURELY DIE; AND THOU GIVEST HIM NOT WARNING, NOR SPEAKEST TO WARN THE WICKED FROM HIS WICKED WAY, TO SAVE HIS LIFE; THE SAME WICKED MAN SHALL DIE IN HIS INIQUITY; BUT HIS BLOOD WILL I REQUIRE AT THINE HAND. YET IF THOU WARN THE WICKED, AND HE TURN NOT FROM HIS WICKEDNESS, NOR FROM HIS WICKED WAY, HE SHALL DIE IN HIS INIQUITY; BUT THOU HAST DELIVERED THY SOUL. AGAIN, WHEN A RIGHTEOUS MAN DOTH TURN FROM HIS RIGHTEOUSNESS, AND COMMIT INIQUITY, AND I LAY A STUMBLINGBLOCK BEFORE HIM, HE SHALL DIE: BECAUSE THOU HAST NOT GIVEN HIM WARNING, HE SHALL DIE IN HIS SIN, AND HIS RIGHTEOUSNESS WHICH HE HATH DONE SHALL NOT BE REMEMBERED; BUT HIS BLOOD WILL I REQUIRE AT THINE HAND. NEVERTHELESS IF THOU WARN THE RIGHTEOUS MAN, THAT THE RIGHTEOUS SIN NOT, AND HE DOTH NOT SIN, HE SHALL SURELY LIVE, BECAUSE HE IS WARNED; ALSO THOU HAST DELIVERED THY SOUL." This is said to all. Many souls, hear, repeat, and profess false beliefs; and the most deceitful of all professed beliefs being, that a simple profession of belief that JESUS is the SON of GOD, ensures salvation. Any belief that absolves mankind of responsibility and faithfulness to find GOD in love and fear is false. GOD sees our belief and love through our obedience to every word in Scripture; and this is GOD'S standard. All through the Bible JESUS says he loves those who love and obey FATHER. Scripture says the fallen angels know JESUS is FATHER'S only begotten SON. yet they tremble, because they know they do not lovingly obey him. We must resist religious pollutions, as people are deceived, and our hearts break at knowing this. We should not be able to contain our desire to cry out and warn. Many are lost to the deceptions of souls who comes as angels of truth; but they lie. This is not to insult, but to warn; an unlearned child who knows no better, is happier with a bag of gold foil covered chocolate coins, than real gold coins. The unlearned child will trade a bag full of real gold coins, whose value they do not know, for the chocolate coins, whose value they do know. In this same way, unlearned children trade the truth for a lie, especially if the lie says they are meant to have both heaven and earth, loving and enjoying all the things and ways they see. But the truth of GOD says we can have our ruling love *for* GOD'S Kingdom, or our ruling love *for* this polluted kingdom, not both. I have observed for a long moment, that so many professing a belief in GOD, possess a strange power. Being that within themselves, by the desires of their heart, they reject all soundness, in not seeing soundness. They embrace with desired sincerity, unsoundness, in their not seeing unsoundness. I've wondered for decades, asking by what ability or power do they do this. The power of self-deception has many facets, and a lack of a natural fear of GOD is a strong one. Isaiah 29:11-13, "*AND THE VISION OF ALL IS BECOME UNTO YOU AS THE WORDS OF A BOOK THAT IS SEALED, WHICH MEN DELIVERED TO ONE THAT IS LEARNED, SAYING, READ THIS, I PRAY THEE: AND HE SAITH, I CANNOT, FOR IT IS SEALED: AND THE BOOK IS DELIVERED TO HIM THAT IS NOT LEARNED, SAYING, READ THIS, I PRAY THEE: AND HE SAITH, I AM NOT LEARNED. WHEREFORE THE* ***LORD*** *SAID, FORASMUCH AS THE PEOPLE DRAW NEAR ME WITH THEIR MOUTH, AND WITH THEIR LIPS DO HONOR ME, BUT HAVE REMOVED THEIR HEART FAR FROM ME, AND THEIR FEAR TOWARD ME IS TAUGHT BY THE PRECEPT OF MEN:*" The truth of GOD in Scripture is sealed to all, without HOLYSPIRIT to unseal it. And a weak taught fear of GOD, is not a strong natural fear. Like fallen mankind's weak love for GOD, they hold a weak fear of GOD. It is a tragic feeling, to witness the natural fear of GOD disappearing from the earth. Yet, it is part of the final separation and days of purification, and the final preparations for the presentation of the Bride to JESUS. I hope you see with clarity, that Scripture says there are the few that love and fear GOD above all things, and the many who deceive themselves, and believe they love GOD enough. 2nd Corinthians 6:17, "*WHEREFORE COME OUT FROM AMONG THEM, AND BE YE SEPERATE, SAITH THE **LORD**, AND TOUCH NOT THE UNCLEAN THING; AND I WILL RECEIVE YOU, AND WILL BE A **FATHER** TO YOU, AND YE SHALL BE MY SONS AND DAUGHTERS, SAITH THE **LORD ALMIGHTY***". Which am I? Which are you? Are you and I, more of this world, or of GOD? JESUS said in Heaven, we will be as the Angels; so, let us desire and emulate that semblance now. As servants and messengers of GOD, be ever fearful of worldly lust finding an advantage to enter your heart and pollute your love. Elevate your search for the truth and love of GOD, and never relent. How serious are you about GOD, and being eternal? Are you half-hearted, are you fervent or passive? Why is there a great number of souls, who with their lips profess CHRIST, but will not obey his laws of love? We must realize those bound to the flesh, have temporary desires, and little eternal hope. Those in the flesh however do have a fearful sense they will physically die, and live no more. Look at a tiny bug with no eternal soul or hope; yet it flees if you bring a threat near it. Why, because it is flesh and has only a physical instinctive desire to continue. The flesh can sense temporary excitement, pain, comfort, hardship, security, and danger, but it cannot know eternal spiritual love. The flesh is attracted to the things that give it temporary continuity. These, and only these things, can the flesh desire. If a soul, being trapped in the flesh, hears the false teaching of an afterlife providing for their every physical desire, don't be surprised if they are attracted and joyous at their eternal physical provision of comfort and security in a place called heaven. But we see that same fleshly desire, will resist all sacrifice on the part of its pleasure in this world, or emulate JESUS to get there. Those in the flesh will joyfully gather together in houses of worship to sing praise, and share some of their abundance. They can conform to teachings, that bring physical health and prosperity; they can care for those in physical need. But that is where most bound to the flesh, tend to draw their religious line of belief. GOD has called and commanded us to *crucify* the flesh. To walk in the truth and love of GOD, as his beloved Children; for therein is found our life eternal.

THE LAST DAYS OF GREAT FULFILLMENT

What does fulfillment mean in Doctrine and Prophecy? It means that in GOD'S eyes, the final touch of completeness, fullness, and fruition, have come to pass. It means it is known by GOD, to have been truly accomplished. At the fulfillment of GOD'S daily and prophesied work of finding his eternal family, that will bring the final loving perfection to GOD'S eternal needs. And being fulfilled, no more cleansing or trying of souls will be needed. All that remains is every soul's eternal continuation in their chosen perfection, or imperfection. What a moment to imagine and long for, that on a coming day and moment, and even at a certain single thought within FATHER, that his Holy work towards finding his family, will be fulfilled and set in its eternal loving rhythms and purpose. But for yet a little longer, the sands of the many things yet to be fulfilled, remain in that top half of the hour glass. And as each grain of sand passes through the narrow now of reality, that fulfillment grows nearer. We create and perceive all things through our current belief of understanding, and we are aware we exist, only within that very narrow now of reality's hour glass. This is where we are, where we have always been, and will forever be. But at that coming moment of GOD'S fulfillment, with every eye in the heavens watching, as that last Prophesied grain of sand bounces down through that narrow now of the hour glass, and falls through against a background of universal silence, it will make a sound never before heard. A sound to be remembered, and never heard again. That last tiny sound of change, will divide and separate all eternal love, from all irreparable rebellion and corruption of creation forever. A sound of change to a new joyous perfect freely chosen sinless reality for us, and an eternal unfeeling restricted imprisonment for the lost, that never allows them to commit a single physical act of sin again. It is hard to imagine the absolute end of so many old imperfect things and ways, and the beginning of all that is perfect. Do you clearly see the Prophesies that have already been fulfilled? Do you see Prophecy being fulfilled now before your eyes? Do you watch for it with anticipation? Well, only if you truly believe it. How firmly or weakly can a soul believe? Does the strength of our belief truly matter? If we faithfully live with HOLYSPIRIT, we will see reality as it unfolds in detail. Without HOLYSPIRIT, no one can properly comprehend anything, living through the destruction coming upon the Earth. Are we allowing HOLYSPIRIT to prepare us for what will occur? Are we not told in Scripture to study, watch, and prepare our soul for the coming day of the LORD? Scripture says very clearly, if we accept the Mark of the Beast, we have prior to our physical death, made an evil vow of worship that seals us as eternally lost. There exists no remedy, once one takes Satan's mark. The Bible also speaks of the seal of GOD being placed upon many in their forehead in the last days, showing GOD has called, chose, and finds them faithful and fulfilled, prior to their physical death. Is GOD'S seal a tattoo? No. The seal of GOD is in HOLYSPIRIT'S fire of sealed spiritual truth and brightness. To GOD'S Angels, these seals are bright signs of shared spiritual goodness. Satan's physical mark outside, is like branding cattle. Isn't it subtle that in all the things that lead up to this mark, that Satan has been getting the world to embrace tattoos, even on their face. Those sealed first by GOD'S Angels in the Tribulation, are described as first fruits; a hundred and forty-four thousand virgins of all the Tribes of Israel. Then we see a great multitude sealed, too large to number from all the nations. When I was young in CHRIST, a great fear preached in most churches in the early 1970's, was the Mark of the Beast, and the beheading of those who refused that mark. That fearful preaching has in great part left most churches, being replaced with the denial and false hope that those who simply profess

CHRIST'S name, will be raptured out before the Tribulation and that Mark. The fear of this Mark, is not gone from all, but it is rare. Do you fear this coming mark of the damned. A mark of accepting and following all evil. And what is the seal of GOD, is it our Baptism and indwelling of HOLYSPIRIT? No, it is something more and after that. We can turn away and lose HOLYSPIRIT, but if we are sealed, we can never fall away. Revelation 7:2-3, *"AND I SAW ANOTHER ANGEL ASCENDING FROM THE EAST, HAVING THE SEAL OF THE LIVING **GOD**: AND HE CRIED WITH A LOUD VOICE TO THE FOUR ANGELS, TO WHOM IT WAS GIVEN TO HURT THE EARTH AND THE SEA, SAYING, HURT NOT THE EARTH, NEITHER THE SEA, NOR THE TREES, TILL WE HAVE SEALED THE SERVANTS OF OUR **GOD** IN THEIR FOREHEADS."* This seal of GOD is given to those living and going through the Great Tribulation, who have shown their devotion, and swore their life unto death for GOD. The Elders, Prophets, the Sealed, and the Martyrs, are in the first resurrection. All other true Christians who died short of this highest standard, wait until the White Throne Judgment, after the Millennial Reign. Please know for CHRIST'S Millennial Reign, those resurrected there are a small portion of all souls who will be resurrected later to eternal life with

GOD. Are we losing our life for CHRIST as commanded, or on balance do we save and live our life for ourself. We are called to hear HOLYSPIRIT rightly dividing GOD'S words. What does rightly dividing the words of GOD mean? Well in great part, it's properly understanding the History of GOD'S efforts to restore mankind, its knowing GOD'S perfect efforts, and mankind's few perfect, mostly imperfect responses. We must never believe mankind's evil responses, are accepted by GOD, as they do all they can to steer us. We must see all the Apostles acts as perfect, or stumbling. The Apostles made early mistakes along their way; we saw JESUS rebuke Peter, we saw Thomas doubt the resurrection, and we saw Judas fall away at his hardest moment. We must hear every word spoken by all Biblical souls, filtered through GOD'S every word being supreme. Do these few things open your eyes to a few aspects of rightly or wrongly dividing Scripture. What will happen if you wrongly divide Scripture? If wrongly divided, we see great loss in Jews being separated from their Messiah; we see worldly gentile Christians, separated from FATHER'S beliefs, desires, and commands. Within the Old Testament, or Torah, FATHER delivers his one true religion. JESUS, a Jew, tells a Samaritan gentile woman in John, 4:22, *"YE WORSHIP YE KNOW NOW NOT WHAT; WE KNOW WHAT WE*

WORSHIP; FOR SALVATION IS OF THE JEWS." The New Testament is FATHER'S greatest gift of love to the Jews; and by adoption, to the Gentiles as well. Examine your worldly choices, and your choices where you truly put GOD'S ways first. A book of our works is being written, but it may be edited to erase our past sins by true repentance. In true repentance, followed by GOD'S forgiveness, our past created perceptions of sin are appointed to be washed by the Blood of JESUS. Would GOD'S Angel say you are nearing the seal of GOD? Or on balance, see more your desires for this world? At our judgment, the books, including our book, will be opened. Yes, there is a book of each life. Revelation 20:4-6, *"AND I SAW THRONES, AND THEY SAT UPON THEM, AND JUDGMENT WAS GIVEN UNTO THEM: AND I SAW THE SOULS OF THEM THAT WERE BEHEADED FOR THE WITNESS OF* ***JESUS****, AND FOR THE WORD OF* ***GOD****, AND WHICH HAD NOT WORSHIPPED THE BEAST, NEITHER HIS IMAGE, NEITHER HAD RECEIVED HIS MARK UPON THEIR FOREHEADS, OR IN THEIR HANDS; AND THEY LIVED AND REIGNED WITH* ***CHRIST*** *A THOUSAND YEARS. BUT THE REST OF THE DEAD LIVED NOT AGAIN UNTIL THE THOUSAND YEARS WERE FINISHED. THIS IS THE FIRST RESURRECTION.*

BLESSED AND HOLY IS HE THAT HATH PART IN THE FIRST RESURRECTION: ON SUCH THE SECOND DEATH HATH NO POWER, BUT THEY SHALL BE PRIEST OF ***GOD*** *AND OF* ***CHRIST****, AND SHALL REIGN WITH HIM A THOUSAND YEARS."* A few verses later. Revelation 20:11-15, *"AND I SAW A GREAT WHITE THRONE, AND HIM THAT SAT ON IT, FROM WHOSE FACE THE EARTH AND THE HEAVENS FLED AWAY; AND THERE WAS FOUND NO PLACE FOR THEM. AND I SAW THE DEAD, SMALL AND GREAT, STAND BEFORE* ***GOD****; AND THE BOOKS WERE OPENED: AND ANOTHER BOOK WAS OPENED, WHICH IS THE BOOK OF LIFE: AND THE DEAD WERE JUDGED OUT OF THOSE THINGS WRITTEN IN THE BOOKS ACCORDING TO THEIR WORKS. AND THE SEA GAVE UP THE DEAD WHICH WERE IN IT; AND DEATH AND HELL DELIVERED UP THE DEAD WHICH WERE IN THEM: AND THEY WERE JUDGED EVERY MAN ACCORDING TO THEIR WORKS. AND DEATH AND HELL WERE CAST INTO THE LAKE OF FIRE. THIS IS THE SECOND DEATH. AND WHOSOEVER WAS NOT FOUND WRITTEN IN THE BOOK OF LIFE WAS CAST INTO THE LAKE OF FIRE."* We admonish all to know we are each writing our book right now that we started at conception. And we just read Scripture that says at the second resurrection, all the books are opened, including the LAMB'S Book of Life. If all of GOD'S Children were in the First Resurrection, why would the Book of Life be opened again if there are no more names within. Scripture says in the last days, the multitudes of the religious will not hold sound doctrine, but be turned to fables. Most souls will be unstable and in constant changing beliefs. Their beliefs will have no eternal foundation or mooring. After this second final Judgment, we see FATHER Create a New Heaven, and New Earth. Think about the level of love and devotion between GOD and all others required to live together in a perfect peaceful responsible way, where all give their all, for all. We each must ask ourselves; do we have this faith and love? If we don't have it now, can we grow it? Yes. But only if we try and want this above all else, and for the right reasons. We are told our love and faith grow with study, prayer, and loving obedience to live the truth and ways and beliefs of GOD. It has to be understood that this is a two-front war of understanding. Front one is learning the goodness and will of GOD; front two is learning to deny and avoid the evils of Satan and this world. Can we grow spiritually learning of only the goodness of GOD and his ways, without learning what is the evil to be rejected. Hebrews 5:13-14, *"FOR*

EVERYONE THAT USETH MILK IS UNSKILLFUL IN THE WORD OF RIGHTEOUSNESS: FOR HE IS A BABE. BUT STRONG MEAT BELONGETH TO THEM THAT ARE OF FULL AGE, EVEN THOSE WHO BY REASON OF USE HAVE THEIR SENSES EXERCISED TO DISCERN BOTH GOOD AND EVIL." As we grow and learn more and more to know, discern, and recognize both good and evil, as they are revealed to us by HOLYSPIRIT. Then we can see and know the perfect love and ways of GOD that perpetuate life and justice eternal, and know if our hearts are true and one together; this is GOD'S promise. We know by FATHER'S omniscience, that all his creations and Prophecies are part of his eternal plan. Every single smallest detail in all physical reality was omnisciently known to FATHER first, before he consciously and forcefully perceived it to exist. When we start by understanding there is now an eternal perfect loving Triune GOD, we see our own eternal perfect love for GOD, fulfills their true need and desire. And we know the true and perfect definition of eternal love has been defined by FATHER alone. But that defining is falsely remade within the individual minds of mankind beyond any true semblance of its true state. It is an absolute necessity for us to know and accept FATHER'S definition of eternal love, before we can fill our hearts with that Holy love. Holy love lives and thrives in eternal faithful responsibility, above the strongest purest love one could possibly feel towards their own newborn baby. GOD'S love is there for us first in all hope, with no assurance of reciprocity. It is there towards us through their every pain and suffering we cause, up until all hope of reciprocity has been lost. And then that love from GOD towards us, painfully dies. This world sees love as feeling and experiencing physical need, joy, and elation, along with its hurt and loss, as each soul's desired and undesired attributes change. But the love of GOD is a force of perfect life that flows into eternity, bringing joy and peace. The endless loves of this world, always fade to indifference; and with the passing of enough time, all worldly love turns to hate. FATHER defines eternal love, as being minus all hurt, disappointment, and loss; but all worldly forms of love are filled with these. FATHER'S creation is divided by these two opposing loves. Can the true understanding of these two loves always be shared with others? All knowledge of GOD'S love and Creation are to be shared in hope with the world; but to be truly shared as it is, something must be received by the other in the exact way and form it was given. Whether this loving understanding can be properly received by others, is a free-will question of their desire. Now FATHER'S perceived laws of creation, reveal many illuminating details when studied in depth. But deep wisdom never appears if one only sees the surface of a rippling sunlit shimmering pond, blinding them to its deeper good and evil realities below that surface. We must dive down to know its depths, lengths, and breadths. We have heard that GOD can do all things physical, but what can this mean for love. Can we all please know and accept that love is beyond being created by GOD? All highest love for GOD'S ways and nature, must be built between GOD and free-willed conscious souls. So, returned eternal love is beyond the control of GOD, just as all free-willed desires are beyond GOD'S directed will. Does that reduce GOD in any way or measure? No. GOD is still GOD, but FATHER cannot gift out free-will souls, to automatically exhibit spiritual eternal love. If he did, the fulfillment of his needs, would come by his creation, and not free-will choice of love. By all the free-will rebellion we see against GOD, we know love cannot be created by GOD. And now, like GOD, you know something most souls don't. The saddest knowledge that exists, is that eternal love, being the most valued, desired, and treasured, of all things to GOD and to all, comes only at a greatest most painful eternal cost. Do you see that GOD loves, yet cannot create love? But, do you see FATHER did create a physical reality, and a complex layered dynamic process, and gifted billions of eternal free-will souls, that all together, allows eternal spiritual love to be born, grown, and held in the hearts of at least, some souls. In some, a faithful responsible eternal spiritual love is grown to be returned to GOD in the same form it comes from GOD. In this way, GOD does accomplish the greatest of all things in a created free-will love among many; just not among most. This saddest truth, reveals the immense eternal cost of it all, making eternal love the greatest most valuable of all realities. So, eternal love is the most powerful, costly, and treasured, of all things. And the Principles of Eternal Love, govern all GOD'S judgments, who exist in all their want and search for it, as do we. We see in Scripture how GOD has warned, knowing he could not gift to us his omniscient understanding. Each soul, cannot just possess a gift of sudden omniscience. Omniscience is an acquired, built, pure understanding, that comes in seeing all proper layers of GOD'S created truth come together. The understanding of all these highest eternal things, requires we first understand the one true eternal foundation it is built upon; being gifted free-will souls cannot absolutely know anything, and all our knowing comes by faith. So, GOD has extensively warned that we truly need to trust in, and obey them alone, as only they can lead us in love to find GOD'S good ways. GOD will lead us toward all good things, if we follow and flee all evil temptations as we sense and pass by them along our way. GOD gave us their words of warning and remembrance for obedience throughout Scripture. GOD warns, there are great inherent dangers within the growing imbalance in the creation. We see chance, choice, and free-will, are required for love to exist and to flow, so reality must exist in this painful costly way. We see there is truly a bright line separating the vastness of GOD'S true and pure good, from the vast abyss of all individual conscious hells. We further see that since the fall of mankind, we each are born just over and against this line, but on the wrong fallen side. Yet GOD made a way, and says turn and come back over and be loved. We find ourselves in the middle of a vast great sphere of complex emotional processes, as all things around us move and interact. We find we are gifted with a wondrous ability within our soul to sense the physical and spiritual aspects of the conscious and physical reality in fascinating ways. Each soul can form an individual subjective reality, or choose to share in the one objective reality of GOD. Yes, all created reality is subjective to GOD; but it must be seen as objective to us. Now FATHER'S Soul, is maybe trillions of orders of magnitude in size, scope, and forceful reach, as compared to our souls. And we know

FATHER perceived this universe, being the second required half of our existence, making us reliant on him for our loving, peaceful, eternal awareness. Yet, each soul can disagree, rebel, and become an individual and be all alone forever. We are gifted and created by, yet far under FATHER; but we are all gifted with souls sufficient enough in essence, to desire, learn, understand, and believe his truth. And if only good desires lead us, we are led to find, not the destructive, but all the good righteous pleasures. We find ourselves alive, and in our souls we can have desires for the flesh and this physical creation pulling us away from knowing and loving the Creator of it all. We see and hear around us, all manner of deceptions that seek to confuse and weaken our soul. Did the eternal ways to sustain a relationship between GOD and mankind change after the fall? They did not. Can we still fulfill these? We can. There were the perfect conditions and ways with GOD before the fall, and the ongoing imperfect ways that now perpetuate the conditions of that fall. Yet there is repentance and forgiveness, and turning back to those same prior perfect conditions that remain with GOD. What exactly is the situation in which we now truly find ourselves? What do we see between us, and our re-established eternal relationship with GOD'S love and truth? We see the many want to believe GOD allows and forever forgives each soul's endless sins. We see all the temptations, brought by the same free-will desires and curiosities that over powered Satan, 1/3 of the angels, and Eve and then Adam, and so on to most all of mankind. And we see those vile perverse desires that gave rise to Sodom and Gomorrah. If our obedient love for GOD grows weak, our lust for the creation will grow strong, and we will choose. Many souls try, but backslide from the eternal ways of GOD, and fall away. Adam and Eve fell by turning inward and yielding to selfish prideful desires and curiosities. They considered, desired, and knew, individual satisfactions and pleasure, born of their imagination and self-indulgence. Nothing has changed in how to have a relationship with GOD, or in what destroys that relationship. GOD cannot alter the definition of eternal love to fit individual wants. Eternal Love, is a unique and specific force that perpetuates all perfect shared eternal life, requiring a perfect environment and continuous flow to eternally thrive. Our true eternal love, can only be known and thrive in emulating that loves pure examples. If we look deep into this creation, down into the causal level and strength of all temptations. We find we are gifted souls from GOD, adjoined to a most sensitive amazing interfacing physical nervous relay system, that reflect to us, GOD'S great wonders of creation. What powers will these senses have over us, if we do not understand and control them? To see, hear, feel, smell, touch, and be touched, are desirous powerful interactions within GOD'S Creation that tempt us with choices of what we will do with this great power. GOD calls out into this fallen creation, seeking all who might desire their loving spiritual truth and ways enough to overcome all physical temptations. Scripture says that the good spiritual things of GOD are not of this life or this world; but of the next. Yet a weak believer, still in the flesh, reads Scripture and hears of physical rewards and gifts from GOD in both this world and the next. It is amazing how the imperfect balance of forces and senses throughout creation, tempt the free-will of mankind, saying play a god if you will, enjoy, consume, gather, build, or destroy; and forget the truth, love, and names of GOD. While from outside the universe, the will and love of FATHER to sustain it all fades. Mankind's desire to redefine good and evil, and to be individual entities separated from GOD, is at the tip of the root of all evil. And in this fallen world, the power and ability to accomplish this in all its splendor and self-glorification, is best found in the accumulation of wealth, being power. So, by this physical root temptation for money, eternal life and love are destroyed in most free-will creatures. By fallen angels and fallen mankind's physical desires, all creation dies in decay, bringing unimaginable suffering, pain, hell, and death. By free-will, this is all set beyond the righteous controls of GOD. We see GOD, since the fall, leaving those of this world to their own desires and devises, foregone and predestined by the choices within their hearts; even allowing all abominable false religion and worship. What is this perfect truth and love we must find for eternal life. It is simple, and is written within and upon FATHER'S every external creation; being his aether, his universe, and all created physical life. But his gifting of all free-will souls under his purpose of finding a great eternal family, has brought destruction upon it all. We can fully see it from both ways, action to purpose, and purpose back to action. To correctly see GOD'S one truth, we start with the fact that all things are created and gifted by GOD for the sole purpose of finding eternal love. Truth is the truth of GOD'S love, so, any true eternal love for GOD, is a love for their truth; in this we find all we need. When mankind studies Scripture without HOLYSPIRIT, it raises a lot of false worldly questions, answers, and contradictions. One question rarely asked, is centered on the fact that true understanding tells us FATHER knew that with free-will and a real desirable choice, rebellion would probably come. By a true desirable choice, a true probability arises within all souls, yet not a certainty for any soul. FATHER also knew if he showed love, and taught what eternal love is, each soul could choose this, or not, by desire. So, eternal love also became a probability for all, yet not a certainty for any. FATHER further knew that once all souls did freely divide by their differing desire, that they must then be kept eternally separated. So how can this be pure love, and not part cruelty. Yes, GOD knew this could happen, and it's true that in the end, all who forged themselves as evil and contrary, must be separate and imprisoned. Knowing the truth of GOD, we see beyond this painful correlation, to see that by all laws of eternal love and free-will, among many gathered together, it just must be. There is no other way for eternal love and peace among the free-will of many to exist. Yet this cost is found unacceptable by the lost, thinking it cruel, thinking there surely was a way for FATHER to find love without the suffering and separation of those who did not love him. But there is no other way; if there were a better way, it would be known to FATHER and embraced. We all see the cost, loss, wonder and value of a free-willed love and life among each other, by our desire for it; and we see no other way for it to be. We know we cannot be guaranteed that if we love someone, they must love us back. We know the cost of all love is loss, rejection, and heartbreak. And measured against the suffering anguish, and the eternal loneliness of FATHER remaining all alone, we see that a free-will way for choosing eternal love, is worth any cost. Even if most choose to suffer, in throwing eternal love and life away, and choose a temporary illusion. All are warned there is no eternal assurance of life, love, or peace, in choosing this world. All are told the wrong choice brings only death and hell; yet most choose it. So, to me, the cruel souls are those who make their cruel choice against GOD. So, cruelty is also subjective, depending upon what one loves, and where one is eternally headed. The only meaningful question is what do we each choose, and who chooses pure eternal spiritual love. Though it is far beyond our grasp, we must try to see, feel, and understand, the great endless suffering GOD has endured throughout their great work to find our love. If we could see all mankind's combined suffering, GOD'S pain is far, far more. Add seeing we cannot create reality or life, all who eternally love GOD are benefactors. We have all to gain or lose by the desired choices of our hearts. The truth must be known, that it was always GOD'S desire that none should perish, and that all would come to GOD. But GOD'S will in this, is subject to mankind's free-will and all their rejection. It is undeniable, that according to all that has been discussed, GOD has done everything they painfully and righteously could, to bring each soul to find the truth and love of GOD. The world has been given the greatest example of love that could ever be shown. FATHER gifted his creative powers in each of us. He has given all he has to all who ask in truth, love, and faith. Yet this world en mass, rejects any submission to the true loving ways of GOD'S Creation. They choose to be a god of their own destruction. I find it impossible to find the first cruelty on the part of GOD in these ordered facts. I do find on the other side, by fallen mankind, and the fallen angels, only great cruelty, arrogance, and self-worship. Blaming GOD for their own evil, being destroyers of eternal love, truth, and all the righteous desirous creations of GOD. The fact of FATHER knowing rebellion was a probability, is a stumbling block to many, but only in their wanton ignorance, denial, and dishonesty. GOD has said in the last days there will come perilous times. Many religious souls are steeped in and accepting of sin and lust within themselves and the church of their choosing. These are blinded to GOD'S truth in not seeing their theft, not seeing that their physical lives are not their own. These have a firm grasp on the lie that their physical lives are their own; believing if they acknowledge GOD with thanks, they can live out their lives in the ways and joys of this world. But these fulfill the Prophecy of CHRIST, saying he would tell many who thought they were saved, to depart from before him, as he never knew them. This is impossible for most to hear; their ears being shut to this truth. There are many hard aspects to the firm immutable truth of GOD. Another stumbling block, raised by those seeking excuse, is why does one-person love GOD, and another does not? Why do some souls, love GOD more, and some less? What are the reasons for this, is chance involved, is this purely our doing? Some ask how can we help what we love? Are we all born, loving or not loving GOD? These souls ask, which of these wrong questions is correct. These ask, is love predestined, or is love left up to chance and a roll of the dice. It is neither, love is a free-will choice born of our deepest core free-will desires. We see a choice and choose by our inner drawing force of desire. Yes, our deepest core desire is hard for us to declare and define; yet by our soul's free-will, it is there. And so, our desired learning enters our equation. There is only one great true question in all reality that has survived as the supreme question in every soul that has ever existed; being is GOD there? Our deepest core desire is between do we want GOD to be there above us, or do we want to be the god above us? Our answer to this question is what we all are led by, and consciously boil down to, in and by all we do and choose. The answers to all hard questions, are always hard to hear. And the answer to the hardest question of all, can be the hardest to hear. We all are wanted by GOD to come to know GOD'S truth. And in circular fashion, we can never eternally love GOD, without knowing them intimately. From this, we must hold sacred, that from the beginning of our first hearing, we are called to strive to love and know more and more intimate knowledge of GOD. This is our hopeful helpful path of promise for a life set out by GOD. To choose anything else, is to choose the spirit of rebellion. If any say to us, they have a full life, and don't have time for such a search. Remember, this is not your answer to us, but to GOD. There exists no excuse, as this is our life's created purpose. All excuses are made in dishonest selfishness of the flesh. If you would love GOD, seek them out in the awe and passion due such a loving GOD. Can any void their responsibility to pursue this fullest of truths, and instead, give GOD a lesser measure of their time and effort; and yet claim hold to the benefits of GOD'S Children. The answer is, what is the true measure of our love for GOD. And GOD alone, will Judge these things. There is always this epic ongoing battle of definitions for this word evil. And in this world, we all are, or once were, participants in this battle. There is no measured battlefield to avoid, and any sidelines that one tries to stand upon, creates a lost observer. There is only one line drawn by FATHER between his perfect created ways that perpetuate eternal life; and rebellious ways that perpetuate the destruction of the same. FATHER sees us each, as on one side or the other. Many try to straddle this line, which GOD says he hates most. So, arm yourselves with wisdom. We know FATHER created all things during the first six rotations of the earth. And the seventh day, he rested from creating all observed complexity, of which we have only scratched the surface. All things will be fulfilled, all works, and all rests. GOD in six days created the beginning of all things; also, within those same six days, was set the most probable coming end of so many things, it all bringing the fulfillment of eternal love. We pray you receive all Prophetic understandings, and that you be given great revelations from GOD. Great visions born of love, hope, and wonder, that are there to be received. We remind all, Scripture says count and consider the full cost of our desires before going forward. As the fullness of this shared knowledge of Scripture is seen only from GOD'S vantage point. This fullness, is only seen through HOLYSPIRIT and a fervent love to find and behold all of GOD'S wonder. This while all worldly religions have abandoned GOD'S Truth, Prophecy, and his Sabbath; and fallen into this world of covetousness.

In this fallen insane world, we all see objects and people, separating and being separated routinely. It is all a part of the everyday fallen life, passing before our eyes. But this eternal last separating of all souls GOD speaks of, comes by each soul's ever-growing chosen drifting away from and losing the one pure truth authored by GOD. And to the flesh and to this world, this separation of losing GOD is not felt as unusual; it is just a normal aspect of a temporary godless progressive worldly physical life. From before the fall of mankind, there was in Satan, his evil plan to separate every soul away from GOD. It was initiated first in Heaven, where Satan lured away one third of the stars (angels) in GOD'S Celestial realm. Satan is called the father (creator) of all things evil, including the oldest of all military strategies, being to lie to divide and concur. And it works as you get your target audience to believe more and more lies, and this divides them more and more against each other. If two people see and agree a certain evil is good, yet later, one hears something new and no longer sees that evil the same way. There is a new division, and both now see the other as wrong or evil, and needing correction. It is just this simple. Souls just keep playing a god of the truth, in reinterpreting the one truth of FATHER'S created reality, to fit their individual desires; and they share their interpretations to propagate more and more options for others to believe. At this moment, Satan cares little about which lie you believe, as long as you don't believe GOD'S truth. In this world, this is called individualism. If you desire, and become an individual, this immediately starts building personal separation*s* between you, and GOD, and others; and as individualism grows within your soul, you will separate more and more from GOD and all other souls. This leads all lesser souls to envy, strife, hate, and violence. Satan has one goal for mankind, being to turn us all away from GOD, and then to himself. This can never work, as the only possible eternal unification is with GOD. At so many times, FATHER has tried to gather all to be one people under CHRIST, being filled by HOLYSPIRIT. Because FATHER is above all reality, knowing all things, and creating all things. FATHER knew separations could come to any creation built on free-will choice to find love. In FATHER giving all souls a choice between his good and their own evil, he saw all free-will souls become a conscious probability equation, and potentially, self-serving creatures. A great weakness in all lesser creatures with free-will, rests in that they cannot in all, and every way, absolutely know anything as being the purest of fact. We must, we do, and we can only believe things in faith. And so, being given a true choice, we choose to believe what we desire to believe. This inherent and innate free-will choice of desire that defines a soul, has been gifted to us all. So, the things we deeply desire, determine our every choice of belief. FATHER can take us to the ends of creation and back, to show us miraculous things we can never comprehend. But at the end of even that, we still must believe by faith. We still must desire to believe what we see, or we will search for our reasons to doubt. And if after all FATHER has shown a soul, if Satan or his children come with lies, and say there are other better Gods, more accepting Gods, or even no supreme God. All souls then must make a choice of belief, and all will choose by their desire, what or who they believe. In this fallen confused world, is our every choice, between a lie and the truth? Or is it most always between two or more lies? By being gifted and created, you cannot know anything absolutely on your own. In this world, Satan contrives endless multitudes of wrong choices, or lies. But from GOD we are given only one truth, and one choice. We see by the sum of all mankind's correct and incorrect choices, that the purpose and need of GOD, will eventually be fulfilled by those fewer who truly choose GOD. Why would Satan, given great heavenly forceful responsibility, consider or think he is, or could be equal to, and overthrow GOD? Because he deeply desired it to be so. But why would he desire this? Who does he love most? It is himself. In being created and gifted by GOD, Lucifer at first had faith that GOD was the GOD of all things, and served GOD in love, seeing GOD as his Creator and Provider of life. But at some point, Lucifer began to lose one desire, and find another as he questioned GOD'S truth in his heart. And by his own desire and seeing his beauty given him by FATHER, Lucifer began to turn and speak to his reflection in a mirror. And at a certain point, he could no longer see the vast difference between himself, and GOD. Satan no longer saw or felt the same loving reasons and purpose for his life he once did. Lucifer had no knowledge of anything before his gifted creation, and he was taught what he knew by GOD. But at a certain first moment, Lucifer asked himself, is GOD truly my creator holding all power and life, or is GOD just a part of reality as I am? Am I, or can I be equal to GOD? And he fell from GOD, and became Satan the deceived, and the deceiver. All such confusing thoughts push GOD'S prominence from one's belief, and divide and concur ones understanding. Such thoughts open one's self to new desires to accept new godless choices and options. And right there and then by Lucifer, FATHER'S perfect dynamic creation came under attack, and the endless abyss of the lying vanity and pride, born out of love for self, began. The truth is that FATHER gifted and created all we are and sense. And Lucifer conceived the first of endless lies within his desire. And as his confusion eroded his fear of GOD, Satan arose and said, I will rise up to be like GOD. This rebellious confusion was then spread by him to all who would hear it. Every lie is an offer of a false choice of belief. The first lie was created by Lucifer, making him Satan, the father of lies. All free-will souls are vulnerable, and for a liar to try to deceive others, they must have first deceived themselves, and continue to do so. We know Lucifer was first to succumb and believe his own deceptions, which are now become his vast delusions. But will Satan re-know the truth at his end, once FATHER takes away all his forceful creative reach. In Scripture, we see Satan is imprisoned and powerless for a thousand years; yet, when set free again, and his power returned, he immediately rebels again. Satan can never accept GOD'S supremacy or mercy, ever again. And at Satan's last free moment, father will strip him of all forceful reach, and leave him a shell of anguish and hell. What motivates Satan? He does what he does, because in the depths of his confusion and pride and envy, he just does not and cannot ever love or truly know GOD again. Loving and knowing GOD is our required bond, not just knowing alone, or loving alone. We must know what we love, and love what we know; all great eternal love for GOD is grown in knowing GOD. Our love for GOD begins at our hearing of a loving GOD, and our first believing GOD is there. So, we must truly see and know that GOD is a loving GOD. And we can love GOD, only once we know him as such; and love GOD more with all knowing. Satan's disapproval of GOD'S ways, has forever blinded him to GOD'S truth. Isaiah 14:9-16, *"HELL FROM BENEATH IS MOVED FOR THEE TO MEET THEE AT THY COMING: IT STIRRETH UP THE DEAD FOR THEE, EVEN ALL THE CHEIF ONES OF THE EARTH; IT HATH RAISED UP FROM THEIR THRONES ALL THE KINGS OF THE NATIONS. ALL THEY SHALL SPEAK AND SAY UNTO THEE, ART THOU ALSO BECOME WEAK AS WE? ART THOU BECOME LIKE UNTO US? THY POMP IS BROUGHT DOWN TO THE GRAVE, AND THE NOISE OF THY VIOLS: THE WORM IS SPREAD UNDER THEE. AND THE WORMS COVER THEE. HOW ART THOU FALLEN FROM HEAVEN O LUCIFER, SON OF THE MORNING! HOW ART THOU CUT DOWN TO THE GROUND, WHICH DIDST WEAKEN THE NATIONS! FOR THOU HAST SAID IN THINE HEART, I WILL ASCEND INTO HEAVEN, I WILL*

EXALT MY THRONE ABOVE THE STARS OF GOD: I WILL SIT ALSO UPON THE MOUNT OF THE CONGREGATION, IN THE SIDES OF THE NORTH: I WILL ASCEND ABOVE THE HEIGHTS OF THE CLOUDS; I WILL BE LIKE THE ***MOST-HIGH****. YET THOU SHALT BE BROUGHT DOWN TO HELL, TO THE SIDES OF THE PIT. THEY THAT SEE THEE SHALL NARROWLY LOOK UPON THEE, AND CONSIDER THEE SAYING, IS THIS THE MAN THAT MADE THE EARTH TO TREMBLE, THAT DID SHAKE KINGDOMS."* Without the greatest aspect of all sentient awareness, being a responsible faithful love towards GOD, we can never please, love, honor, or worship GOD in any correct sufficient eternal way; nor find our own enduring faithful responsibility. We must see how our level of loving obedience to every word spoken by GOD, reflects and is the exact measure of our loving faithful responsible belief in GOD. We know that to sin, is hateful and divisive, and all sin is against GOD. And to sin against GOD, or his ways, is always born of one's divisive questioning of GOD'S authority. Vanity is love of self, as it rebels, questions, and rejects GOD'S commands. So, by what confused standard can the gifted and created say to FATHER who gifted and created them, that what they now individually love and believe is right and true above GOD'S own loving beliefs? This is the separation between the fallen and GOD, which is called the confusion of hell and death. The Apostle Paul writes of those who battle against the flesh and evil, while not having HOLYSPIRIT. Pointing out how he also did this before his own regeneration. Speaking of his once being in that losing battle with the flesh, having only the law, without CHRIST'S love, or HOLYSPIRIT. Paul says he was always losing that battle. Romans 7:18-25, *"FOR I KNOW THAT IN ME THAT IS IN MY FLESH, DWELLETH NO GOOD THING; FOR TO WILL IS PRESENT WITH ME; BUT HOW TO PERFORM THAT WHICH IS GOOD I FIND NOT. FOR THE GOOD THAT I WOULD I DO NOT: BUT THE EVIL WHICH I WOULD NOT, THAT I DO. NOW IF I DO THAT I WOULD NOT, IT IS NO MORE I THAT DO IT, BUT SIN THAT DWELLETH IN ME. I FIND THEN A LAW, THAT WHEN I WOULD DO GOOD, EVIL IS PRESENT WITH ME. FOR I DELIGHT IN THE LAW OF* ***GOD*** *AFTER THE INWARD MAN: BUT I SEE ANOTHER LAW IN MY MEMBERS. O WRETCHED MAN THAT I AM! WHO SHALL DELIVER ME FROM THE BODY OF THIS DEATH? I THANK* ***GOD*** *THROUGH* ***JESUS CHRIST*** *OUR* ***LORD****. SO THEN WITH THE MIND I MYSELF SERVE THE LAW OF* ***GOD****; BUT WITH THE FLESH THE LAW OF SIN.* We have heard this scripture turned and taken out of context many times and made to sound as accepting defeat and being trapped in the flesh and sin. Many teach that Paul is saying we can serve GOD with our mind, and at the same time, we cannot resist the flesh and sin. This is contrary to all true meaning; being that before CHRIST, Paul could not overcome and win in the flesh with only his mind and the knowledge of GOD'S law. The lesson we are meant to derive from this is we must crucify and bury the flesh in water burial Baptism, with its constant evil desires. If not, we will be trapped and serve sin as long as we remain in the flesh. And so, we must live and walk always in the spirit, emulating CHRIST. To substantiate this correct interpretation, we read the *very next* verses. Being the second half of Paul's overall teaching. Being Paul's (but now this) most Preachers fail to include. Romans 8:1-10, *"THERE IS THEREFOR NOW NO CONDEMNATION TO THEM WHICH ARE IN* ***CHRIST JESUS****, WHO WALK <u>NOT AFTER THE FLESH</u>, BUT AFTER <u>THE SPIRIT</u>. FOR THE LAW OF THE SPIRIT OF LIFE IN* ***CHRIST JESUS*** *HATH MADE ME FREE FROM THE LAW OF SIN AND DEATH. FOR WHAT THE LAW COULD NOT DO, IN THAT IT WAS WEAK THROUGH THE FLESH,* ***GOD*** *SENDING HIS OWN* ***SON*** *IN THE LIKENESS OF SINFUL FLESH, AND FOR SIN*

CONDEMNED SIN, IN THE FLESH: THAT THE RIGHTEOUSNESS OF THE LAW MIGHT BE FULFILLED IN US, WHO WALK NOT AFTER THE FLESH, BUT AFTER THE SPIRIT. FOR THEY THAT ARE AFTER THE FLESH DO MIND THE THINGS OF THE FLESH; BUT THEY THAT ARE AFTER THE SPIRIT THE THINGS OF THE SPIRIT. FOR TO BE CARNALLY MINDED IS DEATH; BUT TO THE SPIRITUALLY MINDED IS LIFE AND PEACE. BECAUSE THE CARNAL MIND IS ENMITY AGAINST ***GOD****: FOR IT IS NOT SUBJECT TO THE LAW OF* ***GOD****, NEITHER INDEED CAN BE. SO THEN THEY THAT ARE IN THE FLESH CANNOT PLEASE* ***GOD****. BUT YE ARE NOT IN THE FLESH, BUT IN THE SPIRIT, IF SO BE THAT THE SPIRIT OF* ***GOD*** *DWELL IN YOU. NOW IF ANY MAN HAVE NOT THE SPIRIT OF* ***CHRIST****, HE IS NONE OF HIS. AND IF* ***CHRIST*** *BE IN YOU, THE BODY IS DEAD BECAUSE OF SIN; BUT THE SPIRIT IS LIFE BECAUSE OF RIGHTEOUSNESS."* Paul says there is a huge separation and enmity between the spirit and the flesh. And we cannot walk in both the flesh and Spirit at the same time. Paul says if we are still in the flesh, we will sin, but if we walk filled with HOLYSPIRIT, we cannot sin. Paul says all who walk in the flesh, are enemies of GOD'S ways, and can never please him. Paul says if HOLYSPIRIT dwells within us, the flesh is dead, so we have the power to not sin. Paul says if any walk in the flesh and sin, they do not have HOLYSPIRIT, and they do not belong to JESUS. JESUS said in Matthew 5:48, *"BE YE THEREFORE PERFECT, EVEN AS YOUR* ***FATHER*** *WHICH IS IN HEAVEN IS PERFECT"*. JESUS forgave, but then always said go and sin no more, lest a worse thing come upon you. We do suffer at our hearing GOD'S immutable words being wrongly divided and twisted by those of this world. There is for us all, our required moments to learn, grow, and overcome; yet most can't or don't learn or grow enough, or at all. In our true love for GOD, our goal is to in the strength of GOD, overcome all vile sins of the flesh, and never claim victimhood to linger in sin. Sin is always weakness and evil, so run to the strong ways of GOD. We see how from many twisted lies, come so many religious divisions. Each lie has its destructive temptation, but lies are easy to see as they all break from the perfection of GOD. By all individual truths of individual perspectives, this world is full of never-ending separations. We see this world is separated by oceans; but in the coming New-Earth, there are no such divisions. We see here, many separation of language, cultures, and religions, and even separation by time itself, day and night. Yet, in the New Earth and Heaven, we read that all these separations do not exist. In all separate Countries, we see divisions narrow into religious and secular political divides. In the New Earth and Heaven these will never exist. All souls united by HOLYSPIRIT, will know and live by GOD'S one shared truth of all things. We know fallen mankind in the flesh is flawed, biased, and constantly tempted by desires, circumstance, chance, and all past and wanted experience. By all of these things, most now desire to believe many lies. And that turns their divisions into a search, not for the truth, but for all earthly power to change society to reflect the ways that please them, and not GOD. This might all be academic if all destructive lies and separations stayed outside the doors of worship and evangelism, but they don't. All lies and disruption brought by Satan have an envious purpose that has nothing to do with the self-destroyed masses, but for turning the few who hold the faith, belief, and love, that pleases GOD. The hard focus of Satan has never been on those who destroy themselves, but on those who do not. We are the focus of his ego; we are what GOD and Satan strain over. What feeling does that raise in your soul, knowing GOD actually fights for each child's love. But only in their Holy Righteous ways. And we who truly love GOD, see the purpose of that coming final separation. In all conflicts, ranging from simple two person disagreements, to world wars, there are always battle lines of separation. These are never just physical lines, these are spiritual philosophical conscious lines of perspective, drawn by desires and beliefs. Factors like a country's borders and treaties, are ways to avoid great physical confrontations; yet the underlying divisions of desire and belief are the cause of all confrontations. Herein lie the bright battle lines, this world does not clearly see. The greatest battle line is now, and has always been, right up against the still open door of the True Church of GOD. And the soon closing of that door, comes at the great final eternal separation. On one side of that door, or line, are all the self-deceived who fail to engage or endure the battle for their soul. On the other, we see the eternal Church. We see this separation began early in the book of Genesis, and continues right up through history, through the tribulation and Millennial Reign, and up to the White Throne Judgment. This world and the fallen dying flesh upon it, are all lost to GOD. It is all corrupt, vile, offensive, destructive, and to be burned. On the side of loss and darkness, there are storms of deception, tragedy, fear, and great lies, hell, and death. Every experience that comes upon all souls, be it good or bad, brings a chosen greater desired closeness to, or separation from GOD. These last days will find and separate out the last of those who believe and love GOD with all their heart mind and soul. By Satan's evil design, there are already countless endless separations in souls everywhere. We see most all souls are separated by beliefs, by likes and dislikes, by age, by wealth, and even physical appearance or beauty. Everywhere pride, envy, weakness, and ignorance are found, people are divided into ungodly groups, and further into billions of ungodly souls, by individual internal lies. If one truly considers what it is, to be one with GOD; then one must consider what it is to be an individual. One must find within these two groups, that one group is looking to a true external GOD, as the other looks to a false internal god of self. In school as a child, a teacher, filled with excitement and conviction, spoke to the class of the wonderful aspects of us all being individuals. She went on about how we should embrace, seek out and experience all the variety, diversity, and endless differences in people and all the things that surround us. She said these were life's true joys, and only these could make life rewarding and worth living. I was not yet exposed to GOD'S truth, yet to me, this felt wrong. Though I did not know GOD'S truth, I loved GOD, and my weak unknowing love alone, gave me a discerning sense of wanting my greatest closeness with him. And not any desire to sense and wander through endless individual differences.

Something did not ring true in her words, and later with my growth, I saw how all feelings of individuality bring strife. Strife exists because various ways of life divide those that want, from those that don't want those ways. As some want this, it conflicts with others who don't want that. Embracing individualized group desires, raises disparities in all things of acceptance, comfort, security, and wealth. The illusory joy of having all the things of this world, is a waring temptation to those in the flesh. This coming greatest of separations is about perfection, and not separating all gold bearing ore, from all other ores. It is all the fiery trials, that separate every impurity, from the purest gold. These are the days of the purification of the Bride of CHRIST. Scripture says there are many impurities and blemishes in the Church needing to be cleansed. Isaiah 30: 7-12, *"FOR THE EGYPTIANS SHALL HELP IN VAIN, AND TO NO PURPOSE: THEREFORE HAVE I CRIED CONCERNING THIS, THEIR STRENGTH IS TO SIT STILL. NOW GO, WRITE IT BEFORE THEM IN A TABLE, AND NOTE IT IN A BOOK, THAT IT MAY BE <u>FOR THE TIME TO COME</u> FOR EVER AND EVER: THAT THIS IS A REBELLIOUS PEOPLE, LYING CHILDREN, CHILDREN THAT WILL NOT HEAR THE LAW OF THE* ***LORD****: WHICH SAY TO THE SEERS, SEE NOT; AND TO THE PROPHETS,*

PROPHECY NOT UNTO US RIGHT THINGS, SPEAK UNTO US SMOOTH THINGS, PROPHECY DECEITS: GET YOU OUT OF THE WAY, TURN ASIDE OUT OF THE PATH, CAUSE THE ***HOLY ONE*** *OF ISRAEL TO CEASE FROM BEFORE US."* There is a fast-growing power of the flesh in today's apostate Church, filled with preachers having a spirit of great error who condone more and more of the perverse desires for the flesh and this world. It is a spirit of idolatry, rebellion, self-rule, and antichrist. This ravenous spirit runs the hallways of worship centers looking for the weak, new, and old, to separate them from the true flock. This weakness controls the many who only sit still in the WORD and make no progress or growth in knowing or loving GOD. These are falsely taught that GOD walks everywhere with them in their constant sin. This is part of the "Prophesied" falling away from the truth. We know that Noah, his wife, and six children, were all that were redeemed from the earth's cleansing flood. And only Lot and his family were redeemed from Sodom, yet in leaving, his wife looked back in her desire, and was lost. GOD has seen the whole of the people corrupt before, and cleansed the earth thoroughly. Leaving in hope, just the barest of a true remnant to rebuild. Ezekiel 14:11-23, *"THAT THE HOUSE OF IS'RA-EL MAY GO NO MORE ASTRAY FROM ME,*

NEITHER BE POLLUTED ANY MORE WITH ALL THEIR TRANSGRESSIONS: BUT THAT THEY MAY BE MY PEOPLE, AND I MAY BE THEIR ***GOD****, SAITH THE* ***LORD GOD****. THE WORD OF THE* ***LORD*** *CAME AGAIN TO ME, SAYING, SON OF MAN, WHEN THE LAND SINNETH AGAINST ME BY TRESPASSING GRIEVIOUSLY, THEN WILL I STRETCH OUT MINE HAND UPON IT, AND WILL BREAK THE STAFF OF THE BREAD THEREOF, AND WILL SEND FAMINE UPON IT, AND WILL CUT OFF MAN AND BEAST FROM IT: THOUGH THESE THREE MEN, NO'AH, DAN'IEL, AND JOB, WERE IN IT, THEY SHOULD DELIVER BUT THEIR OWN SOULS BY THEIR RIGHTEOUSNESS, SAITH THE* ***LORD GOD****. IF I CAUSE NOISOME BEASTS TO PASS THROUGH THE LAND, AND THEY SPOIL IT, SO THAT IT BE DESOLATE, THAT NO MAN MAY PASS THROUGH BECAUSE OF THE BEAST: THOUGH THESE THREE MEN WERE IN IT, AS I LIVE, SAITH THE* ***LORD GOD****, THEY SHALL DELIVER NEITHER SONS NOR DAUGHTERS; THEY ONLY SHALL BE DELIVERED, BUT THE LAND SHALL BE DESOLATE. OR IF I BRING A SWORD UPON THE LAND, AND SAY, SWORD, GO THROUGH THE LAND; SO THAT I CUT OFF MAN AND BEAST FROM IT: THOUGH THESE THREE*

MEN WERE IN IT, AS I LIVE, SAITH THE ***LORD GOD****, THEY SHALL DELIVER NEITHER SONS NOR DAUGHTERS, BUT THEY ONLY SHALL BE DELIVERED THEMSELVES. OR IF I SEND A PESTILENCE INTO THE LAND, AND POUR OUT MY FURY UPON IT IN BLOOD, TO CUT OFF FROM IT MAN AND BEAST: THOUGH NO'AH, DAN'IEL, AND JOB WERE IN IT, AS I LIVE, SAITH THE* ***LORD GOD****, THEY SHALL DELIVER NEITHER SON OR DAUGHTER; THEY SHALL BUT DELIVER THEIR OWN SOULS BY THEIR RIGHTEOUSNESS. FOR THUS SAITH THE* ***LORD GOD****; HOW MUCH MORE WHEN I SEND MY FOUR SORE JUDGMENTS UPON JE-RU'SA-LEM, THE SWORD, AND THE FAMINE, AND THE NOISOME BEAST, AND THE PESTILENCE, TO CUT OFF FROM IT MAN AND BEAST? YET, BEHOLD THERIN SHALL BE LEFT A REMNANT THAT SHALL BE BROUGHT FORTH, BOTH SONS AND DAUGHTERS: BEHOLD, THEY SHALL COME FORTH UNTO YOU, AND YE SHALL SEE THEIR WAY AND THEIR DOINGS: AND YE SHALL BE COMFORTED CONCERNING THE EVIL THAT I HAVE BROUGHT UPON JE-RU'SA-LEM, EVEN CONCERNING ALL THAT I HAVE BROUGHT UPON IT. AND THEY SHALL COMFORT YOU, WHEN YE SEE THEIR WAYS AND THEIR DOINGS: AND*

YE SHALL KNOW THAT I HAVE NOT DONE WITHOUT CAUSE ALL THAT I HAVE DONE IN IT, SAITH THE ***LORD GOD****."* The eternal cause of FATHER is always his love for his family in his cleansing and purification of the Bride of the LORD. Strive, cleanse, and purify yourselves and your family, so when the LORD comes, he will find you separate from the world, unspotted, and watching for his return. Do we seek his purity in all things, or in a few things. We need to protect our children and ourselves from all lies. If you believe you are safe and saved in your ongoing false beliefs; you may falsely rest easy, believing your children are safe in theirs. All gifted souls are at risk of being deceived by the lies that find their way into our eyes, ears, and souls. Do you truly see the level of Satan's work in all school systems, the sciences, the entertainment industry, politics, and religion. These evil weapons give rise to themselves, as they have an evil magnetism and buoyancy in this fallen world. Even one single powerful lie, can destroy many souls. One of the most powerful weapons used against the Church, is the word tolerance. Just a few decades ago, tolerance meant that even though you rightfully know and disagree and speak out against something, it being evil and against GOD'S ways. We must allow this its dark place, out of sight; and not kill or drive it from the face of the whole earth. This is the true definition of tolerance or to tolerate. But the world now says tolerance means to accept as equally good without reservation, to receive with open loving arms in every way, and not speak or show disagreement. There is a rapid satanic changing of meanings for words with each generation, and what fertile tools these are for Satan. Each day, children hear hours of false science, new meanings for words, new social engineering, peer pressure, false cultural political and gender labeling, and Country and Christian shaming while in school and from all forms of media. All designed to divide the young from trusting in GOD and Godly parents; and towards trusting in a liberal godless government. Do you think your children can endure 12 plus years of that, and not become indoctrinated into the exact culture Satan pushes. The Government now acts like our children belong to them. Few adults, having a firm fundamental belief in GOD, could endure this gauntlet of a twelve plus years of bait and switch and misdirection, aimed at discrediting the truth of GOD. Will we each be asked before the throne of GOD; why did we allow the lies against GOD to be taught to his little ones. Have no doubt, that when GOD does asks this, he will say they were his little children, and not ours. These are his gifted, created creatures, entrusted into our responsible faithful care, to raise them for GOD. We need to speak out against the newly created power in the words tolerant and intolerant, now having altered and expanded interpretations. No souls, minus HOLYSPIRIT, agree completely on what it means to be tolerant or intolerant in every situation. In a fallen world, this is lost to individual insane interpretation. The Ancient City of Sodom was said to be extremely tolerant of sodomites, but violently intolerant of those who were not; forcing sodomy upon male travelers who passed through town in the night. Immoral deviants are never tolerant of those who are not. The Holy Bible speaks to all these aspects of life, and any soul not having HOLYSPIRIT, can find many things in the Bible to be intolerant. Beliefs of people in the flesh, are molded by desire, experience, chance, and circumstance, and not by FATHER'S commanded created purpose. Scripture tells mankind, that Judgment is coming for all souls who live against GOD'S created purpose. And these will scream about GOD'S intolerance, in all their absurd vile intolerance towards GOD. And the worst part is the faithless *non*-response, from those who say they love GOD, yet change their doctrine to appease evil and not GOD. Make no mistake, political correctness is evil godless authoritarianism, and it has crept into the Church pulpits, turning preachers into magicians and contortionist, bending Scripture into incomprehensible forms. And this new evil form of tolerance, always moves to allow more sin against GOD, making GOD'S warnings, laws, statutes, and remedies in Scripture, disappear altogether. If the Church separates themselves from the world as Scripture demands, we would have few internal problem; but most churches are separated from GOD and immersed in the world. If we are entangled in the affairs and ways of Babylon (the world), we are surely separated from the ways and Kingdom of GOD. No one can be entangled in fleshly sinful worldly ways, and also be entangled in GOD'S spiritual ways, it is impossible. I repeat a short story, to shine a brighter light on this Principle. There was in our modern day, an old man living in an old shack, minus electricity or running water. He lived deep in the woods, far from everyone. He had no phone, radio, television, computer, or any other modern convenience. He had little food, clothing, and fire wood; but always had enough to live. But this man was forever walking with, talking with, and thanking GOD for every spiritual and physical thing he had. He was loving, happy, and at peace, as he communed, trusted, and walked with GOD all the day long. But this world learned of his poverty, and marveled at his fortitude and ability to remain thankful in abject poverty. They put his story on the evening news. And on Sunday, as worldly preachers saw the report, they told his story as part of their message to inspire their physical charity, as many felt their physical worldly compassions for this man. For they saw the man, as in dire need. In their worldly minds, this man was not blessed by GOD with worldly abundance. They fell upon him, with news cameras catching every moment, as they built him a new home with modern conveniences and comforts. They filled a pantry and his life with all he could physically use, and promised to keep him supplied, so he would not go without worldly things again. They returned next month to re-supply him, expecting to see his joy and comfort for all they had done, but they were perplexed. The old man was miserable and void of joy. And when asked, the old man could not explain his feelings. As they left, the old man realized he had seemed ungrateful, and dropped to his knees crying and praying; LORD tell me why I am miserable and cold in my Spirit? And forgive me for not calling on you much lately, or being thankful, I don't know why. LORD, what has happened to my joyous Spirit? Amen. We all need basic, minimal physical things to sustain our walk with GOD; but our far greatest need, is our constant personal needful relationship with, and reliance on FATHER as our parent and provider. We must see there is an ardent immutable line, between actual physical need, and today's wanton excess.

Abundance above the actual basic needs of the flesh to live, comes at a price. This man had been spiritually and physically provided for in a correct way by GOD all along. The lesson is that to see and want, is not to need. To see something you think will improve or ease your physical existence, does not mean you need it, or should have it. Knowing where to need ends, and wanton lust begins, is a lesson souls ruled by the flesh will struggle to learn. The flesh has no desire to know such things, which are spiritually discerned. Many ask in response to this teaching, what worldly sense can be found in denying oneself physical comfort, pleasure, or security. We refer all who ask this godless question, to Scripture, to seek the true meaning of the afore-mentioned term "worldly sense" which must mean a godless physical sense. For we seek an other-worldly and Godly sense. Also, we hear many say our preaching is harsh, and not at all inclusive, and even very exclusive. How do we respond to this political foolishness? Do we back down, or do we stand and defend GOD'S own words? I ask those who preach a message that is very inclusive and tolerant of sin. Do you at least preach that thou shalt not murder or rape children? And if so, is that not an intolerant message to sinners known as child murderers and rapist. GOD alone draws the line between good and evil. Together, we see HOLYSPIRIT and

Scripture reveal what a true Godly compassion looks like. A false compassion, preaches a false path to GOD, where all souls remain sinners and are accepted as such. Rebellion is seen in how little of GOD'S true words and meanings each Preacher preaches, and how much ongoing sin, they say GOD accepts. If you simply want a physical Church, full of people who don't kill, rape, or rob, if that is the measure and standard you apply to your Church for walking with GOD? And if you measure a church's success by the head count, or buildings, or acreage, or dollars given and shared? We hold no part with you in that. We will never compromise or neglect one word; nor stand idle, silent, or tolerant as others do. We would rather be in a tiny shanty Church with just two poor souls who share eternal love in faithful obedience to every word of GOD; and run from a gilded mega church of fifty thousand souls that does not. We profess above all things, the fullness, power, and strength of GOD'S every word, and we rebuke all worldliness. All the world may have, walk, and preach a broad inclusive path, and count heads and dollars. We can only walk and preach the narrow strait path, and count only hearts and souls in GOD'S Kingdom. Many Church leaders now teach their followers to pray for worldly prosperity. How does Scripture describe such preachers, so we may know and reject them. 2nd

Peter 2:1-3, *"BUT THERE WERE FALSE PROPHETS ALSO AMONG THE PEOPLE, EVEN AS THERE SHALL BE FALSE TEACHERS AMONG YOU, WHO PRIVILY SHALL BRING IN DAMNABLE HERESIES, EVEN DENYING THE **LORD** THAT BOUGHT THEM, AND BRING UPON THEMSELVES SWIFT DESTRUCTION. AND MANY SHALL FOLLOW THEIR PERNICIOUS WAYS; BY REASON OF WHOM THE WAY OF THE TRUTH SHALL BE EVIL SPOKEN OF. AND THROUGH COVETOUSNESS SHALL THEY WITH FEIGNED WORDS MAKE MERCHANDISE OF YOU: WHOSE JUDGMENT NOW OF A LONG TIME LINGERETH NOT, AND THEIR DAMNATION SLUMBERETH NOT."* When Scripture says, *"EVEN DENYING THE **LORD**"* it is not speaking of those who verbally deny JESUS is the SON of GOD. It speaks of those who deny JESUS, by denying who he is and his ways, denying his fullness found in every word he spoke. If we deny who CHRIST is in any single small part, we then deny CHRIST'S perfection, period. Another Scripture on preachers who preach worldly prosperity is 2nd Peter 2:17-18, *"THESE ARE WELLS WITHOUT WATER, CLOUDS THAT ARE CARRIED WITH A TEMPEST; TO WHOM THE MIST OF DARKNESS IS RESERVED FOR EVER. FOR WHEN THEY SPEAK*

GREAT SWELLING WORDS OF VANITY, THEY ALLURE THROUGH THE LUST OF THE FLESH, THROUGH MUCH WANTONNESS, THOSE THAT WERE CLEAN ESCAPED FROM THEM WHO LIVE IN ERROR: WHILE THEY PROMISE THEM LIBERTY, THEY THEMSELVES ARE THE SERVANTS OF CORRUPTION:" We teach all to pray only for the spiritual things and the coming Kingdom of GOD. And let GOD define and provide our physical needs, and be content with his provision of simple food and raiment. Scriptures say often and clearly that GOD does not hear self-serving worldly prayers. Do not follow any who teach that prayers for worldly prosperity are answered by GOD. James 4:1-5, *"FROM WHENCE COME WARS AND FIGHTING AMONG YOU? COME THEY NOT HENCE, EVEN OF YOUR LUSTS THAT WAR IN YOUR MEMBERS? YE LUST, AND HAVE NOT: YE KILL, AND DESIRE TO HAVE, AND CANNOT OBTAIN: YE FIGHT AND WAR, YET YE HAVE NOT, BECAUSE YE ASK NOT. YE ASK AND RECEIVE NOT, BECAUSE YE ASK AMISS, THAT YE MAY CONSUME IT UPON YOUR LUST. YE ADULTERERS AND ADULTERESSES, KNOW YE NOT THAT THE FRIENDSHIP OF THE WORLD IS ENMITY WITH* ***GOD****? WHOSOEVER THERFORE WILL BE A FRIEND OF THE WORLD IS THE ENEMY OF* ***GOD****. DO*

YOU THINK THAT THE SCRIPTURE SAITH IN VAIN, THE SPIRIT THAT DWELLETH IN US LUSTETH TO ENVY?" Consider how wants and desires of the flesh are the cause of all verbal and physical confrontations. Consider that in all Scriptural warnings of correction, when we see the term adultery used by GOD, we know FATHER is speaking of those who claim they are espoused to JESUS, which includes CHRIST'S ways. But FATHER finds and calls them unfaithful by their desire for the ways of the world. Also in this Scripture, when it says, "yet ye have not, because ye ask not. ye ask and receive not, because ye ask amiss. that ye may consume it upon your lust." The true interpretation of this is; you do not have the things of GOD, because you don't pray for the things of GOD. And you don't have the things that you do pray for, at least not from GOD, because you pray for the things of the world and your destruction, being things of the flesh. While we wait on our LORD'S return, let's live the things of love we know are Holy, True, and Righteous. Let's know and love our true chosen and faithful brothers and sisters, according to CHRIST. Not as individuals, but as one in all belief and purpose. Let us labor in GOD'S eternal love, and rest in that love, and suffer in that same love, and sing in and of that love. The last being as important as the first. What a wondrous gift to mankind is GOD'S spirit of music, being so emotional. And though souls in the flesh, love music also, music was not created for the godless passing of time, or dancing. There are forceful patterned musical gateways into the deep reaches of most souls that bring a deeper higher communication of feeling and desire by strumming the chords of our strongest emotions. We see an unlocking of exponential power and an amplification of forces deep within our soul, when certain music comes into the heart. We see there is a communion of awareness, which can be towards GOD, or towards evil and this world. If towards GOD, we can sense a greater presence with GOD, and share our true loving desires and praise for them. In melodious worship, these can be very powerful. The Spirituality of music can be either of love or hate, of light or darkness, as its lyrics and chords reveal and open our soul. Our worshipful music before GOD, can unlock enlightenment and strength and set us at the very Throne of Mercy. Or the music of darkness can drag one down into hell, and push souls to bang their heads against the sides of the pit and the abyss, blinding that soul and searing the conscious in pure rebellion and hate. Music by design is not just a gift, but a directional force upon our heart. And it is also a great part of the pulling apart, and separating the hearts of mankind; it is a pure emotional conveyance. Now, though music is subjective, never be afraid to judge certain music as being evil, being music we are to turn away from. Scripture says the far greatest part of the Church, in the last days will fall away. And the true Church remnant will flee into the wilderness to a place that has been prepared. How will this happen, how will all filled with HOLYSPIRIT find one another? This is an untrue question; for how can we not find each other, being led by GOD? Must there be a final end time gathering of us together in hidden places around the world? Or are we all called to come home to Israel? We surely cannot find each other if we are entangled in this world. And if we refuse the mark, and can neither buy nor sell, Prophecy says even our worldly family will betray us, for the penalty is death. So, we will be reported and hunted. This is foretold, just as our being gathered together by HOLYSPIRIT is foretold. We will have one strong love in unity, suffering, understanding and *martyrdom*. If we are of one belief and one purpose with GOD, we are one in the strength of GOD. Though we never seek death, we also do not fear it, for it is our new beginning. It is in the heart of GOD'S Children, that we are prepared to come out of this world now, not wanting to hang on to this physical bondage as long as possible. If we are taken out of this world, it is a part of our final separating. If we wonder how will the Children of GOD find each other, we see we can't help but find each other, being driven together by GOD. It is in the design of GOD'S fire that separates the pure gold from the impurities. We cannot be held by this world, having no shared continuity with it. Put a feather in one hand and chunk of iron in the other and close your eyes. If we seek the truth, being filled with HOLYSPIRIT, we will discern the differences between all good and evil, easier than the weight and firmness of iron and a feather. A Godly Spirit, and the flesh, are in all ways contrary, having opposing needs, desires, and wanting to go in opposite directions. The flesh gathers, accumulates, and gains, its desired worldly advantage, while the Spirit discerns and disseminates the eternal pure truth and eternal love. Can we be both, and go in two directions? GOD says no. In that we could have no eternal purpose. How much of the world's likeness do we each hold? How much of the world have we put into our children? Have we tried to give them every physical advantage to succeed in this world? Why would we do such a thing? Or have we given them every Spiritual strength to come out of this world, and into the Kingdom of GOD? If you are strong in the comforts of the flesh, and weak in the ways of GOD; and raise your children this way, you tie the same millstone around their neck, you have put around your own neck. Now most souls will have, in great disagreement, put this book down, long before finishing. But, if you have read this book to this point, you must sense a cause to read and consider these words. If you have felt the higher need to study GOD'S words in a deeper way, and seek more, even all of GOD'S ways in your life. Then this book is alive in your hands, and has lived and found its purpose in you. We pray you find GOD'S strength and continue in all growing faith till we meet face to face. Never look back, keep your gaze upon GOD, seeking and living in their ways and commands. Forever become one with them and us in your soul. FATHER in omniscience and power, has perceived and created all things to work in unison for the one good eternal purpose of those who desire him and his ways. We praise him in this. Amen. When you are hurting, and under persecution by the world, and feeling all alone. Remember and hang onto this understanding tightly, for no matter where you are, no matter how you suffer, no matter what is felt, be you rich in knowledge or poor, be you old or young, no matter any occurrence or any distress. Know that you can never be closer to, or know and love GOD more, than when you are in your most dire straits, and completely alone with GOD and no other soul. Say to GOD, it's just you and I now, we are all alone together, and we win no matter what comes. Accept the way, and treasure those times, no matter how hard or how long it takes, no matter the cost or the pain. Endure all things, and all experience for evermore. Accept, allow, and love your times of suffering and being alone with GOD, these are the unique eternally treasured moments to be remembered between you and GOD alone forever. No one else can share in these memories but you and GOD. Do you see what a treasure they truly are? When in Heaven, GOD can look into your eyes and say; remember when *we*... By the words of GOD, all are commanded to learn the truth that FATHER made all things to work perfect and eternal in his love. And FATHER gifted all souls, from JESUS, to all the angels, and mankind, all having pure free-will and able to rebel. Yet 1/3 of the lower angels, and the parents of all mankind, lived perfect according to GOD, only up until their fall. GOD commands us with all their love and power, to restore ourselves to their perfect order. Once we receive this truth, we are born a New Spiritual Being of GOD, yet surrounded by a world of flesh. We are then held to the knowledge and truth of our purpose by GOD. Hebrews chapter 10: 26-31, *"FOR IF WE SIN WILLFULLY AFTER THAT WE HAVE RECEIVED THE KNOWLEDGE OF THE TRUTH,*

THERE REMAINETH NO MORE SACRIFICE FOR SINS, BUT A CERTAIN FEARFUL LOOKING FOR OF JUDGMENT AND FIERY INDIGNATION, WHICH SHALL DEVOUR THE ADVERSARIES. HE THAT DESPISED MOSES LAW DIED WITHOUT MERCY UNDER TWO OR THREE WITNESSES: OF HOW MUCH SORER PUNISHMENT, SUPPOSE YE, SHALL BE THOUGHT WORTHY, WHO HATH TRODDEN UNDER FOOT THE ***SON*** *OF* ***GOD****, AND HATH COUNTED THE BLOOD OF THE COVENANT, WHEREWITH HE WAS SANCTIFIED, AN HOLY THING, AND HATH DONE DESPITE UNTO THE SPIRIT OF GRACE? FOR WE KNOW HIM THAT HATH SAID, VENGEANCE BELONGETH UNTO ME, I WILL RECOMPENSE, SAITH THE* ***LORD****. AND AGAIN, THE LORD SHALL JUDGE THE PEOPLE. IT IS A FEARFUL THING TO FALL INTO THE HANDS OF THE LIVING* ***GOD****."* This clearly states, those who have accepted the Blood Sacrifice, and then trampled on the same by continuing to willfully live in sin, will fall into the hands of an angry GOD. I admonish you to not follow those who say it is in our own power to bow and profess our love for CHRIST with our lips, and that by this simple physical leagalistic work, obtain the grace of salvation. This is not in our power, and this is a false pure legalism. We can surely ask for forgiveness, but GOD considers and chooses who gets their forgiveness. We once again point out that Scripture declares and explains clearly, there are three refining purifying elements, or steps, that bring this final eternal separation of GOD'S children from the lost. The first element, being we experience the many elements and chances filling this vast wondrous creation, created to provide us all, the truest of all free-will choices. These are true temptations to the flesh to play a god of self, that reveal deep worldly desires in many. These temptations quickly separate out all souls who greatly desire such things. This first element of choice under temptation, is temporary to this creation, and necessary to fulfill the last two elements. The second element rests in souls that spring up in hearing about GOD, having a desire for there to be something greater than a temporary dying love, flesh, and creation. These find only love, comfort, and peace in knowing there is a Creator of it all. And the fewer of these, see the temporary ways and wants of this world as being contrary to their eternal hope and desire. These see no eternal pleasure in wealth, or the physical beauty of this world. These see worldly possessions lure souls into a blind bondage and false worldly security. By their choice for a greater hope, joy, and love, these fewer perform their good works to learn

GOD'S perfect ways and separate themselves from this world, and adjoin themselves to GOD. This stronger element of desired self-separation, is declared to not yet be of GOD'S purest nature. This second element must be tried further until found pure, true, and enduring. The third element, is in GOD'S final trying and Judgment of these, and in GOD sealing those he has found to remain faithful. We are judged in these three steps, where we see those that can hear and have heard, are the called. And from all the called, the fewer are the chosen. And from the chosen, there are the fewer still, found in the end to have been faithful and eternally responsible at the final Judgment of GOD. Remember, JESUS said to his Disciples, have not I chosen you twelve out from all those who have been called, yet one of you is still a devil in the end. We are all astounded, that Judas could turn away, after beholding the loving presence of CHRIST. At the last, being unable to love and accept all CHRIST'S commands to go on *without* JESUS there in the flesh, and change who he was in his deepest desire. And we see Judas finally fell away. GOD saw Judas's greatest weakness to be tried, being his desire for wealth. So, CHRIST had him carry the bag with the contributions to their ministry. He had this money in his possession at all times, calling to him, tempting him to run away with it, but he resisted. But then at hearing JESUS say he would soon be crucified, Judas was tempted by 30 pieces of silver, a vast amount above anything he had ever seen. And there he sadly lost his fight. Let this and the fallen angels be warnings to us all to strive in love to know and endure; flee all temptation, and watch in fervent prayer and wait. Galatians 1:6-12, *"I MARVEL THAT YE ARE SO SOON REMOVED FROM HIM THAT CALLED YOU INTO THE GRACE OF* ***CHRIST*** *UNTO ANOTHER GOSPEL: WHICH IS NOT ANOTHER; BUT THERE BE SOME THAT TROUBLE YOU, AND WOULD PERVERT THE GOSPEL OF* ***CHRIST****. BUT THOUGH WE, OR AN ANGEL FROM HEAVEN, PREACH ANY OTHER GOSPEL UNTO YOU THAN THAT YOU HAVE RECEIVED, LET HIM BE ACCURSED. FOR DO I NOW PERSUADE MEN, OR* ***GOD****? OR DO I SEEK TO PLEASE MEN? FOR IF I YET PLEASED MEN, I SHOULD NOT BE THE SERVANT OF* ***CHRIST****. BUT I CERTIFY YOU, BRETHREN, THAT THE GOSPEL WHICH WAS PREACHED OF ME IS NOT AFTER MAN. FOR I NEITHER RECEIVED IT OF MAN, NEITHER WAS I TAUGHT IT, BUT BY THE REVELATION OF* ***JESUS CHRIST****."* I too, stand in this self-same affirmation with all my knowing heart. In our LORD'S kindness, he reminds us, and we do remember that all the realities we experience in our awareness, come only by our sensed patterns and forceful desires projected upon FATHER'S most basic physical points of reality, being his created aether. And in seeing how all conscious forceful physical creations, also create energy; we see energy has no original initial standing. Energy only secondarily appears in the wave momentum, amplitude, and frequency, of all moving created realities. Being caused by GOD above, and all lesser creative souls. What does "now you see it, and now you don't" mean. Scientifically it means all worldly physical experiences are for a short moment, so they are an illusion in need of constant re-experience. All Godly spiritual ways, by their supreme nature and perspective, are eternal and contrary to all temporary physical ways. Pure Dualist know that our minute conscious soul, and a substantive living sensing body, only exist in tandem. And for true followers of CHRIST, our soul must always control the flesh, being a Scriptural wisdom we have received in love. And again, for Materialists who believe, or state, or imply, that the sum of all unseen quantum aspects combined, cannot support, and do not constitute the perimeters and parameters of a quantum ethereal realm of Heaven, capable of sustaining immense ethereal particle/wave-based bodies of angelic life. Then please describe for me the perimeters, parameters, properties, and complexities, that constitute the realm you believe allows you to physically exist. Define why a physical realm, can and does support a physical form of interface/relay for our soul, and why a quantum realm cannot support an ethereal form of the same. Does this firmer realm support a firmer life, because it is entangled, cohesive, and reactive in every needed way in which mankind can perceive. This is also true for the quantum ethereal arena. Can any single quantum aspect of this physical realm, be absolutely separate and untangled from the rest of reality? No. So, all the perimeters, parameters, qualities, and entangled reactive complexities that form the quantum realm itself, must also have no single aspects that can be completely separate from reality. This quantum realm has every needed entangled aspect, filled with the endless interactions and reactions within it to sustain the firmer lower realm of life. Yes, when seen as a differing realm, it obeys differing laws, but there can exist no understanding that can disqualify the quantum realm as a vast arena that supports immense ethereal life. Now by opening up to the truth that there are GOD'S many loving mercies woven throughout reality, we see a rarer certain mercy, that as with all mercies, we must experience it, to comprehend it. Like with HOLYSPIRIT, until we are indwelled by him, we can't truly understand his purpose and wonder. Throughout Scripture, we see we are called, each and every soul to our accountability, and to submit ourselves to GOD'S ways for approval. In this enduring wisdom, is where our heart should always walk and dwell, unwavering and steadfast. Yet we find in 1st Cor. 7:13-16, a miraculous gift offered to us for our loving obedience to GOD; that our children are thereby made Holy offerings. This unusual mercy is not given by GOD as an absolute promise made lightly to every parent having the LORD'S name on their lips. This loving gift is towards the fervent work, that an obedient parent has in the hope and need for the children they raise. That their good loving work towards these children should hopefully not be lost. This is, as is our first mercy of forgiveness, a mercy we cry out for, and a mercy we know only upon GOD granting it. We see by our faith and unending efforts to deliver to GOD, the tiny new souls we raise, and in them we have planted the truth of GOD. And by our firm relentless grip on GOD'S WORD, and our loving obedience to the same. We find GOD'S love can have this rarer extended mercy upon our as yet, imperfect children. This gift of spiritual propagation towards us, is for our steadfast unwavering hope in GOD, that the children we lovingly raise may also be saved. But how far can this mercy extend. As with all mercy, this mercy is granted to whomever GOD righteously wills. We cannot take it, or claim it, we can only pray for it. For each soul, the circumstance by which it will be given are under GOD'S Holy Judgments. Let us strive and endure, that we may receive all his mercies; and if we fail ourselves in this, we fail the children we raise and love. GOD asks us all how shall we hold to and care for eternal love? This is the great calling of love that forms GOD'S Family, and that hopefully seals the children we raise. And how shall we not eternally lose this love? It has been put forth by the truth, that eternal love and the Scriptures are the living waters of GOD. But can we hold onto these living waters in weakness of fearing physical death and this evil world? No. Can we be uncertain and wavering with these living waters, holding these waters in gentle open cupped hands in fear of losing or offending others? No. All we will find, is that this world pushes, pulls, and tosses us about, as we watch these living waters spill from our hands all the days of our life, leaving only old open empty dry barren hands. We ask in great fear of GOD, can the promise of these living waters be squeezed tight and held as commanded, to secure the children we raise. When the stronger we squeeze and grip these waters, the more they seem to again, pass through our fingers. I squeeze and hold tightly the words of GOD, yet I buried my son in his imperfections, and was left in unbearable grief. I cried in blurred teary vision and grief to GOD for his promise. And after ten nightmarish days and nights of being at the edge of my own reality, taking no food and staring into my darkest loss, GOD came near, in the moment my greatest ever anguish sorrow and pain had brought me so near to my own grave. At the far outermost boundary of my grief, and the very edge of my ability to understand, GOD came. I sat there with only tears and my tightly clenched hands holding nothing but GOD'S promise. I then saw a vision through my tears, as GOD held out his open palm for me to give him something. And as I opened my clenched hand, I saw at the center of my hand, one bright tiny glowing drop of water, which by my tight grip on GOD'S WORD could not be lost. In silence, I gently rolled that bright tiny drop from my hand, into the loving hand of GOD. In that moment, I experienced GODS most unusual promise of a merciful chance for my son's soul in an experiential knowledge and qualia. I committed in all loving trust and joy, the eternal soul of my son to the care of GOD. I stood, dried my eyes, and returned to GOD'S work.

THE LAST DAY OF GREAT PERMANENCE

We know that all things conscious and physical, are at present, existing in constant flux. All physical things and most conscious beliefs are also now temporary, confused, unsettled, and blurred. But soon comes a great permanence of division, created by GOD. Let Messianic Jews and Judaean Christians, praise his name. Let those called and chosen and faithful praise the LORD, who by his Covenant of redemption and adoption, are raised the true children of Judah and Israel. Let the SON in his great mercy and desire towards us as our KING, reign over the breadth of free-willed reality forever, and all having HOLYSPIRIT, who adjoins us. Dear FATHER, most all your Creation has lost its way and is turned aside, blinded to all your true gifts that are and will be. We cry to your SON in his love eternal, for his righteousness to harbor our souls as his brothers and sisters. Amen. From the greatest of the Holy Angels, to the weakest Saints of mankind, we know that only by living in all GOD'S ways and commands, can we convey to the world the one truth and eternal love and light of GOD. Are not all Spirits of GOD, righteous and holy. Scripture tells us to try all spirits, and to try our spirit, to see if they, and we, are truly of GOD. In this search, we look only for GOD'S Righteousness and Holiness. When seeking out the Sons and Daughters of GOD from among the living on earth, we look for Holy and Righteous evidence, and never any semblance or associations with this dark Kingdom. We must hear the WORD being rightly divided, knowing GOD'S Holiness and Righteousness do not dwell strictly within their existence, but in our LORD'S grace and mercy towards the faithful gifted creations of life. And our Righteous Holiness dwells not only in our search and finding the oneness and perfection of GOD, but in our pure loving desire to eternally share and remain in the same. As long as we strive to overcome all that is in the world, and do confess and *repent* our every offence against GODS's eternal ways, and are not reticent, but openly ashamed, humbled, and active in GOD'S work. Then, we walk in GOD'S designed Righteous Holiness. We are created to stand perfect as a Spirit before GOD. If any accept a doctrine that does not call you to your perfection, but allows your endless imperfection; or fails to teach you what is your perfection, such doctrines are false. We do always look into the mirror of our soul to see where we truly are. Are we guilty of not seeking the full measure of truth and wisdoms given to us in Scripture? Are there GOD'S deep lessons and wisdoms we do not seek, and neglect to find? Harder wisdoms we might not want to see, and so never find. What about the Prophecy of GOD; does one read once, twice, three times in the flesh, and say it is too confusing. Or can we each be clearly shown GOD'S coming plans. If we know we are in the last of the last days, is Prophecy then a staggering knowledge, deeper and fuller than the oceans with meaning and wonder. But as great as the wisdom of these coming trying events is, it is followed by the far greater wisdom and understanding of the final great division of permanence. The cleansing moment of GOD'S final corrections and divisions and permanence, are the moment when GOD can go no further for mankind; a moment when GOD will have done all they can, for as long as they could. And yet remain True, Righteous, and Holy, in protecting their true family found along the way. In reading of GOD'S Prophesied permanent eternal divide, we see GOD remains eternally loving. But after the last moment of our LORD'S Grace, for the unrepentant it will be too late; their moment of being able to find and accept mercy, will have passed. For only in their desire, and during this third current era of Grace, is CHRIST'S gift there to be received. And with the passing of its last moment, this era of MESSIAH'S sacrificial grace will stop, and never again start. There has never been a sound so loud in all of reality, as GOD setting in stone, that freely chosen permanent line between souls, who are forever divided by an eternal gulf. Souls we have loved, will be on the other side from us, and what will be our regrets for them or ourselves? Did we do all we could? If we say we are called by GOD, do we just rest in that statement? Is our true rest not found in answering FATHER'S call to live in all his created ways? We can all feel and believe we are currently vested in love with GOD; but none can see what we will be at that last moment of our life. Souls fall in and out of love all the time, and I have seen this occur even with love for GOD. Not one Scripture says all of mankind or all the angels love GOD equally. Not one Scripture says GOD loves every angel, man, woman, or child, equally. But an abundance of Scripture says GOD loves those who love them; and that those who seek them earnestly, will find GOD. Scriptures say some love GOD more than others. and GOD loves some, more than others. There are Scriptures that say GOD hates sin and those who commit sin. There is an abundance of Scripture, describing the evident obedient ways, acts, and works of life, seen in all who truly love GOD. Ways, works, and acts, found in our turning from this world, and living as GOD knows our name and has called us under one name. We all have witnessed how a love being neglected or treated without proper respect, is injured and even lost. Love can be nurtured and grown, or it can be destroyed. It abides in our free-will desire and effort, and does not eternally exist of its own force. This coming permanence was destined and told of old, and set out as being inevitable and immutable by, and at the fall of all creation. Not by any arbitrary desire within GOD, but by all our needs, hopes, and requirements, found in eternal love, truth, righteousness, and holiness; and by every other soul's rebellion to the same. We see a good loving purpose that GOD can never alter or waver from. GOD has never changed, and cannot change from this. GOD is set in pure eternal perfection. There is only a loving eternal permanence in GOD'S purpose, whose perfection is bound by their omniscience, omnipotence, and omnipresence forever. GOD has always been perfect, and by all that is, GOD will eternally be so. In his omniscience, FATHER cannot err, he cannot be deceived, confused, or tempted, as all reality is born of him. But in our lacking his full omniscience, we are susceptible to temptation, by confusion through the senses of the flesh, and the desires of the soul. There are many wondrous lessons in this life to learn, but the one we seek to share is the great beauty that lies in the purpose of the will of GOD, and the wisdom within their desires. Just these two aspects of GOD are immense books unto themselves. GOD formed all things for finding perfect free-will eternal love, and a perfect eternal family. And in all this, we still hear GOD calling us into their perfection, which is our eternal protection from destruction. Yet by most gifted souls not desiring GOD'S omniscient love, we see great spiritual imperfection, to which many Angels in

Heaven fell prey. Many souls say the imperfections of pain and death we see in creation, are evidence that GOD does not care, or exist; but I tell you that this waring imperfection, is the greatest evidence that a loving perfect GOD, and all lesser free-will entities do exist. And that GOD did create all we see for a perfect purpose. It is only by pure free-will choice, and the loving and hateful choices made, that a true love for GOD, born of eternal desire might arise in some. Being our response to the pure eternal love of GOD. FATHER'S perfect love is best seen against the striking contrast of all individual chosen imperfections and hate. In omniscience, FATHER perceived a perfect vast complex dynamic designed Creation, being a glorious labor and wondrous work. This all for finding eternal love, as the laws of eternal love, set and order every direction, way, and intention of FATHER. The whole of Creation, down to its smallest unimaginable complex aspect, is designed and formed for one purpose. To bring FATHER'S eternal family, from his SON, and HOLYSPIRIT, down to the least who would love without reservation and trust FATHER. All such souls have come through, and are purified by love and truth's fire. Being led by eternal love, to seek, learn, and desire all eternal righteous perfection and holiness. To never do as Satan or Eve or Adam; and this pleasing FATHER in all his ways.

We are all born and meant to be in part, as one with GOD, through the Covenant of CHRIST, pouring out his life's Blood upon the Earth for the redemption of the fewer who would see and know and find him. By the greatest gift of love, strength, patience, and the baptism of HOLYSPIRIT, and every lawful word of love and righteousness, and by our strongest desire, this covenant is now written by GOD upon our hearts and minds. Live to be loved by FATHER, and as commanded and called, walk perfect before him, for this is our love for GOD. Over decades of searching and asking, we learn many things. One such thing is seen as a mountainous rock, permanent, immovable, and hardened as granite. Being that if we neglect and do not always search for more of GOD'S ways and truth, this world will not cease in trying compromise us, and our path will be compromised. The more we read, pray, live, walk, and even write, with HOLYSPIRIT, the stronger more assured our exercised loving understanding becomes. Let us all see, find, and secure the promise in GOD'S calling, and find rest in answering every call of GOD without fail. We are given ten, and then eleven great commandments. We are given their greater Levitical comprehensive expansion as our foundation to build our soul upon. We have been given all the strongest force and love in Creation. We have been given the greatest love poured out in suffering, shame, and blood. We, are given by HOLYSPIRIT, the knowledge of the whole Creation and all Prophecy. But the one thing we have never been given, is a license or excuse to continue in sin and offence. There is no accepted excuse for our willful walking in disobedience, or holding desires for this lost world. Romans 8:1-14, *"THERE IS THEREFORE NOW NO CONDEMNATION TO THEM WHICH ARE IN* ***CHRIST JESUS****, WHO WALK NOT AFTER THE FLESH, BUT AFTER THE SPIRIT. FOR THE LAW OF THE SPIRIT OF LIFE IN* ***CHRIST JESUS*** *HATH MADE ME FREE FROM THE LAW OF SIN AND DEATH. FOR WHAT THE LAW COULD NOT DO, IN THAT IT WAS WEAK THROUGH THE FLESH,* ***GOD*** *SENDING HIS OWN* ***SON*** *IN THE LIKENESS OF SINFUL FLESH, AND FOR SIN, CONDEMNED SIN IN THE FLESH: THAT THE RIGHTEOUSNESS OF THE LAW MIGHT BE FULLFILLED IN US, WHO WALK NOT AFTER THE FLESH, BUT AFTER THE SPIRIT. FOR THEY THAT ARE AFTER THE FLESH DO MIND THE THINGS OF THE FLESH; BUT THEY THAT ARE AFTER THE SPIRIT THE THINGS OF THE SPIRIT. FOR TO BE CARNALLY MINDED IS DEATH; BUT TO BE SPIRITUALLY MINDED IS LIFE AND PEACE. BECAUSE THE CARNAL MIND IS ENMITY AGAINST* ***GOD****: FOR IT IS*

NOT SUBJECT TO THE LAW OF GOD, NEITHER INDEED CAN BE. SO THEN THEY THAT ARE IN THE FLESH CANNOT PLEASE ***GOD****. BUT YE ARE NOT IN THE FLESH, BUT IN THE SPIRIT, IF SO BE THAT THE SPIRIT OF* ***GOD*** *DWELL IN YOU. NOW IF ANY MAN HAVE NOT THE SPIRIT OF* ***CHRIST****, HE IS NONE OF HIS. AND IF* ***CHRIST*** *BE IN YOU, THE BODY IS DEAD BECAUSE OF SIN; BUT THE SPIRIT IS LIFE BECAUSE OF RIGHTEOUSNESS. BUT IF THE SPIRIT OF HIM THAT RAISED* ***JESUS*** *FROM THE DEAD DWELL IN YOU, HE THAT RAISED UP* ***CHRIST*** *FROM THE DEAD SHALL ALSO QUICKEN YOUR MORTAL BODIES BY HIS SPIRIT THAT DWELLETH IN YOU. THEREFORE, BRETHEREN, WE ARE DEBTORS, NOT TO THE FLESH, TO LIVE AFTER THE FLESH. FOR IF YE LIVE AFTER THE FLESH, YE SHALL DIE: BUT IF YE THROUGH THE SPIRIT DO MORTIFY THE DEEDS OF THE BODY, YE SHALL LIVE. FOR AS MANY AS ARE LED BY THE SPIRIT OF* ***GOD****, THEY ARE THE SONS OF* ***GOD****."* These Scriptures declare that the true Baptism and indwelling of HOLYSPIRIT, centers all good things within us. And that by receiving and using the super-natural spiritual gifts given in this Baptism, the flesh and its desires for this world can all be crucified. And raising up a new freed Spiritual creature who is not a slave to such desires. This physical body we were once enslaved by, is to be a clean and Holy Temple for HOLYSPIRIT and us. By these strong Scriptures, it is evident that the rebels within the Church, do not see the world for the evil it is, and they cannot endure its temptation. And for all who do not crucify the flesh daily, and come fully to GOD, they will all too soon feel and know their permanent separation from the Kingdom. The many religious yet rebellious, do not see or understand that they have chosen a nature desirous of constant change, as they fear an eternity set in continuous perfection. Most souls desire and need things to keep changing, bringing them new experiences, be they bad or good. They don't realize that they need the drama and conflict of differing beliefs, even though some offend them. They need this suspense, vengeance, and opposition in life. They see an eternity of peaceful oneness and continuity of truth, as unfulfilling and intolerable. These need the contrast of death and hate to feel they are loving and alive. They need to sense their own individually defined goodness, as a contrast to evil. And now it is even preached in some churches, that the Angels of GOD, will serve mankind in Heaven. The very words of our LORD, say we will be like the lowest angels. All souls who truly hold to CHRIST, hold to GOD'S unchanging eternal truth and love. Constant chaotic change towards equity in all things, is actually and literally a disordered aspect of spiritual division and entropy. There can exist no eternal rising complexity in that. So, we all must live by GOD'S unchanging ways to be eternal. Do not pray for GOD to in any way heal this sinful world; it is foretold that will not happen before the Millennial Reign. We pray for the current and future Children of GOD. We pray for Israel, Jerusalem, the unborn, for religious liberty, simple provisions, and a public square where we are free to hear and speak the truth, and share the love of GOD. But pray no more for this confused, covetous world to find righteousness. FATHER cannot force them, nor would he if he could. If they cannot want GOD, GOD does not want them. This is how every love eternally works and thrives. The WORD is true, GOD has spoken, and Prophecy always comes to pass. Pray for GOD'S Kingdom and those who might yet be saved. LORD come quickly, as your great evil days are upon, before, and all around us. This World is given over, and lost as Sodom and Gomorrah. We who have never felt any affinity with this lost world, know we don't belong here, in seeing everything in this world is askew and amiss in fearful confusion. This world is a passing vapor, and not the true reality we see or seek. Brothers and Sisters, if you had all the wealth and knowledge of this world, so that the whole of the world turned to you to say all earthly things are yours for the asking, but you received not the love of GOD. You would be on the very precipice of knowing and holding the feelings of eternal emptiness and anguish, and so, being void of all eternal joy and rest. You would possess all things illusory, and no real joy or love. All that is in this world, soon vanishes. And what we need, no matter the denials, is to find, know, and please GOD. But who are we to even seek out GOD? And if we find them, what can we bring and give GOD? Can we give them glory? Yes. But all glory was GODS already. Can we give them praise? Yes. But all praise is already theirs, for no other is above them. Can we exalt CHRIST? We can proclaim he is exalted, but FATHER exalted his SON and HOLYSPIRIT. Do we possess anything that is ours alone to give, that can be our good offering? We have only one thing that is eternally ours alone. If we see ourselves in all honesty, and all truth, all we are given, all we eternally are and can ever be, is our simple free-will choice of who we love and how we live to show that love; being a choice made out of *our* desire. This one small thing is all that we truly are, or can ever be. It is what became us and ours alone, when FATHER by his desire for love, divided our souls out from his. And as a minute portion of his

Soul, he completely let us go, giving us eternal free-will choice. This alone was given eternal, while this current fleshly platform we are loaned, will die once and forever. Everything physical, beyond our conscious created perceptions, are all the complex layered generated physical perceptions of FATHER, including all physical life. For mankind's rebellion, we must surrender all fallen defiled physical things on loan from GOD. So, only by our choice to live in GOD'S ways, in our trust of their love and wisdom do we show that we love GOD. Otherwise, we have chosen to love ourselves and this world more. We cannot love both more; we choose to love GOD more, or parents, spouses, siblings, children, friends, and this world more. Can good ever mix with evil? Is there ever communion between darkness and the light? Can we go to FATHER in worldly physical want and need, and bring no true passioned loving worshipful obedience? But his true children go obedient in their love, washed pure and clean in CHRIST'S Blood, and indwelled by HOLYSPIRIT. In being one with GOD, we fellowship no more with any wanton desires of darkness. And we know that fulfilling GOD'S will, includes receiving his Prophecy. We live to please FATHER, by living in all his created ways. We clearly see how lust, wants us to live in all the temporary ways, pleasures, and provisions of this world; but eternal love wants only to give, share, and live in all the eternal spiritual ways and things of the next world. These are GOD'S laws of love, and the fallen laws of lust. You can make up your own laws of desire to live by; but I assure you, GOD will never consider, allow, or obey, your laws. GOD'S ways, beliefs, and promises, establish eternal love, life, light, and their house eternal. In this world, all the fallen spectacles that surround us with physical sight, sound, taste, touching, smells, and all the strange contrary evil things that people do. These all come by individual false beliefs and perspectives of love and truth, built upon individual desires and hopes. Consider, that if we live in the eternal perspective of GOD, and receive HOLYSPIRIT'S teaching and guidance, we then consider and know reality in much faster, larger, and far more meaningful ways. And when we converse with souls having a slow small temporary physical perspective and false individual understanding of reality; we see they all run in the opposite direction, shrinking, and existing at a faded conscious speed. Our Godly understanding of reality, is the fuel that raises our conscious speed. So, in effect, we see all godless souls live in their our own chosen and desired, slowing relative comprehension of existence, as it unfolds within and around them. Imagine being one in understanding and the conscious speed of the mind of CHRIST, or even a pure but substantial portion thereof. And in that our perspective of perceptions, are relative to momentum, and that all created things are in fact consciously relative, our conscious speed becomes a critical factor. We see all fallen conscious hell, death, and time, falling and slowing in relation to our sum true understanding of GOD'S reality. You can consciously break free from the slow small and temporary, to find GOD'S eternity; not physically, not yet, but consciously and spiritually. Again, GOD'S true knowledge of reality, speeds us in the right direction. So, our added knowledge of true or false belief, increases or slows our conscious speed. True and false, form two literal conscious spiritual affects. One speeding up towards GOD, and one slowing and falling away from GOD. All false understanding and belief, pulls one down into the slowing shrinking timely temporary aspects of conscious hell and death. By this, we see within reality, that a spiritual oneness between all souls occurs only while we live within all GOD'S created and allowed, non-volatile, aspects of deference and choice between all souls. And by sinful rebellion, we see only endless volatile divisive aspects, born of individual choices outside those boundaries. We sense and seek only that oneness of spirit by living within all GOD'S created non-volatile provisions and choices; and we see only division in all chosen personal differences. GOD'S ways work eternally, individual ways do not. Many souls hold to a proximity religion, or proximity belief in GOD. But by GOD'S words and ways, we are called to hit the mark. Meaning GOD alone, has set the eternal standard we are to aim for, and hit, not us. Now when we consider something spiritually profound, as it runs through our soul, say three or more times, as we use more meaningful words each time. We sense our feelings for its spiritual meaning increase each time. So, we are sensing the effect of our conscious speed increase. Each soul's comprehending their own conscious speed and direction, remains subjective in relation to their soul's levels of understanding GOD'S created reality. Yet GOD sees us objectively, above our own sense of who and what we are. We see mankind's false concept of time, consciously and spiritually slows and finally disappears in relation to our Godly understanding of eternal life. So, if we remain adjoined in CHRIST'S mind, through HOLYSPIRIT, we can clearly see and understand reality at, and even beyond the slowest speed of light. Again, the concept of time is found to be relative to conscious momentum; so again, all created things, must in all ways be consciously relative, so we can (not physically) but consciously break free of the physical aspects of time, and sense eternity with GOD. Please remember time is a concept that always runs out, but eternity does not.

THE LAST DAYS OF THE GREAT PROPHECY

A great complete *chronological* Revelation of many certain recent, current, and coming events, was given by JESUS to his beloved Apostle and Prophet, John. And for us to receive it all in its fullness, and the pure way it was given, requires we first lose something; being all false beliefs. A Prophet is a soul who hears the actual Prophetic voice of GOD. Being a soul who is told to record and share GOD'S warnings of things to come with all who can hear. Now to truly understand FATHER, we must know what he knows about who he is, what he has created, and even how he has created it. We must believe what GOD believes, and see and hear what GOD sees and hears. So, for us to understand Prophecy requires those same attributes. So, we will go over a little of that before we begin. Can you imagine being chosen to hear the actual voice of GOD, and how afraid *you* would be. But John had already walked with CHRIST for three years, listening to his voice. We see John, CHRIST'S beloved Apostle, then became a major *Prophet,* in being given this faceted Revelation of a future FATHER will bring. And it fills the last 22 Chapters in the King James Bible. Now why do most preachers see John as only an apostle, when by this Prophecy, he became the second greatest Prophet. The greatest of all Prophets was Elijah, being resurrected as John the Baptist, seeing Elijah had never died, being carried up to Heaven in a Chariot of fire. We must in our earnest search, trade all false beliefs, for the one pure truth of GOD. We must overcome false traditions, cultures, and all false teachings of confused preachers. Scriptures say the truth is for the few who fervently search for it. For some indwelled by HOLYSPIRIT, it is a loving fervent patient thing, for others it's more difficult. But for all not indwelled by HOLYSPIRIT, it is impossible to truly see and hold. Being in the flesh, the spiritual eyes remain darkened by false beliefs and desires for things of this world. LUKE 13:23-24, *"THEN SAID ONE UNTO HIM,* ***LORD,*** *ARE THERE FEW THAT BE SAVED? AND HE SAID UNTO THEM, <u>STRIVE</u> TO ENTER IN AT THE STRAIT GATE: FOR MANY, I SAY UNTO YOU, WILL SEEK TO ENTER IN, AND SHALL NOT BE ABLE."* JESUS says many things here in answering that question "are only a few saved." The first word to focus on is "*strive*" which means to struggle against an opposing force with determination. If you do not strive in a determined level of passion and effort to know GOD, you must ask yourselves; are you currently on a vain, less-impassioned path. Next, we see the phrase "*the strait gate*" meaning the one strait narrow path to salvation and Heaven. It is further described in the Synoptic Gospels, as the narrow gate, and not the wide. This narrow strait gate, means a pure doctrinal truth, interpreted by HOLYSPIRIT. Then we see the term "*the wide gate*" which means a broader weak individual interpretation of Scripture. And finally, we read those very hard to hear words from JESUS "*FOR MANY, I SAY UNTO YOU, WILL SEEK TO ENTER, AND SHALL NOT BE ABLE*" even though they seek to fit through the narrow strait gate. Why? Because of the fleshly worldly desires they cannot overcome, they just can't fit through. Example: Judas strived and held for three years, but at JESUS saying he is leaving, Judas finally fell under his desire for power, privilege, and possessions. We pray you receive the eternal perspective and truth of this detailed study, and know you remain within reach and hope in these last days. When JESUS said, in Luke 5:30-39, that it is hard to add new cloth to old, or put new wine in old bottles; he was saying, once a soul is long invested in a belief of false doctrine, their soul is conformed and hardened against change. So, it is now resistant, or unable to hear the pure ancient truth, it being to them a new and strange quality and quantity. Such conscious states stem from a hard-hearted unwillingness in mankind, to be responsible for what they believe. Instead, many feel they are the epicenter of GOD'S love, and GOD would never allow *them* to be taught lies in the first place. They think GOD'S Truth a simple thing that all can hear in their own way; even as Scripture warns so many times, to guard against this very deception. We are each responsible for *desiring* and finding not our truth, but the truth of GOD. In all this, we see the destructive powers and results of each lie of false belief at work. JESUS'S parable of a cloth and bottle, he leveled at the Scribes and Pharisees, describes how hard to impossible it is for many to admit they have all their life, been taught and believed a false doctrine. So, for their pride and lust for their world order to continue in its ways, they refuse to see their self-imposed deception. They could not tear down their false life, and put their false beliefs away from their heart and mind, to receive CHRIST'S correcting true doctrine, and build a true life. The parable of a vessel or garment, representing the soul, that for most is now worn, stretched, stiff, and so unable to accept and hold a new doctrine. But holding portions of the new mixed with portions of the old, is another evil in itself. Complete denial of the old, and complete acceptance of our correction can be difficult. Mine took several years of desired effort; but with GOD and great desire, all things are possible. It is so important to be patient from the beginning in our search for GOD, and pray and wait for HOLYSPIRIT. Do not rush into promiscuous doctrines, and never see yourself at the epicenter of FATHERS love, where JESUS resides. Many convince their self that GOD would never let them choose what does not please him. Scripture clearly says that only the *fewer*, who through patience and the fire of trying and putting away the imperfect world, find GOD'S truth. Now you must understand, that the Holy interpretation of Prophecy is not the milk of GOD'S truth, it is the meat, sinew, and bone, for a stable strong soul. We will first read a few Scriptures telling us the importance of Prophecy, and how we should seek it. 1st Timothy 4:13-15, *"TILL I COME, GIVE ATTENDANCE TO READING, TO EXORTATION, TO DOCTRINE. NEGLECT NOT THE GIFT THAT IS IN THEE, WHICH WAS GIVEN THEE BY PROPHECY, WITH THE LAYING ON OF HANDS OF THE PRESBYTERY. MEDITATE UPON THESE THINGS; GIVE THYSELF WHOLLY TO THEM; THAT THY PROFITING MAY APPEAR TO ALL."* 1st Corinthians 14:5, *"I WOULD THAT YE ALL SPAKE WITH TONGUES, BUT RATHER THAT YE PROPHESIED: FOR GREATER IS HE THAT PROPHESIETH THAN HE THAT SPEAKETH WITH TONGUES, EXCEPT HE INTERPRET, THAT THE CHURCH MAY RECEIVE EDIFYING."* 1st Corinthians14:29-

33, *"LET THE PROPHETS SPEAK TWO OR THREE, AND LET THE OTHER JUDGE. IF ANY THING BE REVEALED TO ANOTHER THAT SITTETH BY, LET THE FIRST HOLD HIS PEACE. FOR YE MAY ALL PROPHECY ONE BY ONE, THAT ALL MAY LEARN, AND ALL MAY BE COMFORTED."* 2nd Peter 1:19-21, *"WE HAVE ALSO A MORE SURE WORD OF PROPHECY; WHEREUNTO YE DO WELL THAT YE TAKE HEED, AS UNTO A LIGHT THAT SHINETH IN A DARK PLACE, UNTIL THE DAY DAWN, AND THE DAY STAR ARISE IN YOUR HEARTS: KNOWING THIS FIRST, THAT NO PROPHECY OF THE SCRIPTURE IS OF ANY PRIVATE INTERPRETATION. FOR THE PROPHECY CAME NOT IN OLD TIME BY THE WILL OF MAN: BUT HOLY MEN OF GOD SPAKE AS THEY WERE MOVED BY THE* ***HOLY GHOST****."* The last sentence of Rev. 19:10, *"FOR THE TESTIMONY OF* ***JESUS*** *IS THE* ***SPIRIT*** *OF PROPHECY."* Rev. 22:18-19, *"FOR I TESTIFY UNTO EVERY MAN THAT HEARETH THE WORDS OF THE PROPHECY OF THIS BOOK, IF ANY MAN SHALL ADD UNTO THESE THINGS,* ***GOD*** *SHALL ADD UNTO HIM THE PLAGUES THAT ARE WRITTEN IN THIS BOOK: AND IF ANY MAN SHALL TAKE AWAY FROM THE WORDS OF THE BOOK OF THIS PROPHECY,* ***GOD*** *SHALL TAKE*

AWAY HIS PART OUT OF THE BOOK OF LIFE, AND OUT OF THE HOLY CITY, AND FROM THE THINGS WHICH ARE WRITTEN IN THIS BOOK." We hear these verses speak of our dire need to seek the full interpretation of all Prophecy, and the great consequences of weakness that will arise in all who do not hear it. We must know Prophecy is given *after*, and in addition to learning the foundational truths of GOD'S ways and desire; and can often require years of patient effort. Prophecy is given as a truthful account of coming events that all, from Church leadership down, must know. For this reason, there are important basic points we will reiterate to a point of irritation, causing the soul to consider the absolute importance of receiving the one and only true interpretation. We must first have no doubt that this Great Revelation given by CHRIST, to John, is in all ways, the supreme, most complete, and most well-defined of all Prophecy. It is chronologically structured and numbered as an order of events that unfold from the first four seals releasing the Four Horseman, following all the way through to the last vial and the coming New Earth and New Heaven. We repeat, it is all written in the chronological and numerical order in which these things will occur. If you cannot hear and accept this precise order, you will wander through it in false notions, trying to reconcile things that then appear contradictory. We hear many preachers mix and teach a false order of these events. For instance, many preachers mix things from the fifth trumpet with things from the sixth vial as if they are one event. There seems no end to such confused worldly teachings. We cannot change GOD'S order of things, by any private interpretation. The Great Revelation of JESUS, is set as the last true words in the ancient Book of GOD. This shows it is the supreme final chronological Prophecy, and all lesser segments of Old or New Testament last days Prophecy, must be received in the same truth, to properly align and over-lay this supreme Revelation. And in this they enhance and confirm CHRIST'S chronology. JESUS gives this numbered ordered unfolding of every-occurrence, for a most important reason. The first seal is opened, and that event occurs first; then the second. And after the last of 7 seals is opened, we see the first Trumpet sounds, and the others in their order. Then after all Trumpets have sounded, we see the first vial poured out, and the second and so on. If you get irritated at hearing this point repeated, you have probably accepted this important fact, and that is a good thing. But it being such a great problem in the Church, in that many others have not; and with reiteration, they might. This great Revelation from JESUS comes with not only the greatest wisdom, but an abundance of dire warnings.

There are many questions asked about the end of days; but my most frightening question, is why are there so many differing interpretations of these end-time events being taught. We can visit a different church each week, we can turn to TV on Sunday to hear televangelist; and if any each speak on Prophecy, far more often than not, we hear different interpretations of these last day's events. We often hear partial Scripture references, steeped in vagueness, taken out of context, and rendering false conclusions. If you are led by HOLYSPIRIT in following through this full examination, you will see why most do not comprehend it. If you stay with us as we put these many pieces together, at the conclusion you will see a wondrous picture in every detail. HOLYSPIRIT cannot give different souls, interpretations of one event that differ. This would be lying, and you know what I just said must be true. All who truly hear HOLYSPIRIT, hear the exact same thing. Shall not these physical events unfold as GOD has said he shall cause them to unfold? Does Satan want us to know these things? No. So, we know our gained Spiritual strength in receiving this Prophecy, out-ways the cost of our search by orders of magnitude. We must accept that all events leading to, and after CHRIST'S return, and into his millennial reign, and further, are beyond any worldly knowledge. Towards the very end, so much will happen so fast, that those still alive will be in shock as the world experiences so many unimaginable things. And as Prophesied, most souls in the church today, are too weak to cope with the fact that this will happen while they are here. Choosing rather to believe a fantasy, that the weak divided Church will be taken out of the Earth and to Heaven before these events. Many people during these coming events will actually die of fear. Luke 21:26, "*MEN'S HEARTS FAILING THEM FOR FEAR, AND FOR LOOKING AFTER THOSE THINGS WHICH ARE COMING ON THE EARTH: FOR THE POWERS OF HEAVEN ARE SHAKEN.*" Many will go quite insane during this unfolding, rather than believe it is of GOD. We can use this wisdom and understanding to consider and estimate our own level of faith and desires for GOD in these days. Do we see ourselves avoiding a deeper Spiritual life and search for the perfect truth of Scripture, and a true understanding of Prophecy. If you can't want and accept it all, this must be because at this time and stage of your Spiritual growth, it is beyond your faithful desire to see and believe. Many want a belief that just says believe in CHRIST'S name, and GOD will save and grant you eternal life. But these do not desire to know the fulness of GOD'S truth, or be involved beyond that. When GOD searches our minds and tries the reigns of our hearts, and sees a lack of desire for the greater Spiritual things, what does that mean to GOD? Does FATHER see a lack of desire for his beliefs, as a lack of love for him? Does he see the same in our disobedience to his ways and commands? Free will has amazing powers in how it works in the heart and mind. It is nearly impossible for a soul to be convinced of something they desperately do not want to be true. We can tell a soul in rebellion an absolute truth, and they look at us in their strongest sincere convictions, and say we are foolish. It appears, most of mankind and a third of the Angels, are capable of going to their eternal separation, rationalizing whatever they desire to believe. Ask yourself, why and how most souls who read and study the Revelation of CHRIST, while in the flesh and not the spirit, hear many different interpretations of what our LORD says. Surely these souls hear only what they want to hear. GOD gives their Prophecy to the realms in their one true meaning, not endless individual meanings. Why would GOD give their one true meaning, and want everyone to instead, hear their own biased preferred meaning of it all? Where is any good eternal purpose found in that? There is none. The heart of fallen mankind can compromise truth and reality with ease and lack of conviction. All the biblical warnings against allowing one's self to be deceived, are directed at every soul, including me and you. And the truth of Prophecy is surely no exception, being set as the last book of the Bible to be properly known. It is large, yet understandable and knowable *if* heard spiritually and in order. If you are willing to look upon any man, and say you trust your soul to this man's truth; your trust is misplaced. So why then, do we try to teach? Well, we are still so called to speak the truth; and here and there, raise interest in the truth and sow embers of hope. But if any hear my true words, and does not try them through HOLYSPIRIT, they do not obey GOD'S command to try every spirit and guard their soul. A few do speak the truth; yet it *must* be tried by each soul in their accountability to GOD'S belief. We see a few small last-days Prophecies mingled throughout the New Testament, followed by this Supreme Prophetic Revelation to John. So, what are the *very* last days the Bible speaks of. Technically they are the last 25-30 years of this third dispensation, that began with the Crucifixion and Resurrection of our KING. Who being our LORD, again sat down at FATHER'S Right Hand. To date, mankind has experienced three separate rising dispensations; being a strong, stronger, and strongest way to hear and return to GOD. The first came at the fall of mankind, which set their bodies and the creation a certain distance outside of GOD'S perfect physical order, care, and provision; *and* brought their spiritual separation of oneness from GOD. The first of three dispensations lasted around two thousand years, up to the cleansing flood. In the first dispensation, GOD'S Holy Angels still communed directly with mankind, and translated up into Heaven those who lovingly obeyed GOD; example being Enoch. But as mankind propagated, their deception, rebellion, and sin, up to the point of complete saturation, except for eight souls. This pushed GOD away more and more, so GOD had to cleanse the earth. After the earth was cleansed by his flood, we see a *second dispensation* of around two thousand years of GOD, from a greater distance and separation, still trying to save mankind. In the second dispensation, souls were given the Book of the Law and the animal sacrifice, called the Old Testament; being the first Covenant of love. And it having many small segments of past, current, and coming Prophecy. These first two interventions of FATHER at different times, tried more and more to re-establish his truth and mankind's salvation. The second dispensation brought a covenant of obedience, brought under a Priesthood and a Blood Sacrifice. These were added progressively, from the life of Noah, to Abraham, Isaac, and Jacob; and Jacob being renamed Israel by GOD. That second dispensation ended, and the third, started with the greatest historical events to date, being the Birth, Life, Crucifixion, Resurrection, and the Ascension of MESSIAH. This now current, *third dispensation*, also being around two thousand years, and the greatest part of the dispensation of grace and mercy. FATHER'S final efforts flowing down through the suffering blood Sacrifice of his Son JESUS, and the great truth of HOLYSPIRIT, offered to all who truly desire GOD. This third dispensation of grace, runs up through the beginning of sorrows, the Tribulation, and up until CHRIST returns to cleanse the Temple by removing Satan, to begin his Millennial Reign *on* the fallen Earth. This starts the last fourth dispensation of only 1000 yrs, under the iron rule of CHRIST, with his saints and martyrs. We see a total of four distinct ever-increasing approaches, and an escalating effort to restore us to GOD'S perfect ways. In our study, we have to be mindful that there are two literal meanings to the term "the last days". The first meaning in Scripture, is this entire third dispensation of around two thousand years. The second use, being the more applicable here, is the final years of this third dispensation. Many ask, is it for everyone to know this Prophecy? Yes. It is for every true Child of GOD to know. Revelation 1:3, *"BLESSED IS HE THAT READETH, AND THEY THAT HEAR THE WORDS OF THIS PROPHECY, AND KEEP*

THOSE THINGS THAT ARE WRITTEN THEREIN: FOR THE TIME IS AT HAND." Revelation 22:17-20, "*AND THE* ***SPIRIT*** *AND THE BRIDE SAY, COME. AND LET HIM THAT HEARETH SAY, COME. AND LET HIM THAT IS ATHIRST COME. AND WHOSOEVER WILL, LET HIM TAKE THE WATER OF LIFE FREELY. FOR I TESTIFY UNTO EVERY MAN THAT HEARETH THE* ***WORDS*** *OF THE PROPHECY OF THIS BOOK, IF ANY MAN SHALL ADD UNTO THESE THINGS,* ***GOD*** *SHALL ADD UNTO HIM THE PLAGUES THAT ARE WRITTEN IN THIS BOOK: AND IF ANY MAN SHALL TAKE AWAY FROM THE WORDS OF THE BOOK OF THIS PROPHECY,* ***GOD*** *SHALL TAKE AWAY HIS PART OUT OF THE BOOK OF LIFE, AND OUT OF THE HOLY CITY, AND FROM THE THINGS WHICH ARE WRITTEN IN THIS BOOK. HE WHICH TESTIFIETH THESE THINGS SAITH, SURELY I COME QUICKLY. AMEN. EVEN SO COME* ***LORD JESUS****.*" Though this Revelation given to John, gives a complete chronological timeline of many specific events. We also see strategically placed throughout this Revelation, several long running narrations that combine the unfolding of many prior, and often, a few coming events. But within these longer narratives, again, it all remain chronological. Going forward, we refer to these longer narrations, for no great reason as side bars, that JESUS added for a more colored in picture of detail and context for our understanding his enumerated Prophecies. We also properly overlay all concurring Old Testament segments of last days Prophecy upon the complete Revelation given to John. Which miraculously align perfectly to reinforce and affirm the Revelation of JESUS. Now to understand Prophecy and Scripture, we must accept a great foundational truth. Being we were all gifted and created by, from, and for GOD; and for their eternal familial desires; not everyone's; not yours or mine; but GOD'S eternal purpose for *their* eternal family. And by accepting this truth, we find our physical lives, and all that physically exists, are GODS alone. Our loving FATHER of course, has our eternal needs at heart, but he did not create anything we see, to serve anyone's desires as individuals. We were created to live our lives as an eternal Child of GOD, and not as individuals. Another founding truth is FATHER, being all Righteousness and Holiness, is unable to look upon sin without hurt and anger. FATHER has for a long time, as sin grows, been pulling back, turning away, and hiding his face from seeing the evil in the world. And as FATHER pulls away, his sustaining desire for his falling creation, moves with him. Colossians 1:17, *"AND HE IS BEFORE ALL THINGS, AND*

BY HIM ALL THINGS CONSIST". The definition of consist is to hold together, or to be held together, or the manner and force by which all is held together in firmness or consistency. We do see FATHER is pushed away and repelled by sin. This is very important for receiving many things in Prophecy, and understanding how his own purposed elements of wrathful destruction come upon the earth. Sin *is* the only underlying root and cause for all destruction. The final outcome says the current Earth and the Heavens shall melt with fervent heat, as in a *thermonuclear* reaction; and they shall pass away. This very last event, comes at FATHER finding not one more soul who can be saved. And at that point, there remains no reason to allow evil to continue. And so, then and there, FATHER completely loses his forceful desired reasoned purpose, and will for all his fallen physical creations to consist any longer. So, the entire fallen corrupted physical side of reality, dissolves and reverts back to the one-dimensional point singularities of, and within the aether from where they once came. FATHER then perceives/creates a new Heaven and Earth. We must know what it truly means, when Scripture says that by GOD all things consist, meaning that by GOD all things are forcefully held together. We must see how this is the coming reason, that by FATHER, all things will dissolve by the same ways and means he used to create them. Romans 1:18-22, *"FOR THE WRATH OF GOD IS REVEALED FROM HEAVEN AGAINST ALL UNGODLINESS AND UNRIGHTEOUSNESS OF MEN, WHO HOLD THE TRUTH IN UNRIGHTEOUSNESS, BECAUSE THAT WHICH MAY BE KNOWN OF* ***GOD*** *IS MANIFEST IN THEM; FOR* ***GOD*** *HATH SHEWED IT UNTO THEM. FOR THE INVISIBLE THINGS OF HIM FROM THE CREATION OF THE WORLD ARE CLEARLY SEEN, BEING UNDERSTOOD BY THE THINGS THAT ARE MADE, EVEN HIS ETERNAL POWER AND* ***GODHEAD****; SO THAT THEY ARE WITHOUT EXCUSE: BECAUSE THAT, WHEN THEY KNEW* ***GOD****, THEY GLORIFIED HIM NOT AS* ***GOD****, NEITHER WERE THANKFUL; BUT BECAME VAIN IN THEIR IMAGINATIONS, AND THEIR FOOLISH HEART WAS DARKENED. PROFESSING THEMSELVES TO BE WISE, THEY BECAME FOOLS,"* These words say the slow small fallen temporary perspective of mankind can physically see and misinterpret FATHER'S conscious forceful projections of willful forceful desire, he used and uses to align his firmer larger external substantive perceptions to consist and move in their dynamic orders. Mankind can also falsely see and study the ethereal quantum and sub-quantum foundational layers nearer the real and true aether that FATHER created and used and uses to build his physical creations. Or we can repent and receive HOLYSPIRIT to see the forceful layered building order and process by which we see and know GOD and the creation clearly. The consciously projected forceful creative aligning process of GOD, must be accepted to hold the full understanding of Scripture, including CHRIST'S Prophecy. All force, energy, mass, dimensions, complexities, chance occurrences, and experiences, exist within and by FATHER'S conscious forceful projected gifts of creation, and this includes all *our* conscious force projections and creations. All forceful creative laws are found in Scripture, including the 4 nuclear, electromagnetic, and gravitational forces. We see DNA, and quantum mechanics, are in the ancient ways of parlance and understanding, described in Scripture as part of GOD'S plan. All these things are perceived by GOD, by seeing, forcing, and holding reality together in his purpose. This dynamic complex design allows FATHER to let go of his awareness of all its chanceful outcomes to certain extents, thus allowing all our free-will actions. We must go deep into the truth, and hold FATHER'S true understanding of all the created invisible (unseen) things. Most professed Christians can see and agree, that every atom was created by GOD. Why? Because we have seen that atoms exist. And we see atoms are made of smaller particles and wave properties; again, created by GOD. We then cannot dispute, that each atom's consistency is held together by GODS forceful creative will. This willful forceful desire of GOD upon all building blocks is seen by science; yet as those last Scriptural verses declare, they refuse to attribute that force to its Conscious Creator. We can understand all creations, only if we clearly see particles, atoms, and all mass, are formed by FATHER'S projected forceful desires upon the physical points that comprise his "Physical Being". And when both are brought forth by FATHER speaking outward, to be held between his two hands, they exist and remain under his original patterned force, being the physical breath of all his spoken words. Gravity, electromagnetism, the strong and weak nuclear forces, are all just the differing focus of strength and intent of GOD'S conscious forceful desire, projected to align and form his dynamic creation. So, in all *certainty*, it is *only* by FATHER'S diminished projected forceful will, that atoms can come apart. It must require a weakening or loss of FATHER'S forceful desire for all, or *certain* atoms to continue. Fission entered the creation at some point after the fall of mankind. The energy of life is born in the shell of atoms and the fusion of stars. But the fission of the atom's nucleus is about death and destruction, not life. Nothing *physical*, consists of itself beyond the projected aligning force of GOD. So, prior to, and into the tribulation, we see reality physically weakening and altered according to GOD'S projected desire. Our eyes we will see things most say are impossible; and they are physically impossible on their own, but not under GOD. Satan knows this is coming, so he has readied many deceptive cover stories. The world will see ever-increasing catastrophes, and hear it is caused by war, pollution, and man-made and celestial causes. Most will believe these reasons, and ridicule GOD'S truth. We also see a prophesied coming nuclear war. Revelation 6:14, *"AND THE HEAVENS DEPARTED AS A SCROLL WHEN IT IS ROLLED TOGETHER; AND EVERY MOUNTAIN AND ISLAND WERE MOVED OUT OF THEIR PLACE."* We see this perfect ancient description of a scrolling mushroom cloud as it rolls up into the heavens after a nuclear explosion. And every mountain and island being moved. Only nuclear explosions have such a forceful appearance and affect. Many Scriptures speak about the levels of GOD'S wrath or anger, and his greatest wrath is an all-consuming-fire. Now please know, if any soul preaches *anything* that does not in all spiritual and physical ways, agree and align, with both the Old and New Testament, they have unintentionally or intentionally over-emphasized, under-emphasized, misinterpreted, or failed to rightly contextualize Scripture; all being heinous lies. From the beginning of Scripture to the end, every word says JESUS trusts and obeys FATHER'S every loving command, for this is the eternal love and truth authored by GOD. And CHRIST by his every word, called *us* to follow and emulate him in all his ways. At the end of this fallen creation, all physical things beyond GOD'S love and truth are dissolved, passing through FATHER'S purifying fiery fission. We read that the Prophet Daniel's Hebrew friends, Shadrach, Meshach, and Abednego, were cast into a fiery furnace by King Nebuchadnezzar, but the flames did not kindle upon them. So, we see a consciously deterministic explanation, being FATHER in that moment, lifted his veil of control over reality, to humble that worldly King. FATHER perceived by force of will, to not allow loss or alteration to a single cell or atom of their bodies or clothing. Against those boundaries of FATHER'S forceful desire, the destructive aspects of fire are powerless. If not a single cell of the body is in anyway altered, we feel no pain, regardless of surroundings. But if FATHER'S desire is not to protect us, the fire will have its way. 2nd Peter 3:10-12, "*BUT THE DAY OF THE LORD WILL COME AS A THIEF IN THE NIGHT; IN THE WHICH THE HEAVENS SHALL PASS AWAY WITH A*

GREAT NOISE, AND THE ELEMENTS SHALL MELT WITH FERVENT HEAT, THE EARTH ALSO AND THE WORKS THAT ARE THEREIN SHALL BE BURNED UP. SEEING THEN THAT ALL THESE THINGS SHALL BE DISSOLVED, WHAT MANNER OF PERSONS OUGHT YE TO BE IN ALL HOLY CONVERSATION AND GODLINESS, LOOKING FOR AND HASTING UNTO THE COMING OF THE DAY OF GOD, WHEREIN THE HEAVENS BEING ON FIRE SHALL BE DISSOLVED, AND THE ELEMENTS SHALL MELT WITH FERVENT HEAT?" FATHER increasingly turns from love and grace, towards wrath, and bringing his Prophecy to pass. FATHER has already loosed the Four Horsemen of the Apocalypse to bring world war, famine, and disease. Make no mistake, these four souls have been loosed for a while now. FATHER started turning to wrath, long, long ago, and great disasters and storms are a coming occurrence. Luke 21:25, *"AND THERE SHALL BE SIGNS IN THE SUN, AND IN THE MOON, AND IN THE STARS; AND UPON THE EARTH DISTRESS OF NATIONS, WITH PERPLEXITY; THE SEA AND WAVES ROARING;*" The current ever-escalating crises, born of this world's confused rebellious desires, and all their subsequent confrontations, are born of mankind's desires. Yet they are being greatly influenced in very forceful ways by these Four Horsemen, sent to direct this world's chaos to help fulfill all Prophecy. Godly societal structures, norms, and decency, are decimated. But by HOLYSPIRIT and Biblical Prophecy, we can see certain big events as reference points, to reveal where we are in Prophecy's unfolding. Most rebellious souls, will for all these coming changes, desire to blame greed, corruption, governments, religions, corporations, pollution, and extra-terrestrial events. Lies will abound to deny the truth, which will rarely be spoken, and less rarely heard. Mankind's desired blindness, and their many mass psychoses, are used by the four Horsemen to steer mankind's greed, lust, and fear. The raw elements of reality's supply lines and labor control in the earth are fought over. There is a coming great world-war, with global famine and disease. Most will blame these plagues on everything accept the evil in their hearts that separate them and the earth from FATHER'S care. And to truly understand CHRIST'S Revelation, we must never forget that JESUS, in Matthew 24: 6-8, separates the eventful times of the first six seals, he calls the beginning of sorrows; from the approximate ten years of Great Tribulation that begins at the first of Seven Trumpets sounding after the Seventh Seal is opened. Each trumpets events, in their order, become more complex and involved. The Seventh Trumpet itself, consists of many events in heaven and earth, to include the last Seven Vials of wrath being poured out to bring the end. The Beginning of Sorrows, is the Seven Seals, and the Tribulation is the Seven Trumpets. And again, the Seventh Trump, includes the final Seven Vials. The Four Horsemen are well into escalating the beginning of sorrows; and the Tribulation will follow. After all these come to pass, we see CHRIST returns to battle at Armageddon. Let's consider one more truth before we study the Prophecy. We must know that HOLYSPIRIT, and this fallen world have zero in common spiritually. And our responsibility to FATHER'S created order and ways that perpetuate eternal life, rests in our love for GOD. To be a physical person in this world's physical ways of life and possessions, has nothing in common with GOD or their Kingdom. Eternal love as declared by FATHER, is a selfless responsible sacrificial faithful desire, found only between FATHER and his Children who desire to perpetuate FATHER'S eternal Kingdom. The basic physical needs of healing, shelter, simple food, and our spiritual covering from GOD are good. Beyond these, all acts, imaginings, and stored physical provisions, can easily be lust of the flesh and this world. All covetous perceptions are evil works to be burned, along with all unholy pleasures and comforts. Contrary to all expressions that say otherwise, it is always *spiritually* better to physically need something to live, and not have it; than to have physical abundance, and not need it to physically live. Now before we start, we laid out an advent calendar labeled (figure a.) on pg. 486, to help visualize the unfolding of all things. Of course, all events are not in our advent calendar, but enough to see the last days unfolding timeline. We have many time lines to hear and use, and we hope they help you understand what is to come. In our calendar, events are listed horizontally from left to right, in order of occurrence. The top line events, covers *all* that occurs from the Crucifixion and Ascension of CHRIST, up to his Second coming. The second-line is the "beginning of sorrows" being the opening of the Seven Seals JESUS set apart, and called the beginning of sorrows; being a prelude where Satan, mankind, and the Four Horseman increase all world divisions, leading to world war, famine, and pestilence. But in the Tribulation, *all* events are caused by GOD alone. The third line of the calendar covers the Prophecy of Revelation 8:13, being the Three Woes. These last most fearful Fifth, Sixth, and Seventh Trumpets of the Tribulation are quite accentuated, amplified, and broader in what comes to pass. They are dreadfully complex, compared to the first four trumpets. Line four, is the estimated years from opening the first seal to CHRIST'S return. Line five, is the first Seven

Seals timeline. Line six, is the Seven Trumpets timeline. Line seven, is the last Seven Vials timeline, which again, are the last part of the Seventh Trumpet. These all align vertically to reference the placement and occurrence of each event to form one overall timeline; as they align up and down through the graph where Prophecy places them. Further, we divide the graph in half, by a row of forward slash marks. The first line on the bottom half contains reference numbers showing the order and timeline of certain events that Prophetically occur in Heaven. The last line, is the certain ordered prophesied events that occur on Earth. All nine rows align vertically according to the timely placement of their occurrence.

ADVENT CALENDAR

(Figure A)

/- -*The complete Old and New Testament Prophecy of the end of days---* /

/------The beginning of sorrows------/--------The tribulation--------- /

The three woe's /- 1- /---- 2 --- /-- 3--- /

/----------------Time-frame-u/known-/---3.5yrs-----/--------6.3yrs-----/

7SEALS.1___2___3___4____5_____6_____7_______________________

7TRUMPETS1 2 3 4 5___6 ______7______

7VIALS 1234567

//

Heaven events 1, 2,3 (4a----4b----4c-----4d) 5 6, 7

Earth events 1 2 3 4 5 6 (.......... 7.......tribulation...... 8) 9

To use our upcoming event reference numbers, you will be directed to first locate the Earth, or Heaven, event number. Then looking straight up through all reference points to see when or where each event falls. It should be clear to most indwelled believers that we are near or past the middle of the beginning of sorrows. Might FATHER for a brief moment interject a bright light one last time, causing the world to pause in fear about the validity of their own created truths? I pray he will. Before starting this study, please know in your heart and mind, this must be a prayerful earnest study. The planned and ordered redemption of GOD'S Children, and the separation of fallen mankind, and fallen Angels, and the millennial reign, and the destruction of all creation, and a new creation, are not a simple matter to be understood by a half-hearted attempt. This will require a true desire and effort. You will need paper for notes, Bible book marks for quick reference, and the Original Authorized King James Bible. Do not use any marginal foot notes your Bible may have. This study takes commitment, but please don't be discouraged, it brings a great blessing. Now a great wisdom for understanding Prophecy is found in how Scripture uses the terms years, months, and days, without giving any specific dates. Yet Scripture gives us many segments of years, months, days, and verified historical facts, along with Biblical reasons and purpose, that all align to give us certain timeframes. HOLYSPIRIT brings these all together, as they fit together perfect in one astonishing chronological way. Most of these specific time frame references for the last days, appear in the books of Daniel and the Revelation of CHRIST. Now Judean Prophecy often uses the terms: “a time” “times” and “half a time”. These terms in Prophecy indicate a year, or a thousand years; but the terms “one week” or “70 weeks” means seven years, and seven times seven, being 490 years. At the conclusion of discussing this overall Prophetic time-line, you will see how they all fit together perfectly to define and declare something most-important we all need to know. And lastly, *once* within the Book of Revelation, we see the term “for one hour” this term meaning not one literal hour, but a time much shorter than a year. Once we see a very large number of days, being 2,300 days, this is literal, being 6.3 yrs. We begin our study by going through all these related time frames, starting with Daniel 8:13-17. There are many time lines to put together, so please write them down to help visualize and understand. And please do your own math, and I promise you will cherish this revelation. But please do not expect your understanding to form, until we put the many pieces together. Please save your notes and math work for later studies. We start this timeline, reading Daniel 8:13-17. Here we see a discussion between one Saint and another, concerning a very long vision Daniel had just received from GOD. A vision that ends with the last of four Beasts, taking away Israel's animal sacrifice, which is soon restored in these last days. Daniel is shown how long it shall take from the Antichrist moment Antichrist takes away of the Daily Sacrifice in the new last days third Temple, and Satan then enters the Temple and the Idol Image, setting himself up as GOD, up until JESUS returns and drives Satan out, and the Temple is cleansed. Which we read will be 2,300 days or 6.3 years. Please take notes and write all numbers down. Now this 6.3-years ends right *after* the Battle of Armageddon. To visualize this 1st time segment, read Rev. 19:19 through 20:3. Our 2nd time segment is in Daniel 7:24-25, is where the final Third and Fourth Beast, rise up to control the new Ten horn Beast, or new Ten Region World Government. In Prophecy the term "Beast" represents satanic worldly governance. This Third Beast will be allowed power over the Saints for a time, times, and a half time, being 3½ years or 42 months. The 3rd time segment is at the end of the Book of Daniel in Chapter 12:1-9, where one last long vision starting at Daniel 11:1 is given to Daniel that ends *after* CHRIST'S Millennial Reign, at the White Throne Judgment. This is so *vital*, as we see a person asking a Man in white linen, how long it will be, starting from that ancient moment Daniel had this last vision, up until all Prophecy is fulfilled at the White Throne Judgment. They are told by a man in white linen, it will be a time, times, and a half time; and here the measures of time are 1,000, 2000, and 500 years, meaning 3,500 years. Please add this *most* important over-all timeline to your notes, as you will see all of Daniels future Prophecy, fits perfectly within, these 3,500 years. Now in Scripture we see Kings Nebuchadnezzar, Belshazzar, Darius, and Cyrus, all ruled over Babylon or Persia, and Judah during the time of Daniel's life. And so, it was during their reigns these Prophesies were given to Daniel. Modern Archeology estimates that the last of these four Kings reign, ended around 450 before CHRIST. But they could be off by decades; and the year of Daniels death cannot be fixed. So how can we accurately see this Prophetic 3,500-year-time frame our LORD has given to us. Scripture answers this, by working backwards from the White Throne Judgment at the end of CHRIST'S millennial Reign. We just subtract the years we already know from 3500, like the coming Millennial Reign of 1000 years, and the time since CHRIST'S Crucifixion to this writing in 2012, and the approximal 450 years before

CHRIST, back to the last Daniel era King. And so, subtracting backwards, we see 1000 years and 2012 years and 450 years, which equals 3,462 years out of that prophesied 3,500 years. Leaving us just an estimate of 38 yrs between 2012 AD and the Millennial Reign. Which gives us an estimate of 2040 AD for the Tribulation to start, and 2050 AD for when CHRIST'S Millennial Reign might start. So yes, we are in the last days. Now going forward, the greatest part of this study will be about those few decades just ahead. Going on, we see a time-line which says before the birth of Moses, 70 weeks, or 490 years of exile were set upon the people Israel for their rebellion; but only 69 weeks were paid before Moses led them out of Egypt. So, this is the 7-year gap, Bible scholars talk about in the last days. We also read in Revelations, that after CHRIST'S thousand-year reign is completed, Satan is set loose. And he goes out to yet again deceive and gather the world against CHRIST for that final third battle. That could take months between the Millennial Reign, and the White Throne Judgment. We see other small Scriptural Prophetic considerations could subtract or add a few years. But by all these time-lines, the Tribulation should begin after World-War III, approximately 20-28+ years from 2012, or between the years 2032 and 2040. But we should live now, like CHRIST'S return is here. Please see, it could be here sooner than our best estimates. We will know more and more as CHRIST'S Prophecy unfold. We are working with many accurate Scriptural components, yet we see small variables; but in all we do know, we are close. Going forward, all time segments refer to coming events prior to CHRIST'S 1000 yr. reign. Now our 4th time segment is in Daniel 12:11, in reading this, we see that from the time the third Beast, being the embodied Antichrist, takes away the Daily Sacrifice, till he sets up the image and Satan indwells it in the Temple, shall be 1,290 days or 3½ years; or a time, times, and half a time. And the 5th time segment in Daniel 12:12, is related as we read blessed are those who come to see 1,335 days, being 45 days *after* Satan indwells the Idol and claims he is GOD. Is this a day of the first great martyrdom of Saints? Is this the day our LORD begins his battle to remove Satan from the Temple? Is it a day the Seal of GOD is placed upon many Children having HOLYSPIRIT here in the earth? Or is it something else wondrous? All we are told at the time of this writing, is "blessed are they that live to see this day". GOD willing, may we see each other there. Time segment six is in Revelation 11:2; and as we read, we see that in mankind's continued weakness, and knowing Satan's plans, FATHER sees the people Israel will betray him and enter a covenant with the first Beast. Being a new Ten Region World Government that causes Jerusalem, the Temple mount, the Temple's Courts, but not the Temple, to fall under world governance. And being open to the world to tread under foot for the Prophesied seven-year religious treaty; which fulfills that gap to be paid. The Holy of Holies, the Daily Sacrifice inside, and everything inside the Temple proper, will remain secure to the Jews, and only accessible to the Temple Priest for the first 3.5 years of this covenant, until the final Third Beasts come to full power and breaks that seven-year covenant at its midway point. In Rev. 11:3-6 we see the 7th time segment, is aligned to the sixth. In which we see GOD giving his Two Witnesses a Supreme Spiritual Voice in all the Earth for the first 42 months of that seven-year covenant. These Two testify before the Temple of the coming Judgment of this world's betrayals, and against all who gave the land of Israel away from GOD'S protection, for a treaty with the World. These witnesses rebuke Israel, for by GOD'S hand he brought them home, and by his hand they were secure from the world. But they, as always, failed to trust FATHER, and in fearful unfaithfulness, turned back to Egypt (worldly provision). Throughout last days Prophecy, the name Egypt symbolizes a trust in worldly government, over GOD.

And so, Israel owes, and will now serve that last week, or 7 years of exile they still owe GOD. These two witnesses preach against all evil and false religion, and cry out against apostasy, idolatry, and the covetous of this world. They have GOD'S ear, to bring plagues upon the world for its rebellion. These two testify, being untouchable under GOD'S protection, up until the Third Beast comes to full power. Then the Third and Fourth Beast, being the Antichrist and the False Prophet are given "permission" and the Antichrist kills the two witnesses. The 8th time segment has passed, but is relevant as a mirror of the 9th. Before we discuss this 8th time segment, read Rev. Chapter 12, being is a condensed vision of the true Churches battle against Satan and his devils. This battle, started before the birth of JESUS upon the Earth, and rages until the Fifth Trumpet in the Tribulation in Revelation 12:6-7. We read that this 8th time segment speaks of the initial early true Church of CHRIST being helped by FATHER, hidden and fed in the wilderness for forty-two months or 3½ years or a time, times, and half a time. And the 9th time segment in Rev. 12:14-17 is its mirror image in the last days. In this last-days, first forty-two-month period, we see a second Beast come to power and try to move things towards a one-world religion, banning all true public Christian discourse. And GOD again has prepared places in the wilderness for his true Church to flee to and be fed for these first forty-two months; again, hidden from the face of Satan. Now the last specific time segment we will look at *for now*, is Revelation 13:5. Which affirms the power of the Third and Fourth final Beast over the Saints; as FATHER allows the Antichrist to martyr his Two Witnesses and many Saints, and bring desolation to the Temple, and blaspheme all that is Holy for 42 months. There are many time segments spread throughout the Old and New Testaments written by many different souls; yet they all fit together perfectly. This is miraculous, because they were all delivered in small parts, at different times to different writers, from one single Supreme Author. We have abundant context to see how the continuity of GOD'S plan permeates his inerrant words to all who search. So, let's begin our verse-by-verse study, with Rev. 1:1-18. We see prior to the four seals being opened. John the Apostle and Prophet, prays for and blesses those who seek and receive this Prophecy. And being physically translated by FATHER, John is brought to the throne of CHRIST in Heaven. Here John is given a commission to deliver seven written letters, containing admonishments and exhortations to the early Seven Churches who proclaim CHRIST throughout Asia. This will be our first reference in the advent calendar, as *Heaven* event 1 on the chart, on pg. 486. Find the lower line marked *Heaven* events, and go to event 1. Then looking straight up (if needed use a ruler) we see this event, occurred prior to the first seal being opened, and the beginning of sorrows. Follow this procedure for all “Heavenly” or “Earthly” events. Remember some Prophecy events are ceremonial in Heaven, and not seen by, or physically affecting us on earth. Father allows us to know what is going on in Heaven, during the time we witness the coming tragic events on earth. GOD considers our desire and need to know where we are in Prophecy. These bi-realm events can overlap in timing, or duration of their occurrence. This first Heavenly event, of John standing before CHRIST in Heaven, occurred around 30+ years after the Crucifixion and Ascension of JESUS to his Throne in Heaven. Next, read Rev. 1:19 through Chap. 3:22. We see JESUS gives to John seven differing messages to each of those Seven physical Churches. Five include dire warnings, that point out their strengths, and short-comings they must address if they are to be accepted. Warning if their sin is not corrected, CHRIST sees them as enemies. Only the Churches of Philadelphia and Smyrna, who suffer great tribulation and poverty, receive no admonishment at that time. In these verses, we must find that though these are the first seven large physical Judeo-Christian-Churches in all of Asia, these warnings are timeless. The Church then and now, is the one true Church of CHRIST, including the last day's Church. So, we need to rebuke all false interpretations of *these* Scripture. As so many preachers say these Churches, weren't real churches with different levels of obedience; but represent types of Churches, or dispensational Churches. This misinterpretation allows for broad blurred Scriptural meaning. Scripture clearly states for John to send these letters to these seven physical Churches which are in named cities across all of Asia. These warnings were to those Seven Churches about their Spiritual state in following CHRIST; yet these biblical warnings are to all who truly love for JESUS. Five Churches that believed and professed the name of JESUS, receive severe warnings that their profession of belief was just the beginning, and not the finishing of their faith. Our first step of profession alone, does not give us final rest, or eternal life. It is our first step in our journey of growth and understanding. We are called to strive to be one in CHRIST, and we need to continually live and walk in his ways. We are called to infinitely more than a weak believing profession of faith in his name, and sinful living. We are called to purify ourselves as GOD'S temple; to be a Spiritual being, no longer of the fallen flesh and world. And as with these early Churches, we too are warned of the consequences if we will not, or cannot do this. John fulfilled the word given him, to write to those Seven Churches around two thousand years ago. This event started in Heaven, so we will call this Heaven event 2, on pg. 486. Looking straight up, we see it too, occurred before the Beginning of Sorrows, and the Tribulation. Now please read Revelation Chapter 4:1 through Chapter 5:14, which occurs prior to any seals being opened. Here John is brought to the edge of Heaven, and is looking through an open door into the Holy of Holies to see FATHER upon his Throne, and those around his throne. John sees a great Ceremony for CHRIST, attended by all in Heaven. In that Ceremony, they give all honor to JESUS, and declare that the Book with Seven Seals, can be opened by him alone, for only he is worthy. This ceremony, with those last prior events has occurred and been fulfilled, and is Heaven event 3. At this point in Scripture, we point out the following seven seal events are all caused by mankind, and strongly influenced by the Four Horseman. These seals occur prior to, and are of a different nature than all the Seven Trumpets and Vials of the Tribulation, caused by GOD alone. We must keep them in this context, using the fullness of Scripture. The New and Old Testaments shine brightest when seen together as one, and not separately; bringing a fulfilled reality to all Godly life and

Prophecy. We see the fulfillment of these first six seals is a spiritual siren, declaring we are in the days just prior to the Tribulation. Read Chapter 6: 1-8, JESUS opens the first four of Seven Seals, releasing Four Horsemen. These are Heaven events 4a-d. Again, we *could* have also listed these as earth events on pg.486, seeing they were released by CHRIST into the earth; but duplication would crowd the timeline. **Note* JESUS has already loosed and opened these first four seals, releasing these four influential spirits into the earth, who have for years been busy at their work. JESUS in the four Gospels, and the book of the Revelation, tells us these Four Horseman can influence mankind and alter the physical discourse. They stoke mistrust and fear between the powers around the world, bringing division, plagues, war, and famine upon the world. All coming hell and death stem from immorality, energy and monetary pressure, false environmental crisis, false religions, and godless government. Pressures mount by the ongoing influence of these Four Horseman. There soon comes a third world war, using a prophesied great sword, meaning weapons of mass destruction. We must expect nuclear, chemical, and biological war, followed by famine, and disease. And by these three plagues will come the death of about one fourth of all mankind; using future estimates, over 2 billion people. Now in Daniel Chap. 7, we see four great Beasts or Governments or Empires that have risen to rule the known world during mankind's existence. Three are gone, and a fourth ten-region government called the Fourth Beast, is prophesied to come. After World War one caused 8+ million deaths, the world said this can never happen again, so, the League of Nations was formed to work toward one-world unity. That thankfully had no chance. Then World War II, where 52+ million people died. So, the United Nations was formed, calling louder, pushing harder, by Globalist Ideologues. Yet, Global Governance has no chance. Prophecy says *one fourth* of mankind will die by the coming third World War, disease, and famine. Sadly, this time a Global-Government will become a reality. This war will be great, fierce, and almost unstoppable. It will greatly hurt the earth, and cause most all people to say with passion, we all must become one or perish. Now for a broader view of these coming events, read Revelation 12:3, and all of Rev. Chapter 13. We see in Rev. 12:3 an evil alliance and command structure of fallen Angels was first forged in Heaven long ago in wanting to overthrow GOD. And they formed long-term plans. Now again, in Chapter 13, being analogous, it calls this new Global 10 state Government a Beast. Now we read that at first, seven of its ten rulers, have great influence and maybe veto power over these Governors or crowns; each being a regional representative. We see within the last days, one world Government, that four Arab Governors arise within it. Two inconsequential Arab leaders, arise to power early as this Government is being formalized, and then two more, being the Third and Fourth Beast rise before its 3 ½-year mark of rule. So, with in the first 3 ½ years, we see all four kings arise. One of Arab decent, rules over the one most predominately Islamic region, and then another of Arab decent arises to overthrow them and assume their place. Who then quickly takes control of two adjoining heavily Islamic regions; yet then he too, is overthrown. And this 3rd powerful Persian Beast or ruler, is the true embodied fallen angelic Antichrist, who appears as a Persian man, to use an Islamic rebellion to rise to power. We know the last days over-all Beast Government, is a ten-region world Government that quickly falls under the control of one personage, or Beast who overthrows all ten. He is called the Third true Beast in Scripture, being the Antichrist and a fallen angel just under Satan. The Antichrist is joined by a 4th true Beast in Rev. 13:11, also a fallen angel and the False Prophet; also, just under Satan. Under this World Government, the world will never see true peace, until CHRIST'S millennial reign. There will be fights breaking out around the world;

mostly where it currently exists, like the Middle East. But the world-wide war of mass destruction will end. Prophetically, Israel and Jerusalem are central to GOD, so, they are the greatest part of Satan's plan. All who do not love GOD, will turn on Israel in the last days, as the signs of this are lining up now, more and more. We see the current Prophetic enemies of GOD and Israel, chiefly being Islamic Persian or Assyrian countries, have the support of Russia, North Korea, and to a growing extent, China. And many godless antisemitic liberals in Europe and the U.S. will join, or support Islam, as the Four Horseman influence it all until the end. The last days come to pass, as Satan fights to take GOD'S Israel and Jerusalem, and the Temple. Israel is Prophesied to be a burdensome stone in these last days. Please ask yourself why this tiny, tiny country, set in the middle of a desert wasteland, with no oil, water, soil, or natural resources, should be the focus that seems to divide the entire world? Can you think of one good logical worldly reason? Isn't it a great question? I'll tell you why, FATHER has said it is his. So, Satan wants it and will burn down this world to get it. Now in all Prophecy, GOD lets us know if it is for the last days, or for a long-ago time that already occurred. It is so important, to look for these identifiers. For example, we see for Daniel 8:17. GOD says; *"SO HE CAME NEAR*

WHERE I STOOD: AND WHEN HE CAME, I WAS AFRAID, AND FELL UPON MY FACE: BUT HE SAID UNTO ME, UNDERSTAND, O SON OF MAN: FOR AT THE TIME OF THE END SHALL BE THE VISION." We see by GOD'S own words, this Prophecy is for the end of times, being our time. And we see in Daniel 11:6, *"AND IN THE END OF YEARS THEY SHALL JOIN THEMSELVES TOGETHER:"* Again, clearly noted for the last days. When any Prophecy is not clearly identified as being for the last days, the last times, the end times, or the last of something to do with time, it was for a past moment, and occurred long ago. Only a few past fulfilled Prophecies, have any relevance to last day's Prophecy; so, don't blur them together. And remember that all New Testament Prophecy, beyond CHRIST foretelling his Crucifixion, Resurrection, the fall of Jerusalem, and the Temple's destruction, pertain to these last days. Now, by a proper interpretation of specific Scriptures, signs, and events, we see certain last days Prophecies have already come to pass. We've seen great Prophesied cultural and religious changes, and the release of the Four Horseman in Rev. 6:1-8. But what is a large, last days Prophesied earth event we can point to that has occurred; it being a major reference point we have recently seen in Prophecy's chronological unfolding. We have seen the major Prophesied last days earthly event, in Daniel, Chapters 8 and 11. These speak of a major event that occurred just a few decades ago; and they speak of a next major event that will occur. It is good for us to see the last day's earthly events that have already occurred, so HOLYSPIRIT reveals them. Now you should at first expect as with that 3,500-year timeline, that the first pieces of any puzzle, seem obscure, and at first reveal little. Yet by patience and the last pieces, it is undeniable. We start here with the first two of *four* major, *last days* Persian Kings being Saddam Hussein, Ayatollah Khomeini, the true Anichrist, and the false Prophet. These first two Persian Kings are identified as the two horns of a ram. Read Daniel 8:2-3, which we just clearly identified as an end of times Prophecy, *"AND I SAW IN A VISION; AND IT CAME TO PASS, WHEN I SAW, THAT I WAS AT SHUSHAN IN THE PALACE, WHICH IS IN THE PROVINCE OF ELAM; AND I SAW IN A VISION, AND I WAS BY THE RIVER ULAI. THEN LIFTED UP MINE EYES, AND SAW, AND BEHOLD, THERE STOOD BEFORE THE RIVER A RAM WHICH HAD TWO HORNS: AND THE TWO HORNS WERE HIGH; BUT ONE WAS HIGHER THAN THE OTHER, AND THE HIGHER CAME UP LAST."* These few short, vital, rich verses, which worldly Prophecy Teachers give little interest and consideration, are great keys. These two verses are filled with meaning and understanding. Read each word carefully, knowing everything GOD does or says is for a very specific purpose; and having layers of reasons to be revealed. Every word GOD speaks, is an important piece of his whole understanding as it were. GOD does not speak like mankind, using exaggeration or annulling words. We *must* understand each word clearly. We see Daniel was brought in GOD'S Spirit, to the banks of the *river Ulai,* to be given this end of times Prophecy. Ulai is an ancient Hebrew name, for a current tributary at the far north end of the modern-day Gihon River, between Iran and Iraq. Now why would GOD bring Daniel from where he *was*, to that specific place, to receive this Prophecy. It was for great reason and importance. This portion of this river, is not coincidental, it is of modern Prophetic importance. Around 2,450 years ago in Ancient Persia, GOD gives Daniel this clearly labeled, *Last Day's* Prophecy of a ram on the shoreline having two horns. Again, Horns in Scripture represent Kingdoms, or what we call Governed Countries. Here these two horns represent two last days Persian Kingdoms. Is this location insignificant? Did GOD bring Daniel to that specific river location that would border and divide two *last* days Persian kingdoms, to then give him a last day's Prophecy of two future Persian Kings for no specific reason? Of course not. There is no other way GOD could identify for us (not Daniel), who these two *last days* Persian Kingdoms are, except by this specific river's location. It is written so we can identify and understand who these two prophetic *last days* horns will be. So, we see a defined heralding event, revealing that as of this writing in 2012, we are at least several decades into the time JESUS called the beginning of sorrows. Let's read more from Daniel's Prophecy concerning Iraq and Iran in Daniel 8:3-7 *"THEN I LIFTED UP MINE EYES, AND SAW, AND BEHOLD, THERE STOOD BEFORE THE RIVER A RAM WHICH HAD TWO HORNS: AND THE TWO HORNS WERE HIGH; BUT ONE WAS HIGHER THAN THE OTHER, AND THE HIGHER CAME UP LAST. I SAW THE RAM PUSHING WESTWARD, AND NORTHWARD, AND SOUTHWARD; SO THAT NO BEAST MIGHT STAND BEFORE HIM, NEITHER WAS THERE ANY THAT COULD DELIVER OUT OF HIS HAND; BUT HE DID ACCORDING TO HIS WILL, AND BECAME GREAT.* (note; this is Saddam Hussein's Iraq.) cont. *AND AS I WAS CONSIDERING, BEHOLD, AN HE GOAT CAME FROM THE WEST ON THE FACE OF THE WHOLE EARTH, AND TOUCHED NOT THE GROUND;* (all Allied Forces being airlifted in) *AND THE HE GOAT (the west) HAD*

A NOTABLE HORN (the U.S.) *BETWEEN HIS EYES. AND HE CAME TO THE RAM THAT HAD TWO HORNS, WHICH I HAD SEEN STANDING BEFORE THE RIVER, AND RAN UNTO HIM IN THE FURY OF HIS POWER. AND I SAW HIM COME CLOSE UNTO THE RAM, AND HE WAS MOVED WITH CHOLER* (anger) *AGAINST HIM, AND SMOTE THE RAM, AND BRAKE HIS TWO HORNS: AND THERE WAS NO POWER IN THE RAM TO STAND BEFORE HIM, BUT HE CAST HIM DOWN TO THE GROUND AND STAMPED UPON HIM: AND THERE WAS NONE THAT COULD DELIVER THE RAM OUT OF HIS HAND.*" We already witnessed the defeat of the first shorter horn of Iraq. Saddam Hussein is that shorter horn who stood up first. And like Iran will be, Iraq was pushing in all directions and doing according to his own will. I watched Suddam Hussein for years on the evening news stirring up the Middle East against the West and Israel. Cursing the powers of Grecia, meaning all Western Democratic Countries, including Israel and all Judeo-Christian Countries. Suddam Hussein brutalized the region and his own people; including barrel bomb poison gas attacks on villages filled with women and children in the Northern Kurdish Region. And he used most heinous forms of torture on his enemy prisoners. We who were alive then, saw his arrogant self-adulation before the world in the news almost nightly. We saw him do whatever he willed, pushing in every direction, invading Kuwait, and warring against Iran for nine years. We saw as Prophesied, the West, being the Realm of Grecia, form a coalition with a notable horn, being the United States. A coalition in great choler (anger) came across the face of the whole Earth, not touching the ground, to brake and stamp upon him. It took the U.S. and their coalition just three weeks to drive Suddam Hussein into a hole in the ground and decimate his Fourth largest military in the world. The world was in shock and awe. And Iran will be destroyed next. Now we read a more detailed expanded description of this, that again starts with that first Persian horn of Iraq to stand up. Daniel 11:2-4, "*AND NOW WILL I SHEW THEE THE TRUTH. BEHOLD, THERE SHALL STAND UP YET THREE KINGS IN PERSIA; AND THE FOURTH SHALL BE FAR RICHER THAN THEY ALL: AND BY HIS STRENGTH THROUGH HIS RICHES HE SHALL STIR UP ALL AGAINST THE REALM OF GRECIA. AND A MIGHTY KING SHALL STAND UP,* (Saddam Hussein) *THAT SHALL RULE WITH GREAT DOMINION, AND DO ACCORDING TO HIS WILL. AND WHEN HE SHALL STAND UP, HIS KINGDOM SHALL BE BROKEN, AND SHALL BE DIVIDED TOWARD THE*

FOUR WINDS OF HEAVEN; AND NOT TO HIS POSTERITY, NOR ACCORDING TO HIS DOMINION WHICH HE RULED: FOR HIS KINGDOM SHALL BE PLUCKED UP, EVEN FOR OTHERS BESIDES THOSE." Now in ancient Scripture, for a leader to stand, or stand up; means to go to war. As prophesied, Suddam Hussein's army of Iraq/Persia, being that first shorter mighty horn to stand. Met his Prophetic demise; also as Prophesied, hanging not just him in 2003, but his two brutal sons Uday and Qusay Hussein. So, as Prophecy says, Iraq was not given to his *posterity*. And we saw the further final fulfillment of that Prophecy, in that Iraq's new government is a different type of government or (dominion) than Saddam's. The Saddam Hussein Government was Sunni, and his replacement Government is Shiite, which aligns with Iran. This is our first major relevant identified Prophesied last days event on Earth; which we place in the timeline of the advent calendar on pg.486, as Earth event 1. Find *Earth* event 1, look straight up to see it is placed, after the first four seals are opened, and in the beginning of sorrows, yet prior to the Tribulation. Again, this Calendar becomes easier to follow with each use. **Note: at this time of December 2012, concerning all prophesied last days major events; this is where we are.* All Scripture revelation and empirical evidence say the opening of the first four seals, and the Heavenly and Earthly events we've identified, including the defeat of the first Persian Iraqi king, as of this writing in 2012, have already occurred. And at this time, the fifth seal, it being a Heavenly event, may have occurred, but we cannot know at this time. The next Prophesied major events going forward, is the defeat of Iran, and then World War III; And at opening the sixth seal, World War III goes nuclear. So, the next major events, are the defeat of Iran, and a world-war going nuclear, that brings great famine and disease, killing 25% of the people on earth. Of course, many huge things will occur in between, as all things cannot be Prophesied. But big events along the way, the Four Horseman can physically cause, that we can call markers, *are* Prophesied in their chronological order to let us know where we are in Prophecy. There is so much to know, that I sense it may be easier to see it come together if we first give a running narration from beginning to end. Then come back and go Scripture by Scripture to confirm it all. In Daniel 11:2 there are a total of four *last days* Kings to stand up out of an area known as Ancient Persia. For a broad understanding, let's re-read starting at Daniel 10:12 to Daniel 11:6, where the angel says, *"THEN SAID HE UNTO ME, FEAR NOT, DANIEL: FOR FROM THE <u>FIRST DAY THAT THOU DIDST</u>*

SET THINE HEART TO UNDERSTANDING, *AND TO CHASEN THYSELF BEFORE THY* ***GOD***, *THY WORDS WERE HEARD, AND I AM COME FOR THY WORDS. BUT THE PRINCE* (fallen Angel) *OF THE KINGDOM OF PERSIA WITHSTOOD ME ONE AND TWENTY DAYS: BUT, LO, MICHAEL, ONE OF THE CHIEF PRINCES* (Holy Arch Angels), *CAME TO HELP ME; AND I REMAINED THERE WITH THE KINGS OF PERSIA. NOW I AM COME TO MAKE THEE UNDERSTAND WHAT SHALL BEFALL THY PEOPLE IN THE LATTER DAYS: FOR THE VISION IS FOR MANY DAYS, AND WHEN HE HAD SPOKEN SUCH WORDS UNTO ME, I SET MY FACE TOWARDS THE GROUND, AND I BECAME DUMB.* (faint and unable to speak in fear) *AND BEHOLD, ONE LIKE THE SIMILITUDE OF THE SONS OF MEN TOUCHED MY LIPS: THEN I OPENED MY MOUTH, AND SPAKE, AND SAID UNTO HIM THAT STOOD BEFORE ME, O MY* ***LORD***, *BY THE VISION MY SORROWS ARE TURNED UPON ME, AND I HAVE RETAINED NO STRENGTH. FOR HOW CAN THE SERVANT OF THIS MY* ***LORD*** *TALK WITH THIS MY* ***LORD****? THEN THERE CAME AGAIN AND TOUCHED ME ONE LIKE THE APPEARANCE OF A MAN, AND HE STRENGTHENED ME, AND SAID, O MAN GREATLY*

BELOVED, FEAR NOT: PEACE BE UNTO THEE; BE STRONG, YEA, BE STRONG. AND WHEN HE HAD SPOKEN UNTO ME, I WAS STRENGTHENED, AND SAID, LET MY ***LORD*** *SPEAK; FOR THOU HAST STRENGTHENED ME. THEN SAID HE, KNOWEST THOU WHEREFORE I COME UNTO THEE? AND NOW WILL I RETURN TO FIGHT WITH THE PRINCE* (fallen Angel) *OF PERSIA: AND WHEN I AM GONE FORTH, LO, THE PRINCE* (Holy Arch Angel) *OF GRECIA SHALL COME. BUT I WILL SHEW THEE THAT WHICH IS NOTED IN THE SCRIPTURE OF TRUTH: AND THERE IS NONE THAT HOLDETH* (knows this) *WITH ME IN THESE THINGS, BUT MICHAEL YOUR PRINCE* (Holy Arch Angel). *ALSO, I IN THE FIRST YEAR OF DARIUS THE MEDE, EVEN I, STOOD TO CONFIRM AND TO STRENGTHEN HIM. AND NOW WILL I SHEW THEE THE TRUTH. BEHOLD, THERE SHALL STAND UP YET THREE KINGS IN PERSIA; AND THE FOURTH SHALL BE FAR RICHER THAN THEY ALL: AND BY HIS STRENGTH THROUGH HIS RICHES, HE SHALL STIR UP ALL AGAINST THE REALM OF GRECIA. AND A* (The first) *MIGHTY KING* (Saddam Hussein) *SHALL STAND UP, THAT SHALL RULE WITH GREAT DOMINION, AND DO ACCORDING TO HIS WILL. AND WHEN HE*

SHALL STAND UP, HIS KINGDOM SHALL BE BROKEN, AND SHALL BE DIVIDED TOWARD THE FOUR WINDS OF HEAVEN; AND NOT TO HIS POSTERITY, (children) *NOR ACCORDING TO HIS DOMINION* (ruling authority) *WHICH HE RULED: FOR HIS KINGDOM SHALL BE PLUCKED UP, EVEN FOR OTHERS BESIDES THOSE. AND THE KING OF THE SOUTH* (Iran) *SHALL BE STRONG, AND ONE OF HIS PRINCES* (allies like Russia, China, or Turkey); *AND HE* (Iran) *SHALL BE STRONG ABOVE HIM (Iraq), AND HAVE DOMINION; HIS DOMINION SHALL BE A GREAT DOMINION.* (a Nuclear Weapons Country) *AND IN THE END OF YEARS* (last days) *THEY SHALL JOIN THEMSELVES TOGETHER; FOR THE KINGS DAUGHTER* (an Islamic leader) *OF THE SOUTH (Iran) SHALL COME TO THE KING OF THE NORTH (Russia) TO MAKE AN AGREEMENT: BUT SHE* (the Islamic leader) *SHALL NOT RETAIN THE POWER OF THE ARM; NEITHER SHALL HE* (Iran) *STAND, NOR HIS ARM: BUT SHE SHALL BE GIVEN UP, AND THEY THAT BROUGHT HER, AND HE THAT BEGOT HER, AND HE THAT STRENGTHENED HER IN THESE TIMES.*" We read here in Daniel, that the two horns of Iraq, and then Iran are both defeated. As our study progresses, Scripture clarifies that these both occur *before* the first Beast/World-Government. Where *after* this World Government is formed, a third Persian King will rise up to overthrow one, and then two more regions of a coming World Government. This overthrow will incorporate all strongly Islamic regions. This demonic man rises quickly, being the third Beast or Antichrist, who goes on to take control of all ten (the world) with the help from a fourth Beast, being a demonic False Prophet. HOLYSPIRIT shows us throughout Scripture and History, that Kings rules by sitting on a Throne; and it is said in Scripture and throughout History, that when a King *stands*, it signals a fight or battle. We see again that the first horn of Iraq stands, and then later Iran stands; and then the Third and Fourth Beast stand. We see today many Persian Kings in the Middle East fretting among each other. This in-fighting is, and will always be a lesser part of this world's troubles. In these last days, a total of four main Persian Kings will stand, having a great evil hunger for power in their heart. One has stood and fallen, and three to go. Scripture says the King of the South and Persia, has a daughter; this daughter is the Islamic religion. We see this "Daughter" makes deals with Kings of the North, being Russia, and others. Russia and Islamic leaders have always been enemies, yet unreliable allies against Israel and the west. Both wanting control of the whole

Middle East with its oil and warm year-round ports of the Arabian and Mediterranean Seas. For Islam, there is additional religious desire to possess Israel, Temple Mount, and the world, being GOD'S affirmation of their faith. We read in that last Scripture from Daniel, that Iraq, nor Iran, nor Islam, nor Russia, are victorious in their efforts to gain control over Israel; instead, they all suffer sound defeat. In Dan. 11:7-19, we see fighting continues in the Middle East within these former provinces, even after the Peace Treaty that ends this coming World War III. We will see Islam here and there, continually break the agreement as they always do. In Dan. 11:20, we see a second Persian Governor, has a very short reign over three regions in the coming Ten-region World-Government who is quickly overthrown by the rising Third and Fourth Persian Kings, being the Antichrist and False Prophet. They bring the whole world under their control by the end of this World-Government's first 3-1/2 years of existence. In Dan. 11:21 through 11:31, the Antichrist, at assuming full control breaks all its agreements, including the Seven-Year religious security Article of the World-Government Peace Treaty, made with "*many*" (all major religions). The Antichrist takes away the Daily Sacrifice and all control of the Jewish Temple. In Daniel 11:31-45, we see more details about the

Antichrist's reign, which may give well-seasoned students pause, at hearing the Antichrist is described as a Persian, when other Scriptures call him Assyrian. But remember these proper nouns and pronouns are written by different authors at different times under different Empires; but the area maps of the Ancient Assyrian Empire, greatly overlays the map of the Ancient Persian Empire. So, this is a semantic argument, as these Scriptures speak of the same area. We add that today *most* Islamist don't identify as Assyrian or Persian, being ancient references. We are told that this is the geographical area of heritage, from which these four Islamic Kings arise. By applying all relatable Scripture, we see the region of Pergamon in Turkey, towards the Mediterranean Sea, is the area the Third Beast, or Antichrist, will hale from. As we go through the many coming pages of Scripture, they also give details about the many events that occur during this power struggle. Now Islamist know, after seeing Iran fight Iraq for over 8 years with little to show; and see the West decimate Iraq in three weeks. The second higher horn of Iran, learned a great lesson; being they can never win a conventional war against the west. So, Iran seeks nuclear missiles that can reach Israel and further. And in recent years, going back to at least 2007 or further, they have boldly professed that once they have them, they will use them. They believe this will keep the west from entering the battle when they attack Israel. If Israel needs our help, the U.S. will be there, depending entirely upon the Party and President in power. Our help is not the guarantee it once was. The antisemitism that is once again, starting to brew in the liberals of western democracies, is starting to turn in favor of the oppressive Islamist terrorist, over Israel. Iran is that second horn that will soon *stand* up against Israel and the west. But before Iran is totally defeated, they will see victory is not theirs, and seek face-saving ways to withdraw from the fight to rebuild. With or without help from the west, we read GOD in the last days is with Israel, giving them every victory. That is, right up until they once again turn from GOD, and sign that World Peace Treaty at the end of the coming World War. Israel will be under pressure from the entire world, and most of their own population to sign, as all countries surrender their sovereignty to a One-World-Government. Now the Old Testament Prophesies about four powerful Spirits influencing the trouble of Israel and the world in the last days. Yes, these speak of the same Four Horseman in we see in Revelations. Zechariah 6:5-8, "*AND THE ANGEL ANSWERED AND SAID UNTO ME, THESE ARE THE FOUR SPIRITS OF THE HEAVENS, WHICH GO FORTH FROM STANDING*

BEFORE THE LORD OF THE WHOLE EARTH. THE BLACK HORSES WHICH ARE THERIN GO FORTH INTO THE NORTH COUNTRY; AND THE WHITE GO FORTH AFTER THEM; AND THE GRISLED GO FORTH TOWARD THE SOUTH COUNTRY. AND THE BAY WENT FORTH, AND SOUGHT TO GO THAT THEY MIGHT WALK TO AND FRO THROUGH THE EARTH: AND HE SAID, GET YOU HENCE, WALK TO AND FRO THROUGH THE EARTH. THEN CRIED HE UPON ME, AND SPAKE UNTO ME, SAYING, BEHOLD THESE THAT GO TOWARD THE NORTH COUNTRY HAVE QUIETED MY SPIRIT IN THE NORTH COUNTRY." We see more information about these four in the Old Testament, being that in countries to the north of Israel, the influence of two of these Spirits, have quieted HOLYSPIRIT. This Scripture refers to the geographical north where we find Russia, China, Turkey, Iran, Iraq, and other godless countries. As we move faster and deeper into the last days, Scripture says that from the compound effects of world-war, that by the end of the beginning of sorrows, after the World-War, and just before the Great Tribulation begins; one fourth of all mankind will die by war, famine, and disease. This is not in the far distant future; it is nearer than we think. At the soon coming defeat of Iran, the precarious balance of teetering world peace for Russia, China, and others, will be stretching to a breaking point, overloading the management of aggressions against Israel and the west. This World can be set to war at any moment, yet its fragile peace will teeter a while longer. This world's systems, are collapsing under rapidly changing complexities, and no form of Global Government can save it, even as a growing many believe it can. We clearly see how this world's differing cultures, values, religions, and standards of living, prevent this from being possible. There is no path of everyone compromising all they know, love, and need, to unite under one common compromised, spirit killing, misery filled way. It is an empirical fact, that there can never be true enduring peace, minus complete victory and complete surrender by the loser. There can be no compromise. In the coming war against Iran, if needed for victory, it *may* include a few small tactical nukes. See Iran's coming defeat in the advent calendar as Earth event 3. Now once the future war against Iran ends, it will set in motion confusing, complicated, uncertain, and aggressive world-wide realignments. The flow of oil will be highly prioritized in the build-up to this coming World War III, and the world beyond the largest oil producing and farming nations could be reduced to dire status in weeks. The major powers will rush to control the flow of the strategic

Middle East oil; and the world will be one decisive shot away from World-War III. By uncontrollable events, one by one, the countries of the world will take up new sides, and even change sides. The last weeks running up to world war, will push the energy crises over the edge, which pushes this world's existing financial crises over the edge; and by just one small incident, it will all suddenly break down. The United States, Europe, Russia, China, and every naval power will form two alliances, and move their subs, cruisers, and carrier groups, near and into the Mediterranean and Indian Seas. Military Bases across Europe and the Pacific will posture and fill with allied troops and Bombers. Scripture says, Russia will move aggressively into the region as they surely see U.S. efforts to stabilize the region, are an effort to gain total control; which is true and necessary for U.S. national security. We can be sure the U.S. and our allies, like all powerful countries, will want a stability they control, to favor their countries. We see in the book of Revelation, the names Gog and Magog. These names are by most preachers, attributed to Russia. That belief is not without some broad loose reasoning. Yet it is far more accurate to think of these Biblical family names of Gog, and Magog, as all peoples and forces north of Israel, allied with Russia. Magog and Gomer, were Grandsons of Noah. And Gog was born of the early generations of Rueben. In ancient chronicled Biblical history, these families migrated northward to settle and spread to the far north above Israel. Eventually migrating into regions now known as Turkey and the former southern Soviet Union and further. These peoples were all born after the great flood, so their vast descendants fill these northern regions. And so, we get the Biblical references, "the land of Gog, and Magog" and "Gog the land of Magog". The Country of Russia and others, did not exist when these Scriptures were written. These northern lands are now heavily controlled or influenced by Russia, are Prophesied in Ezekiel Chapters 38 and 39, to move south by a need for power, oil, and the warm year-round ports of the Mediterranean, for greater strategic and commercial access to the world. And Scripture says, the King of the North, in all reason, being Russia, and their Shiite Arab allies, along with Libya and other N. African Countries will move against Israel, just prior to this World War III. But because of long-standing hate for one another, and cultural differences, Russia and these Arab Countries are all being deceitful and unable to trust, communicate, or coordinate. And in this mistrust, deception, and confusion, they end up turning and attacking and killing large numbers of each other before they can attack Israel. And when they do come near to Israel and Jerusalem, they suffer defeat by Israel's small Tactical Nuclear-Weapons. This is the plague and fervent wrath that GOD said he will send upon *all* who come against Israel in the last days. Zechariah 14:12, *"AND THIS SHALL BE THE PLAGUE WHEREWITH THE **LORD** WILL SMITE THE PEOPLE THAT HAVE FOUGHT AGAINST JERUSALEM; THEIR FLESH SHALL CONSUME AWAY WHILE THEY STAND UPON THEIR FEET, AND THEIR EYES SHALL CONSUME AWAY IN THEIR HOLES, AND THEIR TONGUE SHALL CONSUME AWAY IN THEIR MOUTH."* Zechariah 2:5, *"FOR I SAITH THE **LORD**, WILL BE UNTO HER A WALL OF FIRE ROUND ABOUT, AND WILL BE THE GLORY IN THE MIDST OF HER."* Ezekiel 38:18 through 39:6, *"AND IT SHALL COME TO PASS AT THE SAME TIME WHEN GOG SHALL COME AGAINST THE LAND OF ISRAEL, SAITH THE **LORD GOD**, THAT MY FURY SHALL COME UP IN MY FACE. FOR IN MY JEALOUSY AND IN THE FIRE OF MY WRATH HAVE I SPOKEN, SURELY IN THAT DAY THERE SHALL BE A GREAT SHAKING IN THE LAND OF ISRAEL; SO THAT THE FISHES OF THE SEA, AND THE FOULS OF THE HEAVEN, AND THE BEAST OF THE FIELD, AND ALL CREEPING THINGS THAT CREEP UPON THE EARTH, AND ALL THE MEN THAT ARE UPON*

THE FACE OF THE EARTH, SHALL SHAKE AT MY PRESENCE, AND THE MOUNTAINS SHALL BE THROWN DOWN, AND THE STEEP PLACES SHALL FALL, AND EVERY WALL SHALL FALL TO THE GROUND. AND I WILL CALL FOR A SWORD AGAINST HIM THROUGHOUT ALL MY MOUNTAINS, SAITH THE ***LORD GOD****: EVERY MANS SWORD SHALL BE AGAINST HIS BROTHER. AND I WILL PLEAD AGAINST HIM WITH PESTILENCE AND WITH BLOOD; AND I WILL RAIN UPON HIM, AND UPON HIS BANDS, AND UPON THE MANY PEOPLE THAT ARE WITH HIM, AN OVERFLOWING RAIN, AND GREAT HAILSTONES, FIRE, AND BRIMSTONE. THUS WILL I MAGNIFY MYSELF; AND I WILL BE KNOWN IN THE EYES OF MANY NATIONS, AND THEY SHALL KNOW THAT I AM THE* ***LORD****. THEREFORE, THOU SON OF MAN, PROPHECY AGAINST GOG, AND SAY, THUS SAITH THE* ***LORD GOD****; BEHOLD, I AM AGAINST THEE, O GOG THE CHIEF PRINCE OF MESHECK AND TUBALL: AND I WILL TURN THEE BACK, AND LEAVE BUT THE SIXTH PART OF THEE, AND WILL CAUSE THEE TO COME UP FROM THE NORTH PARTS, AND WILL BRING THEE UPON THE MOUNTAINS OF ISRAEL:*" We read here how Israel is protected and defended by GOD, as Russia and her allies are waring and beaten and turned back. That coalition retreats, and Russia returns home with just one sixth of their forward attacking forces. This all severely weakens and demoralizes Russia. And during the escalation up to World War III, the King of the South (Iran) as Prophesied, will blame, attack, and make certain advances towards the King of the North along Russia's southern border. But Scripture says this is short lived, until Russia reenforces their southern troops and replenishes their southern arsenal. For as long as Russia has existed, it has wanted the oil and warm water ports of the Middle East. Russia once thought these could be won by a chess move of force, in gaining a strategic foothold and staging ground in the barren desert Country of Afghanistan. This was foiled by the intervention of U.S. arms, *(stinger missiles)* and training given to the Mujahadein resistance in Afghanistan in 1986. After that, Russia knew the U.S. would never allow an advance that could risk the Middle East. Now Prophecy repeatedly states we will see a Ten-Region, World-Government formed after a World-War III. And we can know it will have ten newly drawn essentially equal regions, having ten united world Heads. Now in a heated run-up to this Third World War, we can foresee frantic U.N. negotiations to adopt a current framework for a World Government, trying to stop this World War. And in the early stages of this war, we might see an adoption of a loose basic charter framework by all smaller weaker non-military countries, trying to stay out of the hostilities and destruction. And obviously, those with the least to lose and the most to gain by signing, will sign first. And those with the least to gain and the most to lose, will of course be the last to sign. The last hold outs will be Israel, all the Major Powers, and some Islamic Countries. We now see the weight of more and more world-wide deceptions being piled upon the top, as many other deceptions hold it all up from the sides. Yet it is all twisting in the winds of ever greater uncontrollable confusion, deceit, and chance occurrence blowing against it from all directions. The one pure truth of GOD has been cast out and lost to this world. And now only endlessly changing biased contrary beliefs, rule its waring parts. Every powerful country in the world has insurmountable debt, being beyond repair by any means other than a negotiated debt settlement and forgiveness of certain dollar for dollar Bond and Treasury Note debt, borrowed between countries. Government too citizen debt, is a hard wild card. And all the new borders of a united ten-region One-World-Government must be agreed upon. Now please see, that though this may appear as the only way to save the world, all prior distrust and hatred will not just disappear. Negotiations, will at best, be heated and argued down to produce a basic minimal loose and tenuous draft. At its full and final adoption, this will not be a pleasant world-wide kumbaya moment, but the great world-wide war will stop. There is no path to any form of world unification, other than a devastating world war. At this point, let's discuss the Prophesied *third* rebuilding of GOD'S Temple on Temple Mount. We read the Third Temple upon Temple Mount in Jerusalem, is built in Prophetic Scripture after certain Prophesied events occur, and prior to other Prophesied events occurring; so, we see a defined window for its building. And by it being written, means it will be built. By all these Prophecies, we know at some point after the defeat of both Iraq and Iran, and before Israel signs that Treaty surrendering their sovereignty, and joining a One-World Government. That Israel sensed GOD was with them, and the Temple is rebuilt. Now through all negotiations to end the third world war, Jews, Christian, Muslims, and all other religions stand firm, demanding guaranteed protection for their free religious practice. This guarantee in the Treaty that deals with Religious freedom and practice and sovereign places of worship, which Scripture says is made "*with many*" will be adopted for seven years; and set to be revisited every seven years. We put the Temple's reconstruction, in the advent calendar near to where it must occur as Earth event 4. There is no exact Biblical reference for when the Temple construction starts or finishes. But we read about many things that occur before it is built, and after it is built. So, we know it is built in between these same things. We see that all last days Prophecy points to *our* near future, and clearly says this treaty made with this World Government, has the provision for not a one world-religion, but *many* religions for seven years. So, why do many Preachers teach that it is a treaty with Israel alone. This Treaty brings all Ancient Religions, seven-years of protection, that also includes the Jews Daily Sacrifice in the Temple. We are sure of these things, because Prophecy says the Third Beast will only rise up and take control of the Temple, and take away the Daily Sacrifice, after 3½ years into the seven-year Treaty. Now obviously the Temple Sacrifice cannot be taken away, if it does not exist; and the Daily Sacrifice can't exist without the Temple. Scripture says it will be built, so have no doubt. The interesting question to ask here, is how will the Dome of the Rock, and the nine or so other mosques, now built upon it, be removed. They may be moved, or destroyed by fire, wind, or even Iran or Islamic terrorist trying to blame Israel. All we know is according to Prophecy, these mosques are removed. Such an event will make world headlines in every news outlet that exists. Let's discuss the structure of this coming Ten-Region World-Government called the Beast. We are not given intricate details of all things. And we know by all souls having free-will, even GOD cannot predict everything in detail. But FATHER can tell us how he will *physically* intercede by his own superior will, and what he will cause to physically happen in spite of the free-will desires of mankind. Remember that FATHER has free-will, and can affect, alter, and even control the sum of physical reality in any way he desires. Mankind is in control of their own choices and desires, but they do not control physical reality, FATHER does. So, by HOLYSPIRIT and Prophecy, we know to expect all Prophesied things will occur in their certain order. We also know Satan works against GOD'S plan, and their work is in some ways cataclysmic, and many other ways, quite subtle. This New World Treaty, will in so many ways be subtle and intentionally deceptive. Scripture warns us of many great evils being subtle; meaning something that can on its surface, or first glance, appear to be one thing, but its truth lies buried in its unforeseen future and unrevealed details. This World-Treaty is designed to gradually force a world full of very different people, to relinquish all national identity; being all their Constitutions, Borders, Laws, Currencies, Police, and Military. But subtly hidden in the details, is that in the end, it is intended to force all to surrender what they believe in and worship. This world's hedonism, lawlessness, and hatred, of GOD'S created ways, has grown for a long time, and will continue to grow. Such evils push Judeo-Christianity towards false doctrines and a blending of non-Christian traditions and holidays being introduced mostly through Catholicism. The term "Catholicism" actually means to form a combined universal belief, or a timely accepting and blending of all religious doctrines, traditions, and holidays into one. Now in seeing the complexity of other treaties, we must know a One-World-Constitution will be extremely complex. It must consist of many large components, probably being thousands of pages long. No soul will be able to understand or read it all, to vote on it. It may at first, be flexible as to a Country's initial participation, with introductory levels of participation to pull all in by giving up less at first, and more later. Such a graduated surrender, may then coincide with each country's earned rights and privileges of membership. But beyond our educated speculations, a One-World-Government is truly coming. And Israel, above all countries, will probably be the last to join, after Russia, China, the Islamic Nations and the United States. The pressures upon the last country will be tremendous. This New World Order must try to put parts of all former countries, into ten fairly equal size and wealth regions. This will erase all former great and small countries from the face of the earth. All countries may quickly cease to exist on a new world map, but people's hearts do not change so easy. People are often forced to verbally agree to things for reasons of fear, or advantage. But a change of the heart is another matter. So, the power-hungry men and women chosen to enforce this Treaty, will hold great power. By this Treaty's promise to all major religions, the Jews retain their religious rights to their Temple and practice. So, we will see pressure from both Israel's religious and secular people, to ratify at least an entry participation. But at the very *moment* Israel signs away the sovereignty of GOD'S land to Satan in that treaty. They will have turned their back on GOD and chosen the world (Egypt) again. They will have sold their Birthright to the Land of GOD'S Children, for a promise of peace this world cannot ever give them. They will have diminished GOD'S Holy Land in this world to exist only within the walls of the Temple; and Satan comes for that in 3 ½ years. The true GOD of Israel who stood with them, keeping them victorious, will be betrayed again by unfaithful children. The Jewish people will at that time of joining, give up sovereign control over their laws, borders, security, trade, currency, military, policing, and all weapons. And the Seven remaining years they owe GOD, will begin. A great debate must arise, as to which of the ten provinces, are they going to put the former land of Israel and Jerusalem. The former Arab Nations will want it in their most populous region, this in their mind, fulfilling their raging religious visions. Israel will never sign under this condition, and GOD and his will not allow this. Israel will be placed in a region with much of former Eastern and Western Mediterranean Europe, or the Indo-pacific, or what was once Africa. Now though the Islamic religious leaders will sign this treaty; but their belief says they are allowed to lie to and deceive their enemies. We know the Islamic faithful can never let go of their vision to rule the world, and GOD'S land known as Israel. So, we see the reason, fervor, and catalyst, for the Third and Fourth Persian Beast to rise out from among the most Muslim regions. The religious exemption in this World-Treaty, is not made between the Third and Fourth Beast, being the Antichrist and False Prophet, and Israel. It is between the First Beast World-Government, and all religions. This Seven-Year Treaty with *many*, that allows the Daily Sacrifice, is described in Revelation 13:14-15, Matt. 24:15, Daniel 9:27 and Daniel 11:31. And this treaty, once it is signed by Israel, triggers the so-called prophetic gap. Bringing the remaining seven years owed, of Israel's determined 496-year exile from GOD for a prior betrayal. These years of exile were appointed to Israel the last time they as a nation betrayed GOD, and all but a small remnant was taken and enslaved by Egypt. Now we must know that this treaty has nothing to do with any Israel Palestinian treaty. And at signing this agreement by enough countries to stop the world at war and exert some coordinated pressure of authority, the world will officially enter into and start the 3½ years of the First Beast. Mark the date down and remember it when it occurs as it is important to know that day. We know by reading Rev. Chapter 13, and Daniel 7:7-21, that Scripture reveals several different ways to use the term "The-Beast". Now these four Islamic leaders, or beast that arise within the first 3 ½ yrs. of this future World Government, are not the four Prophetic Historical Beast of Antiquity we read about in Daniels visions. But this new World-Government, having ten regions and Governors; being Ten-Horns with Ten-Crowns, is itself Daniels Fourth Beast of historical antiquity. And its Second leader, or horn or beast, is an Islamic leader who rises up to take control of one and then three regions. After this Islamic uprising is started, it is taken over by another leader, being the Antichrist who claims the government and the world are his. By force, he becomes the Beast ruler of all ten; but only after he is released from the pit at the Fifth Trumpet of the Tribulation. This Third Beast, being the second in command under Satan, quickly enters the political arena, followed by the Fourth Beast being his false Prophet. Scripture says Antichrist holds nothing with Islam or any religion; yet the Third and Fourth Beasts, being from Persian areas will use Islam to concur and control in the early parts of Satan's plan. We read the Antichrist rises up from one high Muslim-Region, using Muslim revolt. The first three regions concurred will contain most of what was formally known as the Arab Middle East and two others Muslim dense regions. This Third Persian Beast or Horn or King, being the actual embodiment of the real fallen Angelic Antichrist, will receive a fatal head wound from an enemy. But still being an angel who cannot physically die, recently released from the bottomless pit by Satan at the Fifth Trumpet, he appears to miraculously heal, and is feared and worshipped. He continues to take control by force while promising peace and prosperity, until he has a firm hold over all ten regions. All this occurs by the end of the first 42 months, or 3½ years, of the World Treaty. At this 42-month point, the Antichrist is then allowed by FATHER, according to Scripture, to have control for the second half or 42 months of the World treaty. Upon taking control, the Antichrist takes the Jerusalem Temple from the Jews, and all religious places of worship, and declares he is preparing the way for the true God, and outlaws all worldly religions. The Antichrist, with FATHER'S permission, kills GOD'S two witnesses who preached Judgment of the world for unfaithfulness outside the Temple. Antichrist has the continual help of a very powerful fourth fallen angel, called the False Prophet. We see the Antichrist and False Prophet rule the entire world, and bring about the completion of Satan's task, and prepare for the abomination of desolation. They create and place a Great Image/Idol in the Temple of GOD in Jerusalem. And Satan then indwells and causes the Image/Idol to speak in the Temple on the Holy Temple Mount. Satan then and there declares he is God, ruler, and owner of the Earth. This is the Prophesied *Abomination of Desolation*, meaning everything in GOD'S earthly creation, including the Holy of Holies, has been destroyed and polluted. Many will think this image in the Temple is a super A.I., superior to mankind, who must rule over mankind. Remember, this all happens after the Fifth Trumpet sounds. And only after this desolation occurs, does Antichrist call for and implement his mark of the beast. A call to the world to bow and worship the Image of the Beast that speaks. Scripture says Satan is then standing where he ought not be. And at *that* time, Scripture says Satan installs Ten new Demonic Leaders over all Ten World Regions, who enforce his new World Religion under the eventual penalty of death. Now many ask if this final compulsory religion is Islam or some form of it? No. We read in Dan. 11:37, that the Antichrist does not honor the GOD of Islamic fathers or any others fathers. He honors only force, power, and Satan. We see the World-Government that Israel signed a treaty with, is now the Government of Satan. Satan subtly used this Ten-Region, Ten Ruler approach to get the major sovereign powers and peoples of the world to give up their sovereignty and power. All people would resist giving one leader control, but ten is the lowest magic number of leaders the world can accept. This World Charter results in ten regions drawn to equalize natural resources, stabilize trade, and bring equity of rights and wealth and a single currency to all regions. But in its subtlety, it spreads all policing powers equally, it removes all advantage, leaving the whole world equally weak, tired, and fearful of war. So, after Antichrist unites three largely Muslim regions, the rest quickly fall. Now, about the Prophesied Great Falling Away, we see Antichrist and the False Prophet are accepted by the world at large, and seen to be directly below and sent by GOD. A Trinity is seen and accepted, as Satan duplicates

GOD'S Governing structure and miracles, but not GOD'S laws. By legalizing Hedonism and perverse physical desires, the world follows en-masse. By false beliefs, rebellion, and weakness, the world falls away to follow Satan, whether they realize it or not. Many will be given authority to bring to justice, or if necessary, kill any person proven to be in direct opposition to Satan's ruling religion; who will be labeled as enemies of world peace. The propaganda campaign will be relentlessly focused towards the remaining True-Children of GOD. So, fulfilling the Prophetic words of our LORD in John 16:2, saying, in the last days; when they kill you, they think they do GOD a service. And Daniel 8:25 says that "through peace he will destroy many". Meaning that by all Nations of the World signing this Peace Treaty, and a mass psychosis created and led by Satan towards world peace, the saints of GOD will be hunted by the world, and martyred for refusing to bow. Yes, the true Church will still be here for the Mark of the Beast and is not raptured out. The false weak Christian-Church and all other religions, will have in mass, fallen away to Satan. The world is then seen as eternally lost to GOD, in its bowing to Satan *and* receiving his mark of ownership and worship. So, now after that long narration, we return to our verse-by-verse study of the Revelation, to confirm everything we just laid out. We should quickly make the point, that each new Earth-Event, will in many ways be combined to the former, which contribute and add to all coming events. What we mean, is these events will compound, as all the destruction accumulates. Let's start again at Rev. 6:9 and bring this bright Prophecy to life. By all received understanding, we know the first four, and maybe five seals, have been opened as of 2012. And we saw Iraq defeated, and now await Iran's defeat. And in many ways, these will lead up to opening the sixth seal. This world knew or felt nothing at the opening of the first four seals, other than a gradual world-wide division and tension. Prophecy says these four horsemen went forth with a power to accomplish certain things in the earth. Not immediately, or instantly, but over an unspecified period. Most souls, over this same time, are unaware of these four spirits identity, and their works purpose as world pressures increase. And again, these first six seals are not part of the Tribulation; they are the beginning of sorrows. JESUS in Matthew 24:6-8 says, *"AND YE SHALL HEAR OF WARS AND RUMORS OF WARS: SEE THAT YE BE NOT TROUBLED: FOR ALL THESE THINGS MUST COME TO PASS, <u>BUT THE END IS NOT YET</u>. FOR NATION SHALL RISE AGAINST NATION, AND KINGDOM AGAINST KINGDOM: AND THERE SHALL BE FAMINES,*

AND PESTILENCES, AND EARTHQUAKES, IN DIVERS PLACES. ***ALL THESE ARE THE BEGINNING OF SORROWS.***" Rev. 6: 9-11. Upon opening the Fifth Seal, we see a *side-bar*, to hear certain wisdom and inspiration from GOD to help us understand all things. We see in Heaven, all souls who to date, were martyred on Earth for GOD, prior to the soon opening of the Sixth Seal; and they are all safe under GOD'S Throne. During this Heavenly moment, these Martyrs are instructed to have patience, for more brethren are to be martyred as they were. So again, the Church is still on Earth, and not yet raptured out. This Ceremony may have occurred already, we can't know until we see the defeat of Iran and World War III, which may occur close together. Read Rev. 6:12-17. As the Sixth Seal is opened, we see the heights of this Third World War, going nuclear. Scripture says the Earth is shaken, and the upper atmosphere is darkened with debris and fallout. This coming nuclear war is *Earth* Event 5, in the timeline. In Prophecy we see these bombs described as stars falling from heaven because of their great fire. We read the skies are filled with the debris from bombs and many great volcanoes and fires, darkening the sun and moon. We have all seen nuclear blast clouds, rolling up together, as rolling up a scroll, with every mountain and island shaken and moved. We know Presidents, Generals, and the powerful, will hide in nuclear bunkers deep inside mountains of rock praying to be spared. Though GOD'S Prophetic timeline is clear, many Preachers confuse and mix this next war, with the following two wars. You *must* keep Prophecies Third World-War, and the Battle at Armageddon, and the last war after the Millennial Reign, separate and in proper order, or you falsify Prophecy. Scripture says this next war's destruction, combined with its famine and disease, will close out the beginning of sorrows with the death of one fourth of mankind. Satan knows, *only* by bringing it all to the brink of annihilation, can he get control of this world and Israel through a new ten region government. Next read Rev. 7: 1. This is still part of the Sixth Seal as we see four Angels holding back the winds. By these Angels, the wind and ocean currents become lifeless, coming to a stop. They are held as Angels prepare the first Four Trumpets of the Tribulation. We read in Rev. 7:2-8, of a ceremony in Heaven, as JESUS is given his pure first fruits still in the earth, prior to the Tribulation. We see an Angel place FATHER'S seal upon 144,000 elect Prophets in the earth. All spotless and without fault from the Twelve Tribes of Israel. These 144,000 are not mentioned anywhere outside CHRIST'S Revelation, Chap. 7, and 14. See this on the calendar as *Heaven* Event 5. Next read

Rev. 7: 9-17; still part of the Sixth Seal. We see FATHER'S Throne in Heaven, and a ceremony declaring the approaching Millennial Reign of JESUS, and the promises given to these 144,000 Saints. We also see the far greater over-all number of souls that will be saved on that coming day who washed their robes, and endured great tribulation in the world. What is the purpose for this future ceremony, placed right after sealing the 144,000. GOD in love, wants to make clear for everyone, that there will be redemption for a great multitude, other than these 144,000. Example; I heard back in the 1970's by a Jehovah's Witnesses, that only 144,00 people will go to Heaven. So, we see why GOD clarifies this point. By the end of the sixth seal, we have had world war, poisoning, starvation, disease, many natural disasters, and a 10 region Government is formed. Now many preachers misinterpret Daniel's vision in Daniel, 2:31-45, of the Fourth Beast with ten toes of iron mixed with clay, to represent Ten European Countries as a form of a revived Roman Empire; this is so wrong. And a related Prophecy in Matthew 24:37-38, that many Preachers interpret incorrectly which talks of marrying and being given in marriage, as it was in the time of Noah; speaks about the clay (mankind) being mixed with iron (angels) to produce powerful children as it was in the time of Noah. Scripture tells us to remember this.

Read carefully Genesis 6:1-7. We are warned in his Revelation from the lips of CHRIST, that in the last of days, it will once again be as in the days of Noah, as demons cast down to earth will have children through women; iron mixing with clay. We know without doubt we will see Ten World rulers, not Ten European Rulers. European countries are totally reliant on the U.S for protection and can't even defend their self, so, how can they conquer the world. At the very last, we *will* see ten powerful demons, being the iron of the angelic power, being mixed with clay, being women. Demons taking earthly wives, are then given those Ten Ruling Seats by Satan, to help control the world. We are warned by JESUS, and it is Prophesied in Daniel 2:43-44, *"AND WHEREAS THOU SAWEST IRON MIXED WITH MIREY CLAY, THEY SHALL MINGLE THEMSELVES WITH THE SEED OF MEN: BUT THEY SHALL NOT CLEAVE ONE TO ANOTHER, EVEN AS IRON IS NOT MIXED WITH CLAY. AND IN THE DAYS OF THESE KINGS SHALL THE* ***GOD*** *OF HEAVEN SET UP A KINGDOM THAT SHALL NEVER BE DESTROYED:*" It happened in the days of Noah, and it will happen again. These ten, take the daughters/seed of men for brides to bare mighty unholy children. This is Satan's plan to enslave mankind and combine women with fallen angels to populate the earth with evil hybrid offspring. But CHRIST prevails at the last. The Antichrist, False Prophet, and all the fallen angels will obey Satan as their GOD, as they battle to the end for him. Under Satan's leadership, the world will come to worship this unholy trinity as being gods, and pledge obedience in all things. Even hunting down and turning in family members who are true Children of GOD; again, the Church is still in the Earth. So, who is the embodied Antichrist, and how will we know him. Well, a blind man should be able to see this guy by the time he arrives on the world stage. We have Scriptural references, such as he is said to be a stout man in Daniel 7:20. Stout being an adjective and a noun; meaning both proud and physically bulky in stature. We are have his name and description in Revelation 9:11, 11:7, and 17:8. So, physically we should look for a stout Middle Eastern man, rising out of what was once the kingdom of Assyria and Persia. And adding Rev 2:13 we see he arises out of modern western Turkey, on the north side of the Bakircay River, northwest of Bergama. We read the Antichrist is not let out of the bottomless pit into the earth by Satan, until the Fifth Trumpet of the Tribulation. So, he will not be on the earth until 2,030 at the earliest. He then attains an earthly identity as a Persian man who rises to power trying to subdue the coming Ten Government regions. He receives a deadly head wound, and is healed after appearing dead, to then take control of all ten regions. And after 3 ½ years he sets up an Image for Satan to indwell in the Jerusalem Temple, and the Antichrist demands all souls take a mark, and worship the image. In Rev. 13:18, we see the *number* of the Antichrist's name is 666. Now we see JESUS, HOLYSPIRIT, the 7 Arch-Angels, and the 4 Seraphim who occupy the Highest Heaven, or Holy of Holies, are FATHER'S first 13 gifted souls. Then fallen Satan, the anointed Cherub, would be the 14th soul FATHER gifted, numbered, and named Lucifer. And that then means the Antichrist, being the second to rebel, was the 666th Cherubim soul that FATHER gifted, numbered, and named, among all the Angels. We're not given the number of, or the name given the False Prophet at his soul's gifting; but seeing how rebellion skipped from the 14th to 666th soul, we expect the False Prophet's number of his name to be large. So, we have plenty of info to identify the embodied Antichrist. Now we read Lucifer did not rebel till much later, and after many experiences in his existence. So, we know sin first appeared a certain distance and moment from FATHER'S gifted order of souls in Heaven. We see the fallen angels rarely added to at first, but began to accelerate more and more. And by the time rebellion reached the Earth and the weakness of mankind, we have the near total loss we see today. But beyond all doubt, we know rebellion never arose in JESUS or HOLYSPIRIT or the first 11 far greater Angels. Now let's read what we call the 7th seal of many wonders, Rev. 8: 1-6 (see fig. a. *Heaven* event 6). Here the Seventh Seal is opened with an amazing thing, being a half hour of total silence throughout Heaven. This half hour means a substantial silent moment of wonder and awe. Nothing speaks louder than pure unified silence. This silence is so meaningful as it signals the start of the Tribulation of GOD. Every Holy and fallen Angel, and every Saint in Heaven, are perfectly still, silently watching this momentous event foretold long ago. Around 200 million Holy Angels in Heaven are fixed in prayer on what is about to happen. After watching the continuous grace and mercy of CHRIST being used, abused, turned from, and trampled on for over 2000 years, the stored wrath of FATHER is about to be seen. We see Seven Mighty Angels standing before GOD, receive the Seven Trumpets of the Great Tribulation. We see a golden censer, filled with the smoke of incense, being all the prayers that rose up to the throne of GOD. Then in an answer to those prayers, we see the censor is filled with fire from the Alter in Heaven, and cast into the earth. This has much meaning and significance. So, let's read the mirrored, book-end reflection of this event in Numbers 16:46. We see a mirrored act of this ceremony, called for by Moses at FATHER'S direction to signal his wrath is coming. And placing a line of separation for his wrath to stop at the Children sanctified by blood. This act in both instances, signals FATHER'S coming Judgment. As this censor is cast into the earth, the earth is stretched, groaning, fracturing, and shaken. A stretching of the earth, always comes by FATHER'S wrath and diminished desire to hold it together. The Seven Arch Angels with the Seven Trumpets are prepared to sound. And *this* is the start of the Great Tribulation. These Seven Trumpet events, unlike the Seven Seals destruction caused by mankind's greed, pride, and influence of the Four-Horsemen, are all supernatural, without a single cause by mankind. This wrath as with the great flood, comes only when FATHER sees no hope they might hear, and turn back to GOD. From the First Trumpet, mankind will say these events are caused by mankind, or strange extraterrestrial or natural causes. But with each trumpet, and lastly each vial poured out, it becomes more and more impossible to explain; it all being purely supernatural. By study with HOLYSPIRIT, we gain wisdom that the first *four* trumpets bring GOD'S wrath by striking only the Earth, which mankind severely damaged in World-War III. These events are GOD'S last screaming attempts to get mankind to repent and look to him for their provision. By great mercy, he is removing every other thing they might look to; yet most will not look to GOD. The last *Three* Trumpets, are separated from the first four, and are called the Three Woes. These Fifth, Sixth, and Seventh Trumpets, the Seventh containing the Seven-Vials, are very different from the first four. These cleanse the Earth of all mankind who are sworn to evil. Within these three, comes the Earth's cleansing of three great rebellious groups of people, holding only desires of destruction for GOD'S plan in their hearts. This cleansing is done in preparation for the Millennial Reign of our LORD upon a fallen world. Through Scriptural interpretation under guidance, we receive that these coming last three great woes will start at the end of a Passover, just as that first censor and cleansing death came in the Book of Numbers, passing over Egypt, and sparing only those with the blood of the Lamb on their door post. This second cleansing was set, when MESSIAH died and rose again at the end of his Passover Sacrifice to become our eternal Passover Sacrificial Lamb. Remember, each trumpet releases certain events over an *unspecified* period. We will see months between Trumpets, and as with the seals, the damage from each remains, to add to the next. Read Revelation 8:7; as the First Trumpet sounds, we see FATHER gather millions of tons of atmospheric debris from nuclear bombs, volcanoes, cities, forests, and tens of millions of human and animal bodies burning around the world. We read he sends back to mankind, the fire, radiation, and blood of their war, in burning masses, down upon the remaining green areas of the earth. He gathers the moisture in the atmosphere to form poisonous hail formations that weigh as much as a man, also filled with human remains. The blood of burnt millions, mixed with nuclear debris in hail stones, raining down to destroy the remaining green parts of the Earth. If you desire more contextual background, read Ezek. 32:1-16; being a continuation of Chapter 31. GOD speaks in Ezek. 31 of Satan and the Beast; and in Chapter 32, of Satan's Kingdoms coming desolation. Now these last days Ezekiel Prophesies of, overlay the first Four Trumpets perfectly. Next, we read Revelation 8:8-9. The Second Trumpet sounds, which contains these words; "*AS IT WERE A GREAT MOUNTAIN BURNING WITH FIRE.*" The words "as it were" mean what could amount to, or appearing equal too. We see GOD send mountainous burning masses, far greater than the First Trumpet, to contaminate a third of the remaining living oceans with radiation and toxins. We can surmise, untold burnt bloody radioactive bodies will wind up in rivers and oceans, as plague filled bodies wash up on coastal areas and inlets around the world. Read Rev. 8:10-11. As the Third Trumpet sounds another event, where these same fiery masses fall and contaminate a third of all remaining rivers and lakes beyond recovery. Many more die, drinking these bloody diseased waters. Rev. 8:12-13. As the fourth trumpet sounds, GOD brings a truly miraculous effect. Its description tells us GOD gathers the remaining atmospheric debris around the Earth, *and* creating magnetic fields around the earth, he gathers all remaining debris into a concentrated ring, being flat and wide as it encircles the earth in space. According to Scripture, this band of debris must be at least a thousand miles wide, and thick enough to block all sun light. This band completely encircles the Earth similar to the ring of Saturn, but laid flat and wide like a wedding band, and running north to south over the poles around the Earth, this band being stationary with one side always facing the sun. And as the Earth turns within this stationary ring, everyone on earth will see morning break as they come around to the Sun, and then see a giant black ominous band in space like a curtain, come over every man woman and child for a third of the day bringing darkness. And then after one third of the day, they come out from under the band and see the light of the evening. And then on the other side, as all go into night, they see the stars and moon at first, then again, they go under this band of darkness for one third of the night, and again coming out before dawn. For more descriptions of this, read the corresponding Old Testament segments in Ezekiel 32:7-8, and Zechariah 14:6-7. Read now Rev. 9:1-12, as the Fifth Trumpet sounds, which is the first great woe. These three escalating woes, are no longer against the Earth, these cleanse only GOD'S irredeemable enemies from among mankind. We read that an Angel, is cast out of Heaven and falls to Earth. This angel falling to earth, with a key, is of course Satan. Here, the word "fall" is very important, in that *no* Holy Angels of GOD, ever fall anywhere. Satan and his Angels, at this point in Prophecy, have all been defeated in a war that has raged in Heaven for at least 2000 years. And Satan is cast out and down forever from Heaven, (see fig. a. *Earth* event 6). We hear JESUS, Prophecy of this coming event in Luke, 10:17, "*I BEHELD SATAN AS LIGHTENING FALL FROM HEAVEN.*" We see this Angel named Lucifer cast down to earth with the speed and force of a lightning strike, has only *now* at the fifth Trumpet happened. He is given a *key* (power and permission) to open the bottomless pit. It being a prison of confused dark and deep conscious hell, where all those defeated evil angels have been imprisoned, in that long battle in Heaven. We also see as part of this release, those two most powerful fallen angels under Satan are set free, the Antichrist and False Prophet. Remember, this occurs after World War III, and some time prior to the second forty-two months of the One-World Treaty. So, we know the first Five Trumpets sound within the first 3½ years of the New World Government. Satan is cast down, and he opens that bottomless quantum prison to free his minions, including the Antichrist and false prophet. Read Rev. 17:8, "*THE BEAST THAT THOU SAWEST WAS, AND IS NOT; AND SHALL ASCEND OUT OF THE BOTTOMLESS PIT, AND GO INTO PERDITION: AND THEY THAT DWELL ON THE EARTH SHALL WONDER, WHOSE NAMES WERE NOT WRITTEN IN THE BOOK OF LIFE FROM THE FOUNDATION OF THE WORLD, WHEN THEY BEHOLD THE BEAST THAT WAS, AND IS NOT, AND YET IS.*" Please know the last third of the next segment of Scripture that is underlined, can't be truly interpreted without HOLYSPIRIT. 2nd Thessalonians, 2:1-9. "*NOW WE BESEECH YOU, BRETHREN, BY THE COMING OF OUR LORD JESUS CHRIST, AND BY OUR GATHERING TOGETHER UNTO HIM, THAT YE BE NOT SOON SHAKEN IN MIND, OR BE TROUBLED, NEITHER BY SPIRIT, NOR BY WORD, NOR BY LETTER FROM US, AS THAT THE DAY*

OF CHRIST IS AT HAND. LET NO MAN DECEIVE YOU BY ANY MEANS: FOR THAT DAY SHALL NOT COME, EXCEPT THERE COME A FALLING AWAY FIRST, AND THAT MAN OF SIN REVEALED, THE SON OF PERDITION; WHO OPPOSETH AND EXALTETH HIMSELF ABOVE ALL THAT IS CALLED GOD, OR THAT IS WORSHIPPED; SO THAT HE AS GOD SITTETH IN THE TEMPLE OF GOD. REMEMBER YE NOT, THAT WHEN I WAS YET WITH YOU, I TOLD YOU THESE THINGS? AND NOW YE KNOW WHAT WITHHOLDETH THAT HE MIGHT BE REVEALED IN HIS TIME. FOR THE MYSTERY OF INIQUITY DOTH ALREADY WORK: ONLY HE WHO NOW LETTETH WILL LET, UNTIL HE BE TAKEN OUT OF THE WAY. AND THEN SHALL THAT WICKED BE REVEALED, WHOM THE LORD SHALL CONSUME WITH THE SPIRIT OF HIS MOUTH, AND SHALL DESTROY WITH THE BRIGHTNESS OF HIS COMING: EVEN HIM WHOSE COMING IS AFTER THE WORK OF SATAN WITH ALL POWER AND SIGNS AND LYING WONDERS." We read here, how we shall not be gathered (or raptured) to CHRIST, until all things come to pass. We read that only when Christ's Prophecy says it is time; only then can the Beast be loosed who is withheld in the bottomless pit. So, we know who it is that "withholdeth" the Beast. It is CHRIST'S ordered events of the Prophecy that "withholdeth". We read that Satan is cast down out of Heaven; but again, *only* at CHRIST'S appointed Prophesied time does Satan "letteth" the Beast out. This released fallen angel, is the Antichrist who rises from the bottomless pit, who gains his seat of rule upon the Earth by the end of the first forty-two months of Global Governance. And we read that only the Angel that falls from Heaven, (Satan) is *given* the key to open that pit and facilitate that evil. And then at CHRIST'S final appointed Prophesied time, all these evil angels, are again *taken out* of the way by the brightness of the coming of our LORD. So again, it is CHRIST and his Prophecy that "withholdeth" Satan and the Beast; *and* at the proper time of CHRIST'S Prophecy, that Satan "letteth" out the Beast; and lastly, at the appointed time, CHRIST comes and Satan and the Beast are taken out of the way, revealing them as evil imposters. Again, exactly when in CHRIST'S order of Prophecy, does Satan release this Beast from the pit. Rev 9:1-2, "*AND THE <u>FIFTH ANGEL SOUNDED</u>, AND I SAW A STAR FALL FROM HEAVEN UNTO THE EARTH: AND TO HIM WAS GIVEN THE KEY OF THE BOTTOMLESS PIT. AND HE OPENED THE BOTTOMLESS PIT; AND THERE AROSE A SMOKE OUT OF THE PIT, AS*

THE SMOKE OF A GREAT FURNACE; AND THE SUN AND THE AIR WERE DARKENED BY REASON OF THE SMOKE OF THE PIT." This is CHRIST'S Prophecy being fulfilled; but how will FATHER assure this. At the beginning of all knowledge is this great truth, that GOD has *said*. And from this statement alone, we see that it is by what GOD has said, and that GOD is all powerful and cannot lie. So, FATHER brings this about by *his* wrathful will, upon *his* perceived physical creations. By FATHER'S desire, here and there, he can steer all physical events, being mixed in among all chance and free-will choice occurrence. By this, GOD can topple governments, countries, and steer the physical courses of all ordered Prophesied events, without ever once over-ruling the conscious free-will desires and choices of outcome for a single soul. Conscious free-will is seen in our wanting something, not in getting that something. If a person freely desires to cross the street, and in the middle, they are physically hit by a car. Their *conscious free-will* to cross, was not overruled, only the physical outcome. When any two free-will souls, conflict over a physical issue, neither can perfectly prevail. The physical outcome of each soul's life is constantly subverted by chance and all other free wills, and physically by GOD. And though mankind's free-will desire can be tempted and influenced, they cannot be controlled by anyone else, even GOD. His conscious force of desire, that governs all the physical spins, layers, and orbits in FATHER'S perceived physical creation, can force every physical change he desires upon this raging, confused, deceived, political and falsely religious landscape; yet never once override mankind's free-will desires. GOD also sends his Holy Angels, and has prepared a few souls here and there, who appear to be of mankind. But they are agents of GOD'S will. So, we see all the true separate souls of mankind are gifted to desire and choose what and whom they will love, forever having free-will. A free-will belief among mankind, is not proven to be real, by one soul or the another being right. If two souls are not one with GOD, they are by their free-will, both wrong. This is the eternal chosen essence of division, called individualism, being a great divisive evil. Now consider how Israel is central throughout Prophecy. And how fervent Islam is, even to the death, to take Israel and destroy GOD'S chosen people. Islam is a demonic force that can't be reasoned with by mankind. It is a tool used by 3 of the 4 Horsemen. It is a demonic force that can only by GOD'S constant efforts, be held at bay. But know that with each day, the confusion of the world and the costs and pain of this fight, weakens the western world's resolve to stay engaged and protect GOD'S

Holy Land. This weakening is now in much of Europe, and is seeking entry into the Democrat Party in America. When the balance in the free west, turns against Israel; Islam will explode in power raising insane fervor within the Muslim world. Once this box is opened, it can never be closed by mankind; only by GOD. And later in the Sixth Trumpet, FATHER closes it forever. Know this truth, mankind was not created with the properties or abilities to govern each other, or themselves. Mankind by their inherent inabilities, will always fail in their stolen power. Ever-rising crisis management, can never constitute good governance. Now, reading after the Fifth Trumpet sounds in Rev. 9: 1-12, the bottomless quantum realm pit is opened by Satan, and around 100,000,000 fallen angels are released into the earth. Many released as FATHER creates for them, a physical body of demonic locust, having tails like a scorpion. Remember, all physicality and bodies in this reality, are perceived and created by the forceful will of FATHER. These demon locusts are told (knowing they can be physically reimprisoned) not to eat any grass or leaves, but instead, sting and torment only all fallen rebellious mankind for five months. And these stings leave agonizing sores that will not heal. These locusts are commanded to *not* touch any soul having the seal of GOD in their foreheads. Again, this shows the Church is still here, well into the Tribulation. I have heard Preachers teach the detailed description of these locust, represents some type of modern helicopter gun-ship. That is ridiculous. Weak preachers try to force GOD'S words to mean something they have seen in the world, something they can easily believe, in being unable to see the truth. This Scripture clearly says these come out of a quantum pit of darkness, and are commanded to not eat grass, or trees, or any green thing. Why would GOD release helicopters from the dark confusion of hell, and command them not to eat grass or trees; that's absurd. And why would GOD command gun-ships to hurt, but not kill mankind for five months. These creatures will be as described, being strange looking locust, with scorpion tails. These will actually fly around the earth stinging people for five months that don't have GOD'S seal in their foreheads. Will mankind claim these are a mutated cross breed of locust and scorpion caused by genetic engineering or nuclear war's radiation? Probably. And in reading Rev. 9:3 and Rev. 11:7 together; we see the Angel over the locust comes out of the Bottomless Pit, and did not open the pit, and is a ruler over the Demons from the Pit; not a King of helicopters. This King is the embodied Antichrist. So, *only* after the seals of the beginning of sorrows, and the first five Trumpet events, does the actual embodied Antichrist come to earth, to rise to power. Until then, never think anyone is the Antichrist. At this point of the Fifth Trumpet, we do not know how long the First Beast, or New World Government, has existed when Antichrist is released from the pit. We only know the Antichrist is released, and rises to full control of the earth, by forty-two-months, or 3 ½ years of that new Governments first Seven Year Religious guarantee. Going forward, these and many other Scripture and facts, validate GOD'S prophesied timeline as it unfolds before our knowing waiting eyes. We hope you have at least started to see a bigger picture unfold in how all Old Testament last days Prophecies, in their own chronological precision, overlay CHRIST'S Great Revelation. And together, they reveal the continuity of a precise narrative. Jumping a little forward, we are given even more detail in reading Rev. 13:15-16. Where we see that a false image or idol of GOD is placed in the Temple, prior to the implementation of the Mark of the Beast. The placement of that Image, or Idol, and Satan's indwelling it, is the evil abomination of desolation, spoken of by JESUS in Matthew 24:15, Mark 13:14, and Prophesied by Daniel in Dan. 9:27, 11:31, and 12:11. So, only after Satan himself, by force of deception, assumes the place of GOD, in GOD'S Temple, towards the end of the tribulation. Only then, do we see the mark of the beast implemented and accepted by the world. Satan knows he must duplicate as much of Prophecy and the Holy Trinity as possible, trying to deceive GOD'S True Children who are in this world till the end. The phrase abomination of desolation, means that the very last footprint of GOD'S Holy of Holies and presence upon the earth has been remove. So, by that last act of taking GOD out, the earth has become an abominable desolate place. We dispel all false teachings we hear, that say the Mark of the Beast, comes *at* the formation of a One World Government. We also know the Mark of the Beast can only exist upon the earth, for at most, the last 2¾ years of the approximate ten-year period of GOD'S Tribulation and cleansing. It can only come after Satan's direct and open rule in the Temple, on Temple mount, who the world will accept as being God. And this mark is truly a mark of Satan being worshipped, loved, and obeyed, by the world, it is not forced. Read now, Revelation 9: 13-21. The Sixth Trumpet sounds, being the second greater woe. Again, having many components, we see each of these last three woes grows in scope, complexity, and duration. The first component of this Second Woe, or Sixth Trumpet, is an instruction from Heaven to let loose four powerful Angels and the entire Host, or Army of GOD. These have been prepared to wait until that Prophetic moment. We read about their *literal* one year, one month, one day, and one hour of preparation for their coming attack upon the worst enemies of Israel; notice there was no term of week or weeks mentioned. This army is Sabaoth; it is the Host of the Army of GOD. This Scripture, is *only* about the Army of GOD, being 200,000,000 Holy Angelic Spiritual Horsemen. Now it is clearly stated, they have gone through a time of preparation, and are led by four powerful Holy Angels. Are these the 4 Seraphim? This can make perfect sense. This Army of Horseman is so often confused with the much later Prophesied Kings of the East, coming to the battle at Armageddon. Why can't preachers leave all Prophesied things in GOD'S Prophesied order. The 10 demon Kings, and the mortal armies of the East, come to Battle at *Armageddon* much later, after the Euphrates is dried up by the Angel who pours out the Sixth Vial near the end of the Seventh Trumpet. *Please*, allow all events to remain in the perfect order set by GOD, or you spread Satan's confusion. This Sixth Trumpet Prophecy is not part of any war being fought, by mankind. Everything that happens from the fifth trumpet forward, is like the great flood. It is GOD cleansing the Earth of all who resist his truth to their death, preparing for CHRIST'S return to rule for a thousand years. This cleansing by the Holy Angelic Army of

GOD is clearly stated to be upon those who desired and are responsible for the diminishing and loss of FATHER'S set borders for Israel. The land that was promised his people in perpetuity. Other Old and New Testament references, aligning to this cleansing by GOD'S army, are in Joel 2:1-18, 2nd Kings 6:16-18, Rom. 9:29, and James 5:4. This fiery judgment, comes upon *all* souls who do not in their hearts and words, accept and defend the borders of Israel, as set in Scripture by GOD. These souls, by all such beliefs are enemies of Israel; and thereby enemies of GOD. We know the coming World War III, famine, and pestilence, kill one fourth of mankind. Then those who die from the effects of the first four trumpets, being many millions for sure. And now we see the Army of GOD, preparing for CHRIST'S Millennial Reign, to cleans a third of all remaining mankind, being all who deny Israel as defined by GOD. All these deaths together, reduce the earth's population by over fifty percent, and we have far to go. Please remember, that *as* all of this happens, not one true Child *of* GOD, is hurt or killed by Holy hands. Many of GOD'S Children will be martyred by evil souls along the way, but not by Holy hands. We hope that through these truths, set by Scripture, we undo all confusion and harm set upon many, by many. In returning to Scripture, we see in Revelation 9:13-21, this army cleanses one third of all mankind on earth; being mostly from all former Islamic regions. This Army destroys specifically those people who desired to take, divide, scatter, and trade the land of Israel, Judah, and Jerusalem. No reasons for doing so, are acceptable, including the hope or desire for world peace. GOD has drawn the borders of Israel for the strength and security of his people; mankind has since that time, attacked, altered, redrawn, and destroyed GOD'S lines and his people's safety. At CHRIST'S return to rule for 1000 years on the nearly destroyed earth, Israel's boarders will be re-established. And *all* who in any way or for any reason did not agree, are no longer alive. Let's read all other Prophesies that support and confirm this Sixth Trumpet cleansing. Pay great attention to the physical description of the Horseman in Rev. 9:16-17 and 2nd Kings 6:16-18; where we get another glimpse of GOD'S Angelic Army coming to Earth. We also read that if in the flesh and no the spirit, that mankind can't see this Army. Read Zechariah 1:18-21, where we see an Old Testament Prophecy of these last days Four Angels, leading this same Army of GOD to accomplish this exact same last day's task. But here they are described as Four Carpenters, yet these are the same Four Angels that wait in the Euphrates. We read they come to fray, which means to battle and destroy, as in he, bravely ran into the fray. And the Four Horns in this Scripture, are four Kingdoms or regions of people. This Scripture says these Four Horns, are those that have used their power to scatter not just the people of the Tribe of Judah, but *the lands* of Judah, drawn and given them by GOD. These four Horns, are the last days, four strongly Islamic Regions. Read Joel, chapter 2:1-11, which overlays the 6th trumpet in perfect detail. We see Horseman, marching side by side, unstoppable and indestructible, devouring all souls in their path that hold evil in their heart to divide GOD'S lands. All such individuals are killed with fire, brimstone, and smoke; this without doubt is the Army of GOD. Read Obadiah, as this cleansing, does and will include not just Islamist; but all peoples, including Jews who would in any tiny way, re-draw Israel's borders for a hope and supposition of peace. The reason does not matter. GOD said in Ezekiel 9:6, start the cleansing at my Sanctuary, being GOD'S Temple and its High Priest. If any Priest, within his heart, holds any desire to compromise the borders of Israel for any reason, he is the first to be killed. This 6th Trumpet cleansing will continue out from there, till every man, woman, and child, who hold this specific rebellion, are removed from the face of the earth. Being around one third of all people left alive at that time. Will these be able to see this Spiritual Army of fire breathing Horseman coming at them, bringing the wrath of GOD. We can't know if GOD opens their eyes, like Elisha's servant's eyes were opened to see GOD'S Fiery Army. This one third being killed, from among all remaining mankind, would be nearly 2 billion people; bringing total deaths, to way over half of all mankind. This number may be far higher due to constant deaths from the accumulated conditions of disease and famine in the world. This is all prior to six more coming plagues *and* the battle of Armageddon, which will at the last to cleanse the earth of all alive, having the mark of the Beast. We keep in mind that Scripture says, there will be a remnant of people scattered around the earth who survive all these coming plagues. These will be GOD'S Children, and some fallen souls who are not reprobates and may yet hear the truth of reality, and turn to GOD. After all further cleansing is accomplished, the remaining, will live, build, and re-populate the Earth for a thousand years under CHRIST and his Priests and Princes. All these wars, famines, and plagues are in effect, just like the great flood. Cleansing the world of all unrepentant rebellious reprobate enemies. Read Revelation 10:1-11, where we see the second component of this complex Sixth Trumpet. We find the second component is full of wonder, intrigue, and understanding. We see a Mighty Angel, with a Little Open

Book in his hand. We see this Angel has one foot on the sea and one on the land, and cries out as when a lion roars; which is then answered by Seven Thunders. We see this Angel is mighty, meaning a weaker Angel could not accomplish this, as it requires a spiritual knowledge and strength, held only by a higher forceful Angel. There are differing responsibilities and abilities among all the Holy Angels. In Daniel 10:12-21. We see a *mighty* Angel, comes through a battle with fallen Angels to bring information and truth asked for by Daniel in prayer. And Daniel is told by this mighty Angel that there are none *"who holds with him"* in this truth, save for he and Michael. Meaning *only* these two Angels, know and received it directly from GOD. In Rev. 10:1-11, other amazing aspects are that one foot of this Angel is on the sea and one on land, and the Book is little, and is open. These mean it is knowledge not great in volume, but of great importance. It being open, means it will come only in the last days, for all who can hear it. Now one foot placed on both the land and sea, means it is meant to be taken around the world. The words written in this little book, were not given to us in Scripture at that time; so, we know it is for a future moment. And until that moment, it is only for certain eyes; among those being the Apostle and Prophet John; remember this. Another interesting aspect, is after the Seven Thunders, utter their words, and when John goes to write what the Seven Thunders said, he is told not to. So, we see more sensitive vital knowledge and wisdom to be desired by us all. I want so much, to know what was heard and read by John. But it was given by the Angel to John alone, and held from the ears of the world for a future time to be preached before peoples, nations, tongues, and kings. The mighty Angel declares that *in the last days*, when the Seventh Angel *begins* to sound the Seventh Trump; this signifies a Command from GOD, that by that time, all mysteries of GOD will be told and revealed to us, and these prophesies will have been fulfilled. So, a voice from Heaven tells John to take the little open Book. When John takes the Book, the Sixth Angel tells him to eat the book. This signifies that the words in this little Book will become a part of John. We read that what the Book says, is sweet to read or say, yet hard to hear; and so disturbing when fully absorbed and known it makes one nauseous. So, John alone in all the earth, has the knowledge of what the Seven Thunders said, *and* what was written in the Little Book. But this sixth Trumpet ceremony has even more great importance, as John is given a command from GOD, by way of this Sixth Angel. Remember, John is given this great over-all Prophecy when he is very old and officially exiled to live and die on the

Isle of Patmos. Most scholars, place John in his eighties to nineties when exiled to Patmos. So, he receives and writes this entire Revelation from JESUS at an old age, imprisoned on an island. Yet he is commanded by GOD that he must Prophecy *around* the world, in the last days, these secrets he has just received. What could this mean, seeing this is in CHRIST'S Revelation; which after the warnings to the Seven Churches, deals exclusively with the chronological events of the end of days. By this fact, and according to our firm understanding that certain men of old, have not yet tasted death. This means John will be one of the Two Great Witnesses in Rev. 11:4. The giving of this commission from GOD to John, that he must yet testify again to all the world, those things received from the Seven Thunders and the Little Book, in the Book of Scripture, dedicated *exclusively* to CHRIST'S Prophecy of the last days. And it does appear, just prior to the next Scripture, in which John is shown the last days Temple. John is told to measure this last days Temple in Jerusalem, but leave out the Courts of the Temple. We see the timing of this vision, is where Israel as a Country, has been surrendered to a World Government, but not the Temple; so, this is during the first 3 ½ years. This is so exciting, not just in that it is full of wonder. But it says those who are alive, will hear great additional Testimony and

Revelation from GOD prior to the Seventh Trumpet, and the Second Coming of JESUS. All these things in the Book of the Revelation shall come to pass in the exact order of CHRIST'S Prophecy. So, we with all *certainty,* know the Temple must be completed, prior to this point in John's Sixth Trumpet vision. This is important that John is told to measure only the *Temple*, as everything GOD says or does, has layers of meaning to be revealed. Many Scriptures, point to John being one of the Two Last Days Witnesses. One such verse is at the end of the Gospel of John, in 21:20-23. John wrote; *"THEN PETER, TURNING ABOUT, SEETH THE DECIPLE WHOM* ***JESUS*** *LOVED FOLLOWING; WHICH ALSO LEANED ON HIS BREAST AT SUPPER, AND SAID,* ***LORD****, WHICH IS HE THAT BETRAYETH THEE? PETER SEEING HIM SAITH TO* ***JESUS****,* ***LORD****, AND WHAT SHALL THIS MAN DO?* ***JESUS*** *SAITH UNTO HIM, IF* ***I*** *WILL THAT HE TARRY TILL I COME, WHAT IS THAT TO THEE? FOLLOW THOUGH ME. THEN WENT THIS SAYING ABROAD AMONG THE BRETHREN, THAT THAT DISCIPLE SHOULD NOT DIE: YET JESUS SAID NOT UNTO HIM, HE SHALL NOT DIE; BUT, IF I WILL THAT HE TARRY TILL I COME, WHAT IS THAT TO THEE*?" I have never *once* in GOD'S SPIRIT, believed JESUS ever said anything benign, or without pure

Spiritual purpose. What purpose could be attributed to this saying, could it be just a flippant remark by JESUS to Peter, to mind his own business. Absolutely not, that would be an expression formed by the fallen weak attributes of mankind, beneath the mind of GOD. We must remember that although there are suspect writings, outside of Scripture, saying John was later boiled in oil, yet did not die, and writings intimating other possible executions of John. Yet there is nothing in Scripture about any death of the Apostle John. Add all this to the fact that John is given a great end of days command, that he must prophesy about the Seven Thunders and the Little Book before all nations, tongues, peoples, and Kings. There is no Scripture saying John, being very old, did again travel the world back then to fulfill this command given him directly from CHRIST. If he had, John would have written of it for us in Scripture. While we discuss these needful things, there is a great abundance of Scripture, teaching that the other of these two last days Prophets will be Zerubbabel. Most modern Bible Teachers do not equate the abundance of Prophecy containing the name Zerubbabel, with the last days. Yet the great evidence in Scripture as to this fact, is undeniable. Zerubbabel is of the tribe of Judah, born right in between, and in the direct lineage of King David to our LORD; these three are related by blood. We read a Prophecy in Haggai 2:23, where GOD said; in the *last days* I will take thee Zerubbabel, and make thee as a chosen signet. A signet is a royal seal on communications of command or testimony, sent from a Supreme Ruler to others. So, what is this end of days command or testimony GOD is sending, by using Zerubbabel as his official seal? And to whom is this message sent? We read a second confirmation of this Prophecy in Zechariah 2:1 through 4:14. Again, being clearly defined as another *end of day's* Prophecy, which also deals with the last days Temple in Jerusalem. This Scripture is about Zerubbabel being a force behind rebuilding the last day's Temple. Zech. 2:1-5, "*I LIFTED UP MINE EYES AGAIN, AND LOOKED, AND BEHOLD A MAN WITH A MEASURING LINE IN HIS HAND. THEN SAID I, WITHER GOEST THOU? AND HE SAID UNTO ME, TO MEASURE JERUSALEM, TO SEE WHAT IS THE BREADTH THEREOF, AND WHAT IS THE LENGTH THEREOF. AND BEHOLD, THE ANGEL THAT TALKED WITH ME WENT FORTH, AND ANOTHER ANGEL WENT OUT TO MEET HIM, AND SAID UNTO HIM, RUN, SPEAK TO THIS YOUNG MAN, SAYING JERUSALEM SHALL BE INHABITED AS TOWNS WITHOUT WALLS FOR THE MULTITUDE OF MEN AND CATTLE THEREIN: FOR I, SAITH THE* ***LORD****, WILL BE*

UNTO HER A WALL OF FIRE AROUND ABOUT, AND WILL BE THE GLORY IN THE MIDST OF HER." In this Scripture, GOD says that in the *last* days, Jerusalem will be a big city without perimeter walls. We know by Scripture, that when Zerubbabel *re*built the Temple for the second time, they rebuilt the walls of the city first. So, this is not that Second Temple. To further know these are Last Days Prophesies, we hear CHRIST say he himself will be the glory in the midst of her. Meaning he is the true security of Israel and Jerusalem in the Temple. That is, up until the people of Israel once again betray the LORD. Further we read in Zechariah 2:6 through 3:10, that GOD'S children are called home to Israel, and told JESUS comes to reign and dwell with them in the coming Millennium. We know this only happens in the last days. And all together, this defines beyond doubt that this is a last day's Prophecy. Zechariah 2:10-13, *"SING AND REJOICE, O DAUGHTER OF ZION: FOR LO I COME, AND I WILL DWELL IN THE MIDST OF THEE, SAITH THE **LORD**. AND MANY NATIONS SHALL BE JOINED TO THE **LORD** IN THAT DAY, AND SHALL BE MY PEOPLE: AND I WILL DWELL IN THE MIDST OF THEE, AND THOU SHALT KNOW THAT THE **LORD** OF HOST HATH SENT ME UNTO THEE. AND THE **LORD** SHALL INHERIT JUDA HIS*

PORTION IN THE HOLY LAND, AND SHALL CHOOSE JERUSALEM AGAIN. BE SILENT, O ALL FLESH, BEFORE THE ***LORD****: FOR HE IS RAISED UP OUT OF HIS HOLY HABITATION.*" We read JESUS has come for *his* inheritance, confirming this Scripture is about the last days. And our King who had sat down on his throne beside FATHER, after his crucifixion and ascension; has now *raised up* again out of his Holy Seat. This, our KING, only does in the last days, standing to battle for Israel at his return. If you need more evidence of this, we see in Zech. Chapter 3, a ceremony in the SPIRIT, in which Joshua is ordained as a High Priest during CHRIST'S Reign over a failed Earthly Kingdom. This would be otherwise out of order and place; and we know this even *more* surely by all the Church Elders mentioned in Zechariah 3:8. And it gets even more fascinating as we read Zechariah 4:1-3, *"AND SAID UNTO ME, WHAT SEEIST THOU? AND I SAID, I HAVE LOOKED, AND BEHOLD A CANDLESTICK ALL OF GOLD, WITH A BOWL UPON THE TOP OF IT, AND HIS SEVEN LAMPS THEREON, AND SEVEN PIPES TO THE SEVEN LAMPS, WHICH ARE UPON THE TOP THEREOF. AND TWO OLIVE TREES BY IT, ONE UPON THE RIGHT SIDE OF THE BOWL, AND THE OTHER UPON THE LEFT SIDE THEREOF.*" This describes the coming two last day's

Witnesses; also described in Revelation 11:3-4, *"AND I WILL GIVE POWER UNTO MY TWO WITNESSES, AND THEY SHALL PROPECY A THOUSAND TWO HUNDRED AND THREE SCORE DAYS, CLOTHED IN SACKCLOTH. THESE ARE THE TWO OLIVE TREES, AND THE TWO CANDLESTICKS STANDING BEFORE THE* ***GOD*** *OF THE EARTH."* And then we see Zechariah say he does not know what the vision means, asking for wisdom. But instead of immediately interpreting the vision for him, the Angel must first give him other important Prophetic information, so he may then rightly understand the answer to his question. Zech. 4:6-10, then states who these words are for; *"THEN HE ANSWERED AND SPAKE UNTO ME, SAYING, THIS IS THE* ***WORD*** *OF THE* ***LORD*** *UNTO ZERUBBABEL, SAYING, NOT BY MIGHT, NOR BY POWER, BUT BY MY* ***SPIRIT*** *SAYETH THE* ***LORD*** *OF HOST. WHO ART THOU, O GREAT MOUNTAIN? BEFORE ZERUBBABEL THOU SHALT BECOME A PLAIN: AND HE SHALL BRING FORTH THE HEADSTONE THEREOF WITH SHOUTINGS, CRYING GRACE, GRACE UNTO IT. MOREOVER THE* ***WORD*** *OF THE* ***LORD*** *CAME UNTO ME SAYING, THE HANDS OF ZERUBBABEL HAVE LAID THE FOUNDATION OF THIS HOUSE; HIS HANDS SHALL*

*ALSO FINISH IT; AND THOU SHALT KNOW THAT THE **LORD** OF HOST HATH SENT ME UNTO YOU. FOR WHO HATH DESPISED THE DAY OF SMALL THINGS? FOR THEY SHALL REJOICE, AND SHALL SEE THE PLUMMET IN THE HAND OF ZERUBBABEL WITH THOSE SEVEN; THEY ARE THE EYES OF THE **LORD**, WHICH RUN TO AND FRO THROUGH THE WHOLE EARTH.*" We see clearly, that by the power of GOD, Zerubbabel will be used in the last days. And the Great Mountain, being the World, shall not stop him from *building*, or *Prophesying*. Then we see Zerubbabel with shouting, crying grace, grace; as he heralds and brings forth the Head Stone. The Head Stone is of course JESUS. Psalms 118:22 and 1st Pet. 2:7; being the STONE which the builders refused, is become the Head Stone of the corner. This Scripture about Zerubbabel heralding with crying and shouting grace, grace, is announcing the *second* coming of CHRIST. And it is the *mirror* to the *first* coming Prophecy in Isaiah 40:3, *"THE VOICE OF HIM THAT CRIETH IN THE WILDERNESS, PREPARE YE THE WAY OF THE **LORD**, MAKE STRAIGHT IN THE DESERT A HIGHWAY FOR OUR **GOD**.*" In Zechariah 4:11-14. We now see the Angel answers his original question "who are these two candlesticks" having first given him the necessary knowledge to understand it. The

Angel tells him these are the two anointed ones, that stand before the GOD of the whole Earth. Let's again read Rev. 11:4. We should also quickly go back to read Zechariah 2:4. Where we see the Angel say run, and tell the young man with the *measure line* that the city shall be built this time *without* walls. Who is this young man, is this Zerubbabel? Who else could it be as all of these Scripture are about Zerubbabel. But how is he a young man: well not to us, but from GOD'S, and even the Angel's perspective, he is quite a young man. Before continuing further in the Revelation, reread Zech. 4:2. Let's notice and clarify, as many may have a question at this point, concerning the number of Candlesticks. In this Old Testament Prophecy in Zechariah Chapter 4, concerning these two last day's Witnesses. We do see the two Olive Trees representing two souls, having oneness with the SPIRIT of GOD, and being the two last day's witnesses. But in this first Prophecy we see only one Candlestick. The Candlestick represents the Earthly Temple or physical body or vessel for the Spirit. We see only one, because the second, being John, had not been born yet. In the mirrored New Testament Prophecy of Rev. 11:3-4. There we see the second Candlestick has been added, who had then been born into the Earth. What do we have in total. Scripture tells us Zerubbabel will lead to rebuild the Temple in the end of days, inter-laced with the description of our two last day's Witnesses in Rev. 11:4. And Zechariah 4:7 says, Zerubbabel will with shouting and crying bring forth the Head STONE that the builders rejected. We have much referring to John and Zerubbabel as the two witnesses. But we *also* have a Prophecy in Malachi 4:5-6, which says GOD will send the Spirit of Elijah before the great and dreadful day of the lord, to turn the hearts of the people back to FATHER. But that was about the first coming, and that is the last we read of Elijah. The identity of these two last day's witnesses is intriguing. And we may be prayerfully intrigued by all these facts. It will of course be two chosen by GOD for his purpose, out from all those who live according to GOD'S ways. Another consideration for his two witnesses, is that they must have something additional and powerful to do and speak. Just as John and Zerubbabel are clearly given a command to do. And consider that these Two Witnesses according to Scripture, will be killed after forty-two months at the end of their testimony. So, they cannot have ever tasted physical death before; for Scripture says it is appointed unto men *once* to die, and then comes our judgment. We and many hold the truth, that there are certain men from long, long, ago, who have not yet experienced physical death. And in accordance with that knowledge, we know the two witnesses, must be of those men of old who have not yet put off their old mortal flesh. So, who are all the possibilities: Elijah is considered by many as a last day's witness; who according to 2nd Kings, 2:11, was taken up and did not die and was not buried in the Earth. But in the New Testament, by the spoken words of CHRIST, he is said to have returned in Spirit, to be known as John the Baptist, to prepare the way for the first coming of the LORD. John the Baptist was then beheaded, died, and was buried in the Earth. The Spirit of Elijah may still return again in the last days, but he can't be killed again. There is Enoch, who according to Scripture has not died and been buried in the Earth; and there could be others. I cannot, by any reason see how some believe one will be Moses, who died and was buried in the Earth. And Moses is now one of the Four and Twenty Elders already translated into Heaven. Rev. 4:4, "*AND ROUND ABOUT THE THRONE WERE FOUR AND TWENTY SEATS: AND UPON THE SEATS I SAW FOUR AND TWENTY ELDERS SITTING, CLOTHED IN WHITE RAIMENT; AND THEY HAD ON THEIR HEADS CROWNS OF GOLD.*" According to all that is written, this leaves John and Zerubbabel. As we wait in anticipation, this remains a Prophecy until these two men appear. We also don't know if these will identify themselves by their former names during this time. Elijah's Spirit was translated back a second time, to be called John the Baptist. In Matt. 17:11-13 says "***JESUS*** *ANSWERED AND SAID UNTO THEM, ELIAS TRULY SHALL FIRST COME, AND RESTORE ALL THINGS. BUT I SAY UNTO YOU, THAT ELIAS IS COME ALREADY, AND THEY KNEW HIM NOT, BUT HAVE DONE UNTO HIM WHATSOEVER THEY LISTED. LIKEWISE SHALL ALSO THE SON OF MAN SUFFER OF THEM. THEN THE DISCIPLES UNDERSTOOD THAT HE SPAKE UNTO THEM OF JOHN THE BAPTIST.*" We see, by *all* understanding, given to us as part of the Sixth Trumpet Prophecy, that GOD wants us to know all these earthly events, will happen *after* all Fifth Trumpet Events, and *prior* to the sounding of the Seventh Trumpet. Now in Rev. 11:1-13, we find another long narrative, that will *mostly* unfold before the sounding of the Seventh Trumpet. This Scriptural narration is wisdom, knowledge, and an affirmation of many certain events that occur *after* the last days Temple is built, and in their proper chronological order of occurrence, leading us up to the Seventh Trumpet. Now, it is so important and informative that John is told to measure only the Temple. Yet Zerubbabel is told to *build* the Third Temple, and to *measure* Jerusalem. This tells us, at the time Zerubbabel returns to Israel, the boundaries of Jerusalem are not as in his days on earth; yet Jerusalem and Israel are *still* Sovereign. So, Zerubbabel's building the Third Temple, must at least *begin* preparations, prior to the coming One-World Government. And obviously, John's measuring, is done *after* the Temple is built, and into the first 3½ years of a One-World Government. Meaning at that time, the land and Courts that surround the Temple, have been surrendered to be trampled by the gentiles for forty-two-months. Yet, we know a New World Government has not yet been taken over by Antichrist at this time of John's measurement. Yet GOD'S Holy Land has been reduced to within the HOLY of HOLIES and the Temple walls. And Satan is coming for that next, to steal and assume GOD'S name and place in the Earth completely. We read how at the start of this World Government, FATHER places his Two Witnesses in front of the New Temple to testify. And forty-two months later, both are killed by the Third Beast Antichrist who ascends from the bottomless pit *at* the Fifth Trumpet, and has risen to full power. We know according to Prophecy, the Daily Sacrifice in the Temple, was secured by a Seven Year Religious Right within that treaty. But midway at forty-two months, the Temple and Daily Sacrifice are taken away. This timeline says the Temple construction could start at any time *after* the first Four Seals where opened, meaning now. And prophetically, it must be complete prior to the Fifth Trumpet. And surely many things can occur in between the Fifth and Sixth Trumpets. Meaning many months, to a year or more could pass from the sounding of the Fifth Trumpet until the Sixth. We read that FATHER'S Two Witnesses, once they are killed, their bodies lie in the streets for 3½ *days* as the fallen world celebrates. They celebrate because these Prophets tormented the Earth with plagues for their rebellion. Are these 3½ literal days, or 3½ years; because we see them rise again and ascend up into Heaven in front of the whole world after those 3½ days. The meaning here, is 3½ years. And this must be, to allow the Antichrist to be seen during *his* 42 months, as being sent from GOD. If they stood up and were received into heaven after 3½ literal days, the world would see this as an indication the Antichrist is weak, and not with the GOD above. While on earth, FATHER allows mankind all their desired delusions and the evil in their hearts. So, after 3½ years, these two stand up from being dead before the world, and are carried up to Heaven. And their ascension is followed by a great Earthquake. So, to recap, we know the Third Beast embodied Antichrist was not in power at the beginning of the Two Witnesses Testimony, who testify for 42 months, and the Antichrist kills them upon rising to full power. We see many

Prophesies that have a time-line, of a time, times, and half a time; and others say forty-two months. These are all prophesies of 3½ years, and they are *all*, either the first, or second half of the same "Seven Year Period" just after a World Government Treaty is signed by the World and Israel. Some Prophesies are in the first 3 ½ years and some are in the last 3 ½ years, but they fit together only one way, so, it is easily understood. We also read, that GOD prepares a place for his children to be nourished, and hide from the face of the Serpent for the second forty-two months. Then, after the second forty-two months, Satan enters the Temple to indwell the image, and begin his time of the mark of the Beast for the next 2.8 years. GOD'S Children will then be hunted, and many will be found and martyred by Antichrist, False Prophet, and their followers. When added all together, we see the temple will already be there as the two Witnesses testify in front of it during the *entire* first 3½ years, so the Temple is functioning, if not entirely completed before the World Government is in place. We see the 7-year religious treaty protections put in place by this World Government, endure for only forty-two months. And for the second forty-two months, FATHER prepares a place to hide his true children from the face of the Serpent. GOD is not the author of confusion, it is the fallen of mankind who take the harmony of Scripture and confuses it. Any unintentional or intentional false teaching is a crime of confusion against GOD'S Kingdom. When we speak of the Third Temple, I have hope, and I pray that I might by GOD, be found in Zechariah 6:15, *"AND THEY THAT ARE FAR OFF SHALL COME AND BUILD IN THE TEMPLE OF THE **LORD**, AND YE SHALL KNOW THAT THE **LORD** OF HOST HATH SENT ME UNTO YOU. AND THIS SHALL COME TO PASS, IF YE WILL DILIGENTLY OBEY THE VOICE OF THE **LORD** YOUR **GOD**."* My great desire is to help build the Temple. To wait, see, and worship our GOD with Zerubbabel, John, and other faithful servants. To be there to see John and Zerubbabel sent to prepare the way. To at least have my hand on one stone as it is set in place, and aid, protect, and nourish these two in any way I can. In our work for GOD, this does feel like a selfish desire; but it is there within me, and will not leave. We hope all souls are strong and desirous to continue at this point, for there are over 50 more dreadful pages of the Revelation of CHRIST to detail. There is far to go before we rest in its full understanding. I wonder how John was able to bear the vision of this entire Prophecy before his eyes, knowing he must wait around 2000 years to finish it. It is only a remarkable faithful responsible love for GOD that could bear such great promise, in all endurance. I hope all realize, the Second Wow, contained much more than the First Woe. These three, grow one after the other in such awe and fear. So, we now read about the most fearful Third Woe. The Seventh Trumpet of GOD'S sounds and his final wrath and Tribulation are set upon, not the earth, but mankind. And at its end, this Trumpet has the last quick Seven Vials of wrath poured out upon the earth. This Trumpet, and what follows, filling every page from Revelation 11:15 through 19:21, is dreadful and hard to hold. Read Rev. 11:15-19, where right away, we see another long side bar that starts here on our pg. 582 and ends on 599. Again, Scripture pauses from FATHER'S eventful Judgments to give us some required wisdom, foundation, and guidance, to help us see CHRIST'S Prophecy in its fullness. We see here a Heavenly ceremony, announcing and declaring that by the faith in all Prophecy, the immanent victory of JESUS is at hand; and also declaring what JESUS brings with that victory. We also see the original Ark of the first Covenant, stored in Heaven; so, a search and mystery for this world is over and resolved. We have all heard from documentaries, movies, and unknowing Pulpits about the search for the lost Ark of the Covenant. This has never been a mystery to GOD'S Children. I knew this from the first time I read the Bible all the way through in 1974. We see the Ark of the Covenant is not lost, but was taken up to Heaven. Which GOD foretold, speaking in Jeremiah 3:16, *"AND IT SHALL COME TO PASS, WHEN YE BE MULTIPLIED AND INCREASED IN THE LAND, IN THOSE DAYS SAITH THE* ***LORD****, THEY SHALL SAY NO MORE, THE ARK OF THE COVENANT OF THE* ***LORD****: NEITHER SHALL IT COME TO MIND: NEITHER SHALL THEY REMEMBER IT; NEITHER SHALL THEY VISIT IT; NEITHER SHALL THAT BE DONE ANYMORE.*" We see FATHER took back the first Ark being a vessel made of wood and gold, and filled with Holy Power to enforce that First Lawful Covenant. Being a shadowy forerunner of the fulfilled perfect Covenant and Holy Vessel of CHRIST; the most rare and pure Vessel sent and offered to mankind, because they had not loved or obeyed the First Covenant, or its empowered vessel. FATHER then sent his Purest Greatest Blood Covenant of Grace and life, and then his indwelling HOLYSPIRIT of love, these being the last and greatest of everything FATHER has, and could ever offer. And that we the gifted, could ever know or receive. A love and a life, not to be written on tablets of stone, and held in an Ark of wood overlayed with gold. But written in JESUS'S Blood upon the hearts and minds of all truly repentant obedient souls, being overlaid with MESSIAH'S grace, and indwelled by the power, truth, and understanding of HOLYSPIRIT. Each true Child of GOD becomes an eternal Ark and Temple of this Covenant. During the next three chapters, being Rev. 12-15. GOD gives us an extensive, condensed, prophetic flyover, covering a long period of many events and warnings of many things that occur prior to, and in the last days. These are placed right at this point in Scripture, for GOD'S strong pure Spiritual reasons. These long soliloquys appear in Prophetic Scripture, right where and when our LORD needs to give, condense, and emphasize many things at once. This certain long-expanded truth, is not limited to the chronological events we have so far studied. Yet it aligns and overlays perfectly, and it adds more understanding and enhanced wisdom and clarity. These long narratives, as with every word of GOD, are to be desired, treasured, and understood, as they are the meaningful written promises of GOD. Read Rev. 12:1-17, which reveals that back during the second dispensation of FATHER'S efforts to save mankind, long before CHRIST'S Crucifixion. That as sin and rebellion grew, FATHER'S ever-increasing efforts to find, restore, and separate his children, from all the evil reprobates among mankind. FATHER declared his Family and land, to be called Israel. And going forward, his returning Children, by either *blood* or *adoption*, shall be called and known by one of the Twelve Tribes or Sons of Israel. We Gentiles are called and *adopted* under CHRIST, to come out of this world under the Tribe of JESUS known as Judah; being one of the twelve sons of Jacob, who FATHER renamed Israel. In verse two, we see GOD'S Family's Twelve Names are pained to deliver a Holy Child, to be the King Eternal, born of the Tribe of Judah as the Prophesied Savior. Verse three tells us how in Heaven, a great evil alliance was formed long ago by Satan and his most powerful conspirators, plan to mimic FATHER'S family structure and his plans and appearances to at least a certain degree to lure mankind away from GOD. These demons align, and form the strongest confederation among all fallen angels. They are all of like mind, in their purpose of trying to imitate the Spirits of GOD, and take total control of GOD'S Creation. These want to defeat GOD, by turning the creation against the true GOD, and *to* them. From Rev. 12:1-17, we see an evil alliance and plan, called the Beast, was conceived by these angels, long, long, ago, while in Heaven. This plan was formed of old, before JESUS came to be our Sacrifice, before Michael and his Angels began fighting this long battle to cast them out from Heaven. Satan's early alliance, is the Beast or Dragon, having its initial Seven Heads, and later, its planned Ten Horns of World-Government. Remember that here, it is written there are only Seven demon Crowns upon the First Beast's Ten Regional Heads. We will explain this seven-to-ten meaning in the coming pages, which is very important, as we all are strengthened by this wisdom. We see Satan has been forming a rebellious structured army of his own for at minimum 4-5000 years. In the dark depths of Satan's mind, he denies his efforts are futile; but remember he is confused, delusional, insane, and completely detached from reality. Satan is serious, fearful, and angry, about self-preservation. He has up to around 12,000 years of experience, and extreme power, and is extremely smart. He is doing everything he can think of that might save, and bring him victory; yet in his depths of dark confused beliefs, he must vaguely sense that nothing can save him. We see Satan's envious non-omniscient strategy is to imitate GOD'S ruling structure, to use mankind's lustful desires to introduce greater and greater levels of doctrinal confusion and sexual perversions and deception of reality among believers. The weakest of mankind, will always choose a less lawful and more physical religion, over a more lawful and spiritual one. Satan comes as a false Christ who not only forgives all sin, but allows all sin. And though GOD has warned us of this, it works on all of those souls who do not love GOD'S lawful ways. Satan offers a way of life that pleases and glorifies the flesh, and denies the soul its strength; just like he offered those in Sodom and Gomorrah. A false religion, that allows individuals to decide what is right or wrong and true or false in their life for themselves in the desires of their flesh. Rev.12:4 describes how Satan and his Angels go to Earth, desiring to kill the CHRIST Child upon his Birth, so he can't fulfill the Prophecy. Verse five tells us how, that through the Family of Israel, and more specifically the Tribe of Judah, CHRIST must be born into the flesh of mankind. Meaning JESUS put off his eternal body of being GOD in Heaven, and in his loving trust and obedience to FATHER, CHRIST'S soul was adjoined to the weak flesh formed by HOLYSPIRIT overshadowing and placing the DNA of the loins of King David, in the womb of a virgin maid. The SON of GOD, came down to experience our existence, lower than that of the Angels in the flesh of man; to be Sacrificed upon the Earth. And upon the fulfillment of his great loving obedience, FATHER was pleased in him. Heb. 2:9, "*BUT WE SEE JESUS, WHO WAS MADE A LITTLE LOWER THAN THE ANGELS FOR THE SUFFERING OF DEATH, CROWNED WITH GLORY AND HONOUR; THAT HE BY THE GRACE OF GOD SHOULD TASTE DEATH FOR EVERY MAN.*" After his Crucifixion, CHRIST is raised and received back up to Heaven, and to his Eternal Body, and his Throne of GOD beside FATHER'S. Verse six, shows that back then, FATHER prepared places for the first new early infant Church of CHRIST, to flee the anger of the Ancient Roman Beast, being also ruled by Satan. This protection was to nurture, strengthen, establish, and allow the Early Vine (Church) to root and grow. Verses seven through nine describe how around the time and preaching of John the Baptist, there broke out a protracted war in Heaven. And by this over 2000-year long, ongoing war, comes the eventual total defeat of Satan and his army, as Satan is cast down from Heaven for all time *at* the Fifth Trumpet. JESUS said in Matt 11:12-14, "*AND FROM THE DAYS OF JOHN THE BAPTIST UNTIL NOW THE KINGDOM OF HEAVEN SUFFERETH VIOLENCE, AND THE VIOLENT TAKE IT BY FORCE. FOR ALL THE PROPHETS AND THE LAW PROPHESIED UNTIL JOHN. AND IF YE WILL RECEIVE IT, THIS IS ELIAS, WHICH WAS FOR TO COME.*" We also know that as they were defeated one by one, Satan's Angels were imprisoned in the bottomless pit. And as Satan is cast down last, he is given its key. So, we know this battle, as of this writing, has now raged for over 2012 years, and still rages; but it will soon be over. Remember, JESUS while walking the earth, Prophesied, saying "I saw Satan fall from Heaven". This JESUS said, experiencing a vision of promise from FATHER, for strength to endure his work on earth. Verses 10 through 13, proclaim that final victory of CHRIST, and Heaven being cleansed of Satan and all evil. And it is followed by a dire warning to the Church which is still on Earth, to keep the Commandments and Words of CHRIST, and forever turn from the ways of the world. We then see a further dire warning to us, of the coming fury of Satan, who will be cast down to the Earth at the Fifth Trumpet. Satan's insane fury, after his heavenly defeat, will be turned towards GOD'S Children in the earth. Remember, we see there has been no pre, or mid-Trib rapture mentioned. In Rev. 12:14-17, after Satan is cast down, the last days Vine of Heavens Church on earth, is given the same mirrored, book-end help as that first day's vine of CHRIST'S Church. Again, to hide it from the *direct* wrath of Satan, for the last 42 months of the Peace Treaty; but these are still times of fearful hunting persecuting and martyring the Church. In this long narrative, in Rev. 13:1-18, we see a Prophetic description, showing a broad view of the coming form and stages of a One World Government. And to be thorough in our teaching, yes, we are being repetitive. But so many things are easy to forget while fighting the flesh, so they need repeating. I have read these things many times myself. In Rev. Chapter 13:1-2, we see a coming 10-Horn or 10-Region World Government, known as the First Beast. We see this Ten-Horn-Model was conceived by Satan with his followers, long ago in Heaven. We see the *First* Beast Government, is not of a pure demonic structure, but these ten new regions which erase all former countries and their strengths. These new regions in the first seven years, are ruled by some possessed, and some demon influenced *mortal* men, under a new impossible to grasp Global Constitution. All souls who are alive at the final and complete adoption of this World Government, which occurs before Satan is cast down to earth. Will see Israel with the rest of the world trade their borders, strengths, and realities, for a promise of world peace with religious concessions. As this long narrative, starting on pg. 582 continues, we see this brings GOD'S two witnesses, who stand in front of the Temple in Jerusalem to warn all, and call for certain plagues upon the earth. This seven-year time-frame fulfills the gap of the final seven years of desolations, determined upon Israel for breaking GOD'S First Covenant, and making a covenant with the world. These Two Witnesses will tell us what was in that little book, and what the seven thunders said. Then as all this occurs, we see in verses 13:3-10, in the confusion and dismay of the former and continued world destruction, a new leader arises quickly out of the

Middle East, appearing as a man. He will arise out from the area of Pergamos in Turkey. And yes, he is a fallen Angel, the Antichrist, released by Satan from the bottomless pit at the Fifth Trumpet. He quickly rises and subdues all ten regions. He appears before the world to be assassinated by a mortal head wound. But he is not a man, but an eternal angel, created by GOD to never die. So, by the far higher created ways of all Angel's bodies, he is restored or resurrected by GOD'S pre-set care. And this lying Antichrist is joined by a lying Prophet; as they lie to a world in turmoil and fear, saying they are preparing the world to receive its Supreme Creator, to bring peace to the world. This overthrow process, *starts* at some point in the first forty-two months of the World Government, and is completed by the start of the second forty-two months. During the first forty-two months, the religious leaders of the world, now being more political in having their faith shaken, will work with political leaders to agree on reformations, to move all religion towards one unifying world religion for peace. This was always, *Satan's* subtle ultimate purpose for the seven-year religious addendum, a purpose not seen by the world. The original true and strong written words of GOD, during this time are ridiculed and become hard to find, so write every word on your heart to have it always. The new Bible versions of the very last days, will be altered and hollowed out of all meaning and strength, as we see occurring already. Now, during the first 42 months, the true world power control and reach of the Catholic Church is brought to bear in all political efforts. Many may be offended by this truth, but the wealth, political power, and reach of the Catholic Church, will have a far greater influence in all future negotiations than any other religion. Again, during the first forty-two months, the Jew's Daily Sacrifice remains in practice, as GOD'S two witnesses, Prophecy against, and condemn, and warn all who would make any agreement of covenant with this world and the First Beast Government. The true Children of GOD, and all die-hard radicals from all false religions, will resist every effort to consolidate all religions. But the weak *majority*, being safety in numbers minded, go along to get along types, will be deceived and follow the Beast Antichrist. And here we see CHRIST'S Prophesied great falling away begin. We do *not* know exactly when the Third Beast Antichrist begins, or how quickly he rises to power. But we know he is released upon the earth after the Fifth Trumpet sounds, and he assumes full control by that forty-two-month point. Then this Third Beast, during *his* next forty-two months, with his False Prophet, move to prepare and place that Image within the Temple, for

Satan to indwell as GOD. And though it is not stated, we should assume this image may appear as seated upon a Throne. Then at the end of the second forty-two months, Satan enters the Idol to speak and probably move. Rev. 13:15-18, after which the Antichrist and False Prophet carry out his orders to institute his compulsory One-World *Mark* of religion that signifies worship for the Image as being God. Yes, this created physical image that speaks to the world will be seen as GOD, or a supreme future artificial intelligence. A belief that a future artificial intelligence will be the world's savior is already being put in place, but it will be Satan that speaks. Then Satan appoints ten true Demons as Kings who are loyal to him, over the ten regions to consolidate all rule under Satan for one hour. One hour meaning a short while, in this instance, 2.8 years. With this appointment, these 10 Kings over the world, take human wives to have offspring, as they uproot, outlaw, and consume all religions; including the political power of the Catholic Church. This starts the long over-due Judgment of the original Mother of all Religious Harlots. We will first see Catholicism taken over, and then all Christian Denominations being the daughters of the Great Harlot Mother. And all Sects of Judaism, Islam, Hinduism, and all other religions are taken over. The Third and Fourth Beast, will install Demonic Angels in all religious centers of worship, including "The Great Mystery Babylon" being Vatican City. Scripture says it shall be a cage and hold of foul spirits. Before this, during that second forty-two months, GODS true children are hidden from Satan's face in places prepared by FATHER in the wilderness. In this long narrative, after the Antichrist and False Prophets forty-two-month rule, the image or idol begins to speak, and these two fall to second and third in command in the eyes of the world, completing an unholy trinity. We see Satan, openly, defiantly, before the One True GOD, standing in GOD'S place; and just Prophesied, saying he is GOD for the world to see. The Antichrist and False Prophet will tell the world that GOD can wait no longer to be worshipped by individual choice, and declare all must take a mark of faithful worship and ownership. And they make war against all who resist, saying they are enemies of peace. Those who accept the mark, then have two religious points of focus: one is to destroy GOD'S true Children, the other is to worship Satan as GOD of this world. This is when Scripture says for GOD'S Children to run to the mountains, as a short time and allowance of 2.8 years is given to the Beast Antichrist to find and martyr many of GOD'S true Children. Yes, the Church is here, running, hiding, and in great peril from Satan at this worst possible time. Referring way back to our mathematical adding up of all the timelines on around page 488. We read in Daniel 8:13-14, that there will be 2300 days or 6.3 years, from the taking away of the last days daily sacrifice, until CHRIST removes Satan, and cleanses the Temple at his return. If we subtract the second forty-two-month, rule of the Third Beast Antichrist, who takes away the Sacrifice at coming to power, and remains in power until he places the abomination of desolation bringing Satan into the Temple. This leaves from the time Satan takes control of the Image in the Temple to the time the Temple is cleansed at CHRIST'S return, as about 32 months or 2.8 years of pure Satanic rule. Most Preachers teach that the Mark of the Beast is a mark of World Government control. No, it's a religious mark of worship, and even ownership by Satan. People must freely accept the name or number of the beast to actually be written on their body for all to see they believe the image is GOD. It is in fact Satan putting his seal of ownership upon his property. To even get this Mark, one will have to declare they love and worship the image, no one will be held down and marked as an Authoritarian would do. But once you are caught without it, and sit in prison and refuse it for long enough, you will be made a public example, and be beheaded. And probably on T.V. Satan does not want to behead you, he wants you to say you know he is GOD and accept his mark. We will see the weaker are broken down over time and surrender, and the strong who will never take the mark will be beheaded. Scripture says if *any* accept this mark, it being a mark of worship, love, ownership, and obedience to Satan. There is no possible hope or offer of salvation left to them, ever. They are forever Satan's, and eternally lost to GOD. So, no you can't ask for forgiveness for taking this mark; and we see the Church is still here and will not be raptured out till CHRIST'S return at the Battle of Armageddon. As all Prophecy unfolds and we grow closer to CHRIST'S return, GOD'S true Children may then be able to see a near approximation of the day and hour of CHRIST'S return, but no one can know that moment, but FATHER. CHRIST returns only at his FATHER'S will, to cleanse the Temple and fight the Battle of Armageddon. And we cannot know how long that Battle will last. We are told to watch, understand, wait, and keep our garments clean till our LORD returns, *for* only those with a mark of HOLYSPIRIT, and not the mark of Satan, receive CHRIST'S reward of life; and *only* at that time are we taken up. We read again in 2nd Thessalonians 2:3-4, *"LET NO MAN DECEIVE YOU BY ANY MEANS: <u>FOR THAT DAY SHALL NOT COME, EXCEPT</u> THERE COME A FALLING AWAY FIRST, AND THAT MAN*

OF SIN BE REVEALED, THE SON OF PERDITION; WHO OPPOSETH AND EXALTETH HIMSELF ABOVE ALL THAT IS CALLED ***GOD****, OR THAT IS WORSHIPPED; SO THAT HE AS GOD* *<u>SITTETH</u>* *IN THE TEMPLE OF* ***GOD****, SHEWING HIMSELF THAT HE IS GOD.*" We read, the true Church of CHRIST remains on the earth, until just prior to the battle of Armageddon. We cannot be harvested out, until we see the Third Beast, the False Prophet, and the Abomination of Desolation of Satan in the Jerusalem Temple, claiming he is GOD. The Church must resist the Mark of the Beast for 2.8 years, while many are Martyred. Do not think your beheading may be impossible to accept. After witnessing all of Prophecy unfold as written, seeing GOD'S miracles, knowing GOD is there and that we must put off this fallen body to ascend, we may smile in a pure peace, as they lower our head to their blade. Here we see that blessing of spiritual strength in knowing Prophecy that Scripture speaks of. We see many time-lines throughout Scripture which align only one way; and they all concur perfectly. Knowing and watching this great ordered truth of Prophecy unfold will erase all doubt, and bring unimaginable courage in CHRIST. This is what we are allowed to know for now; for the Holy Angels, and even the SON, will not know the very day and hour of CHRIST'S return, but only FATHER. JESUS said in Mark 13:32, "*BUT OF THAT DAY AND THAT HOUR KNOWETH NO MAN, NO, NOT THE ANGELS WHICH ARE IN HEAVEN, NEITHER THE SON, BUT THE FATHER*." Matt 24:36, JESUS SAYS, "*BUT OF THAT DAY AND HOUR KNOWETH NO MAN, NO, NOT THE ANGELS OF HEAVEN, BUT MY FATHER ONLY.*" At CHRIST'S return, Satan is driven out, and JESUS cleanses the Temple. Satan calls all his rulers, and all who have his Mark, to join him in the long wide Valley of Armageddon that leads to Jerusalem, to fight and reclaim the Temple. JESUS then and only then, gathers his faithful in the air, from all corners of reality, to watch him fight the Battle of Armageddon, which may last days to weeks. And then CHRIST'S Millennial Reign begins. We *note* * as to all time-lines in Scripture, it is not stated plainly that the Seven Great Tribulation Trumpets, and the final formation of the First Beast Government granting the Daily Sacrifice, will start at the exact same time. But we do see Scripture shows both start around the same time. Remember we are still in a long running narrative that started back on pg. 582. And please read Rev 14:1-20, where *after* Heaven was cleansed, and Satan and all evil were cast down to earth, and after the Tribulation is over, and after CHRIST has cleansed the Temple Mount and we see JESUS with his 144,000 upon the Temple Mount, (see fig. a. Heaven event 7). Here JESUS has moved the battle front from Heaven, down to Earth, and is firm and ready for the Battle of Armageddon. Read Rev. 14:6-11. We see Angels flying in the heavens calling all to worship GOD, and warning all souls who have not taken it, to resist the Mark of the Beast. We see an Angel declare the end of all worldly and satanic religions, and their awaiting punishment. In Rev. 14:12-20, we see a voice from Heaven, declare that those who remain and die in CHRIST from this point forward, are very blessed, this being a most difficult time for them to endure. Now right *here,* and not until *here*, is the Church ready to be taken up. We then see a second reaping of the vines, *of the earth,* being all the falsely religious, having the Mark of the Beast. These are killed in the great battle at Armageddon and cast into the press of GOD'S wrath. In this look *forward*, we see *no* sinners are spared. CHRIST has always declared war on *all* sin, *all* falsehoods, *all* false beliefs, and *all* evils over the face of all reality. And that is the end of that long running narrative that started back on pg. 582. We pray you do not have trouble coming way back from the end of these long chronological narrations, back to the confirming specific order of the Seals, Trumpets, and Vials where we left off. Read back in Rev.

15:1-8, where we see a heavenly ceremony, preparing the last Seven Vials of wrath in the Seventh Trumpet to be poured out. These mirror the Old Testament plagues upon Egypt, and upon those who are not GOD'S. Read Rev. 16:1-21 (see fig. a. Heaven event 8), describing how the final wrath of GOD is held by Seven Angels, which are poured out of Seven Vials. We are told often in Scripture that if our love for GOD is true, we are safe from all these plagues. So, if you are true, do not fear the coming of these things. If you are still alive, and standing in the loving obedience to every word of GOD, these plagues will not touch you, you will walk between every evil raindrop. Vial One, brings upon all those in the world who have Satan's mark, a noisome, grievous sore. *Noisome* meaning having a foul odor, and *grievous* meaning life threatening. This is in all aspects, necrosis, or decaying putrefying dead flesh. Vial Two, pours most of the debris of that massive ring in the atmosphere down; bringing the final death to all the oceans, seas, and every single thing that still lives in them. It is staggering to imagine a crust upon the surface of all the oceans, described as like the blood of dead men; and we know blood thickens, stinks, and hardens. As Vial Three is poured out, we see the last remaining debris poured down into all remaining fresh waters, increasing famine and thirst everywhere; don't worry, pure water is provided for the Children. For the lost, drinking water will require great effort and expense, making water, food, and clean air more valuable than anything. In desperate thirst, many of the lost must drink improperly filtered bloody radioactive contaminates and die a horrible death. In Vial Four, we see the protective atmosphere is altered, after that great blanket of contaminates that formed a band around the Earth, was cast down into the waters. We see earth's protective atmosphere is damaged, and the power of the Sun is temporarily increased by an Angel pouring his Vial upon it, causing its surface to temporarily be seven-times hotter. The suns surface is around 10,000° F, so it increases to 70,000° F. This produces so much radiation, it severely burns *all* skin in just a few minutes. From Vial Five, we see something amazing. We know that physical light is GOD'S perceived creation, and his gift of particles and waves that carry to us the complex patterned information that describes for us, our visual representations of GOD'S Creation within our purview. But this Fifth Vial brings a temporary darkness, upon *only* those having the Mark of the Beast. This is not any temporary ocular blindness, and we read about the mirrored act of this by GOD in Ancient Egypt; Exodus 10: 21-29. This is not FATHER altering the ability of their eyes to sense this information, it is his work of altering reality itself. The visual systems in their bodies work correctly, but particles and waves of information carried in physical light, are temporarily repelled around them and withheld from their eyes. This is a darkness that even the born blind have never experienced; and this is why Scripture says it is a darkness that can be felt, bringing great physical distress. If it were just their eyes, they could not sense a lack of sunlight upon their bodies. We read that this plague is temporarily poured out, only upon the Seat and Kingdom of the Beast. And GOD'S Children will see and feel just fine, even as they walk among those who temporarily sense only cold darkness. We know that this holds an additional element, in that all who experience this cold darkness, will know they are found on the side of evil and lost to GOD as they cry out to him. But they cannot be heard, their eternal choice has been made. We must see that as CHRIST comes near to remove Satan from the Temple and Jerusalem, Satan will still claim he is GOD, and call all upon the earth, having his mark, to come and battle against any who has come to fight him. We cannot know exactly when, but prior to the Sixth Vial, our LORD and his Holy Army, have driven Satan and his Trinity from the Temple and the Temple Mount, and are cleansing Jerusalem.

So, CHRIST must then defend Jerusalem from Satan and the world. In the Sixth Vial, please note that the Euphrates River is just now dried up, *and* there is no mention here of the two hundred thousand, thousand Horseman Army of GOD. And the Euphrates was not dried up back at the Sixth Trumpet, as many falsely teach. So, it is now dry, making the way for Demon Kings of the East to come to join the Battle of Armageddon. Also, with this Sixth Vile, three demonic spirits are sent to convince all Ten Demonic Kings who have been given power for one hour, that there is still hope and reason for all who have the Mark, to come and gather together at Jerusalem for a last battle. Rev 17:12-13, "*AND THE TEN HORNS WHICH THOU SAWEST ARE TEN KINGS, WHICH HAVE RECEIVED NO KINGDOM AS YET; BUT RECEIVE POWER AS KINGS ONE HOUR WITH THE BEAST. THESE HAVE ONE MIND, AND SHALL GIVE THEIR POWER AND STRENGTH UNTO THE BEAST.*" This is their last futile act; and their motivation is a desperate self-preservation felt by all around the world who made that wrong eternal decision to follow Satan and receive his Mark. In Rev. 16:19 as the *Seventh* Vial is poured out, FATHER says the very last plague has begun, as by FATHER'S unimaginable force, the greatest earthquake the world has ever felt occurs. This quake breaks

Vatican City into three parts and great buildings all around the world fall to rubble; islands sink below the surface of the oceans, and mountains sink into the ground as hail stones weighing as much as a man, fall around the earth. All Satan's surviving Children gather together against CHRIST, seeing the destruction of this plague is so great. CHRIST suddenly pauses here at the peak of this fierce timeline to remember and relate the events that led up to the destruction of Mystery Babylon, being the Vatican and Catholicism, and tell us how and why total destruction occurred to all unfaithful worldly religions. This in-depth look back at GOD'S wrath upon all false Christianity, and all false religions fills Rev. Chap.17:1 through Chap. 19:3. This wisdom is needed to be shared and heard right here in this moment, and at this very specific point in Scripture to help us see and understand why; and understand GOD'S suffering reasoned wrath. This detailed narration, starts from just before the formation of the One World Beast Government, and it tells us who this Mystery Babylon is, and what she's done. We see the Judgment of this Great Whore, occurs after Satan enters the Temple and installs his Ten Demonic Kings. This next long narration goes through a lot of ancient history and bring us right back to just after the battle of Armageddon. The length of this last narration, declares how important it is for us to know and understand it all. It details how GOD hates all false religions and doctrines formed by men to try and say they worship him, while appeasing the flesh as they falsely claim his Holy Name. Rev. 17:1-18, says, GOD reveals *for us* the details of his Judgment upon this Great Whore, which can only occur after those final Ten Demon Kings are given power for one hour with the Beast. So, it is at this point and place in our study of now having a proper understanding, that GOD reveals this story to us. What can we know about this mother of all harlots. In Rev. Chapter 17, we see a coming Judgment from GOD upon the oldest greatest unfaithful harlot mother, *and* all her religious harlot daughters. The World Government Beast, from its formation, will exert pressures *upon* all religions to compromise towards a single Religion to bring religious peace. Can there ever be a peaceful compromise among religions? No. We must know religious-preservation will be strong. This Great Harlot Mother being judged, is Queen of all unfaithful false religions. And being a mother of harlots means she has unfaithful daughter religions. Will this Great Mother, and most powerful Christ-centric religion on the face of the earth, surrender power without resistance. We know all Judgments from GOD, come for our unfaithfulness, and many may say this Harlot must be Islam.

It cannot be Islam for many reasons. First, in the Prophecy of GOD'S Judgment, Islam is judged at the Sixth Trumpet. Two, a harlot is unfaithful, and one must first be betrothed to be unfaithful. Islam never claimed or received the Sacrifice of CHRIST, and was never betrothed to, or a part of the Bride of CHRIST. So, Islam being a purely antichrist enemy from the start, cannot play a harlot against CHRIST. These harlots are those who took a vow as the bride of CHRIST, and dishonored that betrothal by unfaithfulness through their greater love for this world and the flesh therein. In Rev. 18:7, this Greatest of Harlots sees herself as a Queen. To be a Queen, implies no less than the Bride of a King. No other religion other than Judeo Christianity, has a King and a Bride. We see apostasy reigns to many degrees in *all* denominations claiming the Holy name of CHRIST. Every new weak false Christian denomination, is a daughter of that first Queen, being filled with unfaithfulness. The Roman Catholic Church is the oldest, largest, and the first Church claiming CHRIST to turn to errors. A Church whose age is measured in millennia, so, she is called the mother, meaning first progenitor of all daughters to come after. Roman Catholicism began several centuries after Peter and most all Apostles had died, as leaders within a certain Church, turned to the doctrine of the Nicolaitans; a doctrine GOD hates. This

“Catholic” (meaning universal) Doctrine, formed a Church hierarchy of souls having authority over laymen in the Church body for all things, *including* the interpretation of Scripture. Thus, subverting each soul being taught by HOLYSPIRIT. This started in Rome, by combing Christianity with a universal religion that existed before CHRIST. This false doctrine set in place the authority of one infallible Pope, to interpret Scripture for all members. Each new Pope can change a prior Pope’s interpretations of Scripture, even times and laws; and thereby change GOD’S truth to fit the current society with each new Pope. All who join the Roman Catholic Church, vow to accept each Pope’s interpretation as infallible. The last official Document to use the identity “the Roman Catholic Church” was written by Pope Pius X11 in 1950, as that prefix became a political problem. The Roman Catholic Church has wealth and political influence, beyond our understanding. They own Cathedrals, Churches, Universities, Hospitals, Schools, State Embassies, a gilded scarlet City, and many assets around the world. The Vatican has its own bank and army, and as of as of 2012, Catholic Church holdings could be worth over a trillion dollars. Vatican City, being 100+ acres of gilded history, is worth billions. And remember, true assessed value includes all near-term incomes. Through history, we see written accounts of the Holy Roman Empire using wealth, power, and influence to sway the outcome of wars. We see their power in crowning Kings. The Roman Catholic Church, being a reconfigured of the former Holy Roman Empire, considers itself this world's true universal religion. They abandoned their label of Roman and Empire for political reasons, but know as sure as the sun rises, they are an Empire, and they do not surrender power to other religions. Why do world leaders meet with the Pope? Why do news networks publish every official statement from the Pope? Why do politicians, heads of state, and celebrities seek a Pope's audience, and advice? Why is the Catholic Church so involved in worldly Governance? Do we see this *level* of direct influence from any other religions leader? No. In Rev. Chap. 17, we see the power of Catholicism in *this* One-World-Governments business will be extensive. Does a Pope hold world influence like a President of a Country? In fact, he is the leader of a Country. Do you know Catholicism has their own country and army? Judgment comes for this Queen, this mother, this progenitor of all other religious whores. We see this Queen of Harlots rides upon the First Beast Government as a partner, and sees herself a Queen. But her judgment come quickly by the Antichrist, the False Prophet, and Satan's final Ten Demonic Kings. We read that GOD puts it in their hearts (allows them) to destroy her, as it aligns with GOD'S, and their will. We see Satan's *unholy* alliance utterly consume and erase her presence. Scripture is often misinterpreted, and we see the same for Rev. Chap. 17. This chapter focuses on the Judgment of this greatest oldest whore. The only questions are, *who is* this Mystery Babylon, Mother of harlots; and what is a harlot. But instead, many twist this Babylon harlot clothed in scarlet, *sitting* on seven mountains, to be a revived Roman European Empire as the Beast Government. Having its capital in Vatican City, which *is* called the city on seven hills. This interpretation is so wrong. Yes, the First Beast *is mentioned* in this chapter, but *only* to show that at the start of a One World Government, this harlot is supported by the Beast, revealing her entangled power with the Fourth Beast Government. And the Third Beast Antichrist is mentioned, only because GOD clearly says he and his Ten Demon Kings will bring Judgment upon her. By looking at current facts, the Catholic Church is larger and richer than all Christian Denominations combined; and far more than all other Religions. They are major investors in stocks, banking, and hold a huge portfolio in the insurance industry. They own major interest in thousands of schools, hospitals, universities, and other institutions. The Catholic Church is the oldest, most powerful, most organized, single enterprise in the whole world; they rival the U.S. Government in certain ways. If we could see all the wars they've financed over the last 2 Millennia, and all political coups they financed. And the Crusades their soldiers fought. The Catholic Church has had a standing army for about 1500 yrs, killing who knows how many tens of thousands in the past. They conquered and confiscating wealth and lands over the ancient centuries in the name of GOD. Over the millennia they changed the political landscape. No other religion, claiming CHRIST'S name, has persecuted and killed more Jews, than Catholicism. In many countries, their Kings coronations over the centuries had to be blessed by the Pope, and many were toppled by the same. Many Kings over the centuries have bowed and kissed the Ring of the Pope. Do you know what kissing that ring means? It means fealty to the wearer of that ring. What about Biblical interpretation, which they alter and diminish over time, step by step, line upon line, and precept upon precept. Have we not heard in these last days from each new Pope, how Catholics *must* take a universal posture towards the world, and soften their stand *for* Israel and *against* divorce, adultery, and sexual perversion; even abortion and sin in general. They declare the Catholic Church must change to maintain its domination in the earth. So, do you think that with the formation of a World

Government, this Catholic Church will give up any power? No, they will temporarily be one of the biggest victors. They will compromise, as will all false religions. Catholicism has always compromised with the world to gain financial and influential power. Remember this World-Treaty has a Seven-Year provision made with *many* religions, to protect and allow them to peacefully exist and exercise their faith. But in the second 3 ½ years, this Government turns and works against all religions. Do you think the Antichrist and False Prophet will build all new religious infrastructure? No. Starting from the mark of the Beast, with his ten newly appointed Demon Kings. This Antichrist will confiscate all religious properties and infrastructure, including the Jerusalem Temple, Vatican City, and all cathedrals, mosques, temples, shrines, and every place of worship or spiritual reflection will fall under one name and form of worship. And if their leaders bow, they assume control of them also. This final One-World-Religion, worships Satan as he claims he is GOD in the Jerusalem Temple, with the Antichrist and False Prophet in Vatican City and Mecca, and demons in all religious strongholds. In all this, we don't say any Catholic, or any denominational believer or leader, are *knowingly* in part with the coming Antichrist, but we assume some are. Now to use this great power of the Catholic Church, can the False Prophet pose as a Pope? No. What Revelations talks about in chap. 17, is not any one soul, but the institution and beliefs of Catholicism; and it includes as daughters, all apostate Christian denominations. The Catholic Church, for seventeen centuries has existed in a state of deep apostasy and worldly traditions, contradicting Scripture and being far from the commands of GOD. For examples: one is seen in calling men *Father* (Father means Creator). JESUS never let anyone call him FATHER, he always said he was the SON. JESUS said he always obeyed FATHER, who was and is and always will be above him. This Doctrine of calling a men father, in effect places man in the stead of FATHER, and above JESUS. CHRIST said in Matt. 23:8-10, "*BUT BE NOT YE CALLED RABBI: FOR ONE IS YOUR MASTER, EVEN CHRIST; AND ALL YE ARE BRETHREN. AND CALL NO MAN YOUR FATHER UPON THE EARTH: FOR ONE IS YOUR FATHER, WHICH IS IN HEAVEN. NEITHER BE YE CALLED MASTERS: FOR ONE IS YOUR MASTER, EVEN CHRIST.*" The Catholic Church also turned souls away from the Holy Sabbath of the LORD, in changing the Sabbath from the 7th Day, to the 1st Day. This denies the Church's Holy rest in our FATHER. They changed the day of JESUS'S crucifixion from a Wednesday, to a Friday. John the Apostle wrote it was on a

Tuesday evening, being the day of *preparation* for the *High* Sabbath of Passover, that CHRIST was taken, as CHRIST is the Passover Lamb. Passover fell that year on Wednesday, two days before the Saturday 7th day Sabbath. Catholicism moved the Crucifixion of CHRIST to a *Friday* preparation for a seventh day Sabbath, contradicting Scripture and making our LORD, who rose on the 1st day, a liar. Placing our LORD in his tomb, for less than two days and two nights, contradicting CHRIST'S Prophecy, when he gave us the sign of Jonah, telling us he would be for three days and three nights in the earth. In Ancient Hebrew life, each day started at 6-pm, and CHRIST was placed in a Tomb on Thursday, which starts Wednesday at 6-pm. And he was there thru Sunday, which begins at 6-pm Saturday afternoon. JESUS was in the Tomb, Thursday eve and morning, Friday eve and morning, and Saturday eve and morning; CHRIST rose after 6-pm Saturday eve, being the start of Sunday, or the first day. Mary arose half way through the first day, being Sunday morning, to go to the tomb where she saw the LORD had risen. Catholicism takes away the immersion Baptism, removing our burial of our sinful flesh, and resurrection of our new clean flesh. This act of burial, described by CHRIST to John the Baptist as a part of our fulfilling all righteousness, is replaced with sprinkling or pouring water on the forehead. This does not fulfill all the righteousness CHRIST commanded in Matthew 3:15. If the ordained steps of worship from GOD are not observed, the flesh was not buried, and cannot be raised anew by GOD, or be indwelled by HOLYSPIRIT. Catholicism also falsifies for themselves a multitude of Catholic Saints, and elevated the status of the Apostles and GOD'S most favored handmaid, Mary. They have elevated Mary to a point, that detracts from worshipping JESUS. We see all Angels and Servants of GOD throughout Scripture, warn *all* against giving them adoration, worship, or respect. Instead, they say to give *all* Glory to GOD for the good our soul. Of course, a great many other apostate conditions exist in all Christian Denominations to different and various degrees. Catholicism is just the mother, meaning the first oldest and greatest offender. Making her an unfaithful Mother and unfaithful Queen of unfaithful Daughters. As said, we see from the earliest records, that instead of coming out of the world, and being a separate Holy people. The Catholic Church has been greatly entangled in the world systems and politics so much, that a line of separation at many points is non-existent. At a certain point in history, Popes began to realize as education became a reality to the masses. That to be known to the world as the Holy Roman Empire, was no longer acceptable. So, they moved to more discrete forms of power and influence. They long ago, began relinquishing *apparent* control over the many States of the Holy Roman Empire in a campaign to hide their power. In 1929, by a Treaty for the world to see, they made a deal relinquishing their sovereign control over the last of the former European Papal States. But retained sovereign control over the Vatican as a sovereign country, having their own military and bank. Do any other religions have their own country, military, bank, and Nation Embassies in other countries around the world. We need to understand, there is one pure truth, and *all* Denominations, Sects, or variations entwined with this world, are confused, weak, and apostate. They're deceived and in league with, and aid the contrary spirit called Antichrist. All such rebellious apostasy can only be revealed by shining GOD'S pure light of truth upon it. In its strictest form and definition, this term "Antichrist" means and includes everyone, everything, and every word, that stands just outside the perfect intent, beliefs, and ways of CHRIST. Catholicism is not directly responsible for the new false doctrines of a pre, or mid-tribulation rapture, or once saved always saved, or pre-destination, or the modern prosperity doctrine, or all false doctrines of CHRIST-Centric Denominations. Apostasy will weaken a soul to *fall away* and worship the flesh, the world, the Antichrist, the False Prophet, and finally Satan. We do not say that current denominational leaders, or followers are in *knowing* efforts with Satan. But we point out that most current leaders and followers are in grave error along with Satan. It is Prophesied by CHRIST that many will be deceived, and a *falling away* to Antichrist will occur. Remember, Scripture says Satan comes *brilliantly disguised* as an agent of light. This Judgment of the Great Mystery Babylon, is specifically labeled and clearly speaks only of the Judgment of GOD upon all apostate unfaithful Churches and souls who claim CHRIST, yet are entwined in the flesh and the world for its comforts, pleasures, and provision. CHRIST need not face, or dirty his hands with these whores; he allows the Antichrist to cleanse the earth of these betrayers before his coming rule. Our LORD says that this warning is heard only by the true believers who know what comes. Scripture says this Mystery Babylon Whore, is a great city upon seven hills. At the start of a World Government, we see this City riding upon the Beast, with great influence and arrayed in Royal Scarlet. There is only one such sovereign City in the world, so arrayed in these royal colors, gilded with gold; and that is Vatican City. And it will remain Catholic and sovereign through the first 3½ years of the First Beast. Read Chapter

18:1-4, (see fig. a. *Earth* event 7). Here we see a very powerful Holy Angel, declaring the Judgment of wrath upon the mother of all religious harlots is complete, as this Angel says the Vatican is become a cage of unclean and hateful bird (meaning Demons). Please note the True Church is still here according to Scripture, as they remain in hiding with some souls yet to be martyred before the reaping of the earth. To be martyred is a blessing we may hope to endure. It is our highest gift we can lay at the feet of JESUS. There has been no rapture mentioned in this complete Prophecy of the last days. Would CHRIST leave out such a huge earlier event if it were real? No. At this point, we want to correct *in greater depth*, the false teachings we discussed earlier in 2nd Thessalonians 2:6-7. Which says; *"AND NOW YE KNOW WHAT WITHHOLDETH THAT HE MIGHT BE REVEALED IN HIS TIME. FOR THE MYSTERY OF INIQUITY DOTH ALREADY WORK: ONLY HE WHO NOW LETTETH WILL LET, UNTIL HE BE TAKEN OUT OF THE WAY."* We see the Prophecy of CHRIST in the last days, holding back Satan until his Prophetic time. Then FATHER casting him down and giving him the key to the bottomless pit. We see Satan *letting* out his army upon the Earth, until CHRIST returns and again, and *takes Satan out of the way*. I have heard pre-trib. And mid-trib. Preachers, claim it is the

Church and HOLYSPIRIT who lets, and are taken out. This is such a confused belief, the Church remains here through the Tribulation, and HOLYSPIRIT could never leave the Church. And HOLYSPIRIT never physically battles, and is never withheld be evil, and never lets anyone out of the pit. Satan is withheld, and then Satan lets them out, and then Satan is taken out. All responsibilities and actions of GOD in war and battle, are reserved in CHRIST and Sabaoth; being the Army of Holy Angels and Archangels such as Gabriel and Michael. And at the final Battle at Armageddon, our LORD does join, and lead the battle, but not HOLYSPIRIT in any way. For examples of this, look at Daniel 12:1, which says; *"AND AT THAT TIME SHALL MICHAEL STAND UP, THE GREAT PRINCE WHICH STANDETH FOR THE CHILDREN OF THY PEOPLE: AND THERE SHALL BE A TIME OF TROUBLE, SUCH AS NEVER WAS SINCE THERE WAS A NATION EVEN TO THAT SAME TIME: AND AT THAT TIME THY PEOPLE SHALL BE DELIVERED, EVERYONE THAT SHALL BE FOUND WRITTEN IN THE BOOK OF LIFE."* Here we see that Angels are referred to often as Princes. And Matt 24:30-31, "*AND THEN SHALL APPEAR THE SIGN OF THE SON OF MAN IN HEAVEN: AND THEN SHALL ALL THE TRIBES OF THE EARTH MOURN, AND THEY SHALL*

SEE THE SON OF MAN COMING IN THE CLOUDS OF HEAVEN WITH POWER AND GREAT GLORY. AND HE SHALL SEND HIS ANGELS WITH A GREAT SOUND OF A TRUMPET, AND THEY SHALL GATHER TOGETHER HIS ELECT FROM THE FOUR WINDS, FROM ONE END OF HEAVEN TO THE OTHER." I hope you pray for all Holy Angels, theirs is a battle of Creations epic forces. We know there are Seven Archangels. Read Revelation 1:4, *JOHN TO THE SEVEN CHURCHES WHICH ARE IN ASIA: GRACE BE UNTO YOU, AND PEACE, FROM HIM WHICH IS, AND WHICH WAS, AND WHICH IS TO COME; AND FROM THE SEVEN SPIRITS WHICH ARE BEFORE HIS THRONE;*" Revelation 1:16, *"AND HE HAD IN HIS RIGHT HAND SEVEN STARS:*" Revelation 4:5, "*AND OUT OF THE THRONE PROCEEDED LIGHTNINGS AND THUNDERINGS AND VOICES: AND THERE WERE SEVEN LAMPS OF FIRE BURNING BEFORE THE THRONE, WHICH ARE THE SEVEN SPIRITS OF* ***GOD***." These are the very first seven, highly elevated, most powerful Arch Angels at the right hand of GOD. But a *fighting* HOLYSPIRIT, is never mentioned. We see examples of a creative protective FATHER, and we see his sacrificing, and then fighting, and then ruling SON, But we see HOLYSPIRIT described as a loving, comforting, teaching, spirit, not a feminine mothering spirit, but the pure nurturing spirit of GOD. And these are the three Godly attributes of man. There are immense differences, between the Anointed Cherub Lucifer's fallen Cherubim, and the Angels in the Holy of Holies. Satan and his angels are a lesser order below the Archangels and Seraphim. In Rev. 18:4, we see Holy Angels calling out for souls to resist the call to take the Mark of the Beast, as Catholicism and all Christian Denominations are overthrown. These Holy Angels call all to resist in a love for GOD, and escape the wrath stored up for Satan, his Angels, and all who take that Mark. In Rev. 18:5-24, we see vast economic losses to the millions of souls who were dependent on this great harlot and her holdings for their lively hood, as Satan seizes that vast religious empire. It is an astounding account of the reach of power Catholicism has in this world and how many people depend on that empire. A look of astonishment is upon all those who see it taken in one hour. All her power riches and splendor, being as if it were consumed by fire, leaving only a visual of rising smoke; (rising smoke is symbolic of complete loss and ruin). So, after that last long narration, we read Rev. 19:1-18, where we see certain ceremonial and inspirational words, as the Millennial Reign of our LORD is at the door. All that remains, is the battle of Armageddon. We read our LORD is followed by his Army in triumphant glory, riding his White Steed into battle to take rightful rule of his Kingdom, given him by FATHER for his great responsible faithful Sacrifice. The full Glory of FATHER is placed upon JESUS, as the King of all Kings, and the Lord of all Lords. This great battle Armageddon, like all prior cleansings, started at the Temple in Jerusalem. CHRIST first cleans his own house of all who are not of GOD. This battle and cleansing, then goes out throughout all Jerusalem, and to a place where all those marked armies of the world will gather, and all and any who would fight against, or disobey JESUS. They are gathered into one place for their last great battle and ultimate destruction, which is the final cleansing on earth, of the Spirit of Antichrist *before* CHRIST'S Millennial Reign. The outcome of this battle is pronounced in this Scripture, and it will surely occur. (See *Earth* event 8). When all the deceived and disobedient come to Jerusalem to fight, they are met and destroyed by JESUS in battle, in which Scripture says, they keep coming and coming until there are so many killed, that the blood covered bodies of all those slain, combined, fills this long wide valley to over five feet deep and to a length of a thousand six hundred furlongs. This would require hundreds of millions of slain blood-soaked bodies and horses. We see that after World-War-Three, the world's oil industry is destroyed, and the use of horses will return. In Chap. 19:19 through 20:3, we see JESUS is victorious, and his Angels take the False Prophet and Antichrist, and cast them alive into the eternal prison called the Lake of Fire, to start their *eternal* conscious torment. These have eternal angelic bodies, not created under the weakness of physical destruction like those of mankind. And Satan is bound and imprisoned in the pit for a thousand years, as CHRIST'S Reign on this old destroyed earth begins. Read Chapter 20:4-15; now we have had to correct more than a few false teachings, and yet there are more. Another error is that at this first resurrection of souls, all GOD'S Children are raised to reign with CHRIST, but Scripture says otherwise. We rarely hear it preached that there are two resurrections. We see CHRIST'S Millennial Reign starts with the First Resurrection of Elders, Prophets, and the Martyrs of the Tribulation. These will rule with CHRIST over those who remain alive, and all who will be born in the next 1000 yrs. During this Millennial Reign, these resurrected, are tried and taught to live and rule Holy and Righteously with CHRIST, yet these free-will souls can still fail in this ultimate responsibility. Now all souls throughout history, including the aborted and young, who were killed or died before making their eternal choice, will be born again to make their final choice. But the rest of GOD'S Children, who died being under or beyond GOD'S Holy Covenant and grace, they are not raised until the White Throne Judgment, after the Millennial Reign. We see Prophecy denies any pre, or mid-tribulation rapture. Did you see one I missed? Now this last Millennium, is a time of final teaching and trying of hearts to see who can truly serve as eternal leaders of self and others. If this doesn't sound right to you, read Zech. 2:10-13, and 3:1-10. Here we see Joshua is crowned as *a* High Priest for the Millennial Reign, but he is being admonished by an Angel to be careful to keep himself in the ways of the LORD; meaning if Joshua fails to keep himself, he faces dire consequences. Reading Zechariah, 14:1-21, at the end of CHRIST'S earthly 1000 yr. Reign, we see the earth has been partly healed by JESUS, and is again filled with a multitude of people. And for the last time, Satan is released, who has not repented. And goes out again to deceive those weak in their love for GOD, living beyond the camp of Israel and the LORD. And most, souls once again, by the desires of their flesh, are weak and separated from hearing HOLYSPIRIT. The many again rebel, even living under the very presence of JESUS and his Throne on Earth, just as the fallen angels did in Heaven. After Satan is loosed and riles all who once again don't love or want GOD'S ways; Satan gathers these together at the last battle. And for the first time, every soul in all of history will have *now* made their firm and final choice. And all alive on earth, are firmly with CHRIST, or they gather to surround Israel and fight. So, the flesh of all those coming to fight is consumed by fire coming down from FATHER above. Satan being an Angel whose body is not subject to physical death, is then imprisoned in the lake of fire for his eternal conscious torment. And we see the final White Throne Judgment where all the remaining just and the unjust sleeping in death, are resurrected to their eternal bodies like the Angels. The just are cleansed by the refining fire of GOD, burning away all evil works (perceptions). And all conscious confusion and rebellion called hell and death, being curses; along with all whose names are not found written in the book of life, are cast into the prison known as the lake of fire. Now we correct another misinterpretation about the Millennial Reign on this old defiled earth. We have read that after cleansing all evil and rebellious reprobates from the face of the Earth, our LORD sets up his camp in what will be the cleansed current Jerusalem and Israel for one thousand years. This is not the *New* Jerusalem that comes down from Heaven to the *New* Earth *after* this thousand-years, and after the very last battle that comes after the White Throne Judgment. Now to know the ways of life during CHRIST'S Millennial Reign after that first resurrection, we read about his reign towards the end of the Book of Ezekiel. Ezekiel Chap. 1, to around Chap. 11, are warnings from GOD, pertaining to the life and time of Ezekiel, including the great dispersion. Chaps. 11-17, are a mix of mostly past, and a few future Prophesies. Around chapter 17, we see Ezekiel's Prophecy turns solely to the last days trouble as it builds through chapter 27. In chapters 28 and 29, Ezekiel's Prophecy turns to Satan and the cause and Judgment of his fall found in his confusion, born in his ego. Chapter 30 through 37, continue with warnings of the last days. And around chapter 37, we read of the rebirth of Israel in the last days, which occurred on May 14, 1948. Chapter 38 and 39, start at CHRIST'S Prophesied Beginning of Sorrows, and quickly moves through the Tribulation without detail. But in Chapters 40 through 48, we see in great detail, how after total destruction of the earth by war and the Great Tribulation, that Earth is miraculously and selectively being healed here and there, as life returns in the Millennial Reign under the ruling provision of CHRIST on Earth. We are shown this, *from* the perspective of our LORD over his Government. We read CHRIST'S Government, consists of him as Heavenly King, with a former earthly King, *now* being a Prince under him, being David. We see the Temple Priesthood restored, and much more. We see CHRIST'S Third Dispensation of grace and forgiveness of sins, through his blood sacrifice, has at this point closed, and the Animal Sacrifice is returned during the Millennial Reign, with the Temple Priesthood and all its Holy observances. Now there is little Scriptural teachings about the Millennial Reign and Heaven; so, much of *what is taught* around the world is false. It often seems as if Preachers do not care if they speak or hear the pure truth, as long as it comes from a pulpit with a cross emblazoned on it. I hear preachers and professed Christians say and believe they will be there ruling with CHRIST. I hear many say they can't wait to run to JESUS and hug him. Surely according to Scripture, this will never be allowed in the Millennial Reign. To understand the Millennial Reign, we start in Ezek. chapter 40, as it describes the last days Jerusalem Temple. How do we know this? We need only read that after describing the form and cleansing of the Temple, we see our LORD JESUS enters the Temple through the East Gate as he is Prophesied to do in the last days. We read in Chapter 44, that the East Gate is then shut. Ezekiel is told by JESUS, it is shut, because the LORD has entered by it. And from that moment forward, only David the Prince can come in that Gate to eat bread before the LORD, and go out of that gate; it is known thereafter as Prince David's Gate. For understanding David better, read Ezekiel 34:23-24, and Ezekiel 37:22-25, and also Ezekiel 44:1-3. Now back in Chapter 43:4-10, we see the glory of the LORD has filled the whole house. Glory, meaning an intense cleansing Holy presence of light, heat, and smoke. Then JESUS sets forth the laws of his house. In Chapter 45, we see the division of the Proper Scriptural lands of Israel by our LORD, to the Tribes of Israel. In Chapter 46, we see more laws of CHRIST'S house are revealed. And knowing the period of Messiah's Sacrificial Grace has ended, we see the return of the animal Sacrifice. We read CHRIST reigns through this 1000 yrs. on earth with a rod of iron. After this 1000+ years and the very last war, this earth and the current Heavens dissolve as if they never were. And FATHER says, sees, and consciously forcefully perceives, or creates if you will, a New Earth and Heaven. And CHRIST will rule eternal over that newly created Earth and Jerusalem. By our first seeing JESUS rule here in the old Temple on the old destroyed earth; without doubt, we know this speaks of the Millennial Reign. Also, when we read the last chapters of the Revelation about the New Earth, we see it says there is no Temple in that New Jerusalem, as JESUS is said to be the

Temple in the New Jerusalem on the New Earth. And in Ezekiel Chapter 47 on the old earth, we see a miraculous trickling of living waters coming from under the threshold of the door of *the Temple.* And within a short distance, this healing water is miraculously grown to become ankle deep. As the healing water flows east, it grows and finally increases into an impassable river that after FATHER'S last massive leveling earthquake, is able to flow all the way to the once symbolic Dead Sea. We know, having just studied this, that upon the returning victory of CHRIST, the Earth and all its waters have all been destroyed. We see in chapter 47, verses 8-12, that by these living waters flowing from the Temple, that where these waters go, the banks and the seas are immediately healed. Whether by streams, or by rain, these waters heal the earth they touch locally. We see GOD'S blessings in the immediate return of a multitude of fish, and many trees on the banks of the river whose fruit and leaves never rot that CHRIST brings for food and medicine. Also read Joel 3:13-21. There is abundant evidence in this Scripture showing that this too is the old earth Millennial Reign, and not in the New Earth to come. In Revelation 21:1-27, we see great distinctions between the New Jerusalem come down from Heaven, where there is no Temple, and thereby no threshold for the waters to flow under. But we do see a Temple and a threshold, in Ezekiel Chapter 47. In the New Jerusalem we see all living waters flow from under CHRIST'S Throne, as he is the Temple there. In Ezekiel we read these waters flow to the sea. But in seeing the New Earth in Rev. 31:1, John says one of the first things he realizes, is that there are no seas on the New Earth. Also, though we see healing trees and glorified fruits having great new properties in CHRIST'S Millennial reign on old earth. In the New Earth, we see the return of that additional tree of eternity, called the Tree of Life. The fruit of this tree will be necessary for eternal life, and is not present in the Millennium. Under CHRIST'S rule and presence, the aging process in JESUS'S Millennial Reign, is set back to as it was for Adam and Eve, right after the fall. And as stated in Isaiah 65:20, and Ezekiel Chapter 48, we see the *camp*, meaning temporary home of our LORD, being the lands of Israel on the Old Earth, are divided among the Twelve Tribes for a thousand years. But in the New Earth, we see a New Walled City of Jerusalem and New Earth, but no New Israel. The Great City of Jerusalem has grown to hold all who love GOD. But those who live in the Old Israel will reign and rule under CHRIST for 1000 years over those living outside that camp, who are subject to serve those within the camp, for 1000 years; offering foods, fabric, and the such. From reading so far, about the Millennial Reign, we see only King David and the Holy Priest, are permitted to approach and minister unto JESUS. We read our LORD'S dispensation of Grace is over, and JESUS now rules with not Sacrificial mercy, but a rod of iron (Psalms 2:9 and Rev. 19:15). And he dashes to pieces, all who disrespect or disobey his rule and command. The current third dispensation of forgiveness by JESUS'S suffering, will have at that time passed. This is not a hard thing to grasp and understand. But once we understand CHRIST'S suffering for the sins of others, must eventually end, that JESUS cannot suffer eternally for children who will never cease to hurt him. At a point, his pain and suffering for others sins must and will end. At that point, our LORD cannot be asked to pay for our sins any longer. As for the rest of the sick desolate world, we are not given any complete understanding as to the extent of healing it receives by the Holy waters, *except* we are told that all nations who come up to the feast and worship, will receive these healing waters by way of rain. Those who don't, won't receive it. Let us take a last look at the scope of all the coming Prophesied death. We see a soon coming world war brought by mankind's evil desires, and the influences of the Four Horseman. We are approaching this event now, where conventional, nuclear, and probably chemical and biological weapons, coupled with world-wide famine and disease, will kill a fourth of mankind. We see the Tribulation plagues as they continually increase, and bring the cleansing wrath of GOD up through the Fifth Trumpet. Then with the Sixth Trumpet, we see the Spirit Army of GOD cleanses all who in their hearts disregard the borders of Israel drawn by GOD. Then we have the Judgment of the great whore, being all false Christian Churches, where GOD allows the Beast to make war on the whore to devour her and consume her. This political war on the great whore will kill many people, but most will instead fall away, to take the mark of the beast. Then we have the Battle of Armageddon, killing all souls who have the mark of the Beast, this number will surely be huge, as we see the depth of all the bodies, horses, and blood, fill that narrow valley around JESUS up to the bridal of his horse over a distance of 200 miles. Then as GOD has said, for "the Elects sake" those days of wrath, cleansing, and tribulation, are shortened, or no flesh would be left alive. There may be only a few million souls left alive around the world beyond GOD'S Israel. We must ask ourselves, how is this for the Elects sake? What is GOD'S plan for the souls who survive the Tribulation? It is the same as always, to try and save the faithful from among those who have not made their firm and final choice. There are a few references to this in the latter chapters of Ezekiel, but beyond their required trying and final choice, the plan of GOD concerning the temporarily spared souls is not strongly written. Read Rev. Chapter 21(see fig. a. Earth event 11), where after a thousand years, Satan is released and goes out to deceive again. FATHER, from above, then kills all who gather against his SON. FATHER'S created perception of the Old Earth and Old Heaven, are then forgotten and dissolved. FATHER then perceives/sees/creates the New Heaven and New Earth. All mankind, receive new created bodies like those of the Angels at the White Throne Judgment. After which, all who are of GOD, shall watch as a New Earth is created, and watch the created New Jerusalem, come down from Heaven. A City surrounded by walls of gemstone around 216 feet thick. And being a cubic city 1500 miles wide, 1500 miles long, and 1500 miles high. And all the eternally redeemed, being around 300 million souls; enter and live together in GOD'S peace, love, and Holy ways with JESUS at its center. Though there is much more to the Prophecy of GOD to be known. We end with this Prophecy. Isaiah 66:22-24, *"FOR AS THE NEW HEAVENS AND THE NEW EARTH, WHICH I WILL MAKE, SHALL REMAIN BEFORE ME, SAITH THE* ***LORD****, SO SHALL* <u>*YOUR SEED AND YOUR*</u>

<u>NAME</u> REMAIN. AND IT SHALL COME TO PASS, THAT FROM ONE NEW MOON TO ANOTHER, AND FROM ONE <u>SABBATH</u> TO ANOTHER, SHALL ALL FLESH COME TO WORSHIP BEFORE ME, SAITH THE ***LORD***. *AND <u>THEY SHALL GO FORTH, AND LOOK UPON THE CARCASES</u> OF THE MEN THAT HAVE TRANSGRESSED AGAINST ME: FOR THEIR WORM SHALL NOT DIE, NEITHER SHALL THEIR FIRE BE QUENCHED; AND THEY SHALL BE AN ABHORRING UNTO ALL FLESH*". After the White Throne Judgment, we see two things: the Sabbath is eternal as said in Genesis to Revelation. So, observe it now! And all in the New Heaven will go out the Gate and see the Lake of Fire and all who are there for all eternity. To quickly recap the calendar…

ADVENT CALENDAR

(Figure A)

/- -***The complete Old and New Testament Prophecy of the end of days***--- /

/------The beginning of sorrows------/--------The tribulation--------- /

The three woe's /- 1- /---- 2 --- /-- 3--- /

/----------------Time-frame-u/known-/---3.5yrs-----/-------6.3yrs-----/

7SEALS.1___2___3___4____5_____6_____7________________________

7TRUMPETS1 2 3 4 5___6 ______7______

7VIALS 1234567

Heaven events 1, 2,3 (4a----4b----4c-----4d) 5 6, 7

Earth events 1 2 3 4 5 6 (………… 7…….tribulation….. 8) 9

Heaven Events 1-3 are Holy Ceremonies. Heaven Events 4a-4d are the four-horseman loosed. Heaven Event 5 is the Seventh Seal opened starting the Tribulation. Heaven Events 6 and 7 are the Ceremony of the Seven Vials, and delivering the Seven Vials that lead up to the Battle at Armageddon. **Earth** Event 1, was the Iraq-war. Earth Event 2 will be the coming Iran war. Earth Event 3 is the start of the coming World-war going nuclear. Earth Event 4 is the coming Great famine. Earth Event 5 is the point *after* which the Temple construction must begin. Earth Event 6 is the height of Nuclear World War. Earth Event 7, Satan is cast down to earth. Earth Event 8 is the Battle of Armageddon. Earth Event 9, the Millennial Reign begins. We pray you hear this, and go to GOD and receive and build upon this foundation to share the boundless revelations to be delivered in the near future. Please know, this is just the beginning of the eternal things FATHER shall cause to manifest before our eyes. And lastly we have defined one common path to salvation, which is our growing, learning, loving, and living with GOD in all the created ways of eternal life. And to each, and together, endure *until* we are sealed in our Holy and Righteous fate by GOD. Now within our 4 books, when read in their proper order as a series. We deliver the foundations and beginnings of all the love, reason, purpose, pain, loss, and faith, now found in GOD'S labored reality. We sufficiently explain the broad and basic spiritual and physical *laws* of GOD'S created bicameral reality. We have reconciled the ordered events of the Great Prophecy of CHRIST. So, by all we have received and know and shared, we say eternity is not just something that is always out there in the distance, or the abstract past, or future. The spiritual and physical realities of eternity, are always right here in the eternal now and present. Do not live life your life here, like you are waiting for eternity to arrive; live your life here, like eternity is always now and present. Certain souls, more than anything, want to see, create, and live *now* in each unfolding eternal moment. But most souls have little to no desire to know or live in these eternal ways. Our learning, being our process of building our belief, can be increasingly complex, messy, and unpredictable. There is no guarantee we each and all will build and learn the objective truth of GOD about anything, let alone the sum belief and ways of GOD, and all created reality. The great odds are that most-lesser fallen corrupted confused souls, surrounded by evil choices in this fallen world of great confused temptations, will choose to learn in subjective and predictive ways. And all according to what they are taught and exposed too within their desires, and who they learn and live with, and who force themselves into and upon them. Free-will learning is not in any way mechanical or precise, it is an experiential subjective process of each soul's constant choices to believe whatever ideations they desire to believe and live by. And for all greatly confused souls, each additional thing they try to accept as part of their ongoing complex web of false belief, must first traverse and survive a gauntlet of their built-up ideological walls, gaps, loops, and even great voids, that form the false loose web of their suppositional beliefs and desires. So again, we see each soul gets no guarantee of choosing or desiring to learn the one pure truth of anything. The word lost in Scripture, is huge and does not simply mean someone who has not yet been found. It actually means someone who is spiritually and physically falling, confused, deceived, and in error as to their accepted learning and belief towards GOD and all created reality. And the degree to which each soul falls and is lost is another thing, bringing us to that term reprobate; which means to be lost beyond all hope of return. I pray and hope you are indwelled with all the highest forces and senses that GOD has offered us. If so, and if you have read our four books, I so hope you keep re-reading and know every word in the same one pure truth in which they are known and written. If you have not yet been indwelled by

HOLYSPIRIT, I hope you pray and fast and crucify and purify your soul and receive this most loving wonder who leads us to our LORD'S mind of eternal enlightenment. There is much to know, live, and love in the eternal sense. There are far higher understandings of GOD'S purpose and eternal Holy Righteous qualities that need to be shared. And there is always more we can and should and must always do and give and share as part of GOD'S Kingdom and Family. Please love every true word of GOD'S love and life and light. Please become and be all that you are gifted and created and desired to be from above. Please ascend to your perfect home and place, and never fall into hell again. GOD prepares one place for light, and one for darkness. One for the eternal love of GOD, separated from his wrath. And one separated eternally from his love by his wrath. In joy or sorrow, we all will soon be in one of those two places. We pray that you, in the name of CHRIST, will and can search, and try all these words by Scripture in the HOLYSPIRIT. So, we pray for your Holy Righteous strength, and that you receive ever greater revelations than these from FATHER. GOD have mercy on us. Know that GOD will end this soon. Amen. Now, to share *only* GOD'S love and hope, concerning these words. Or to ask good and true needful questions, please Email to: livetoknowtheGODofProphecy@gmail.com